XXII International Conference on Computational Mechanics and Modern Applied Software Systems (CMMASS 2021)

MATEC Web of Conferences Volume 362 (2022)

Alushta, Crimea
4 - 13 September 2021

Editors:

Natalia Severina
Marine Mikilyan

ISBN: 978-1-7138-6061-7

Printed from e-media with permission by:

Curran Associates, Inc.
57 Morehouse Lane
Red Hook, NY 12571

Some format issues inherent in the e-media version may also appear in this print version.

For additional information, please contact EDP Sciences – Web of Conferences at the address below.

EDP Sciences – Web of Conferences
17, Avenue du Hoggar
Parc d'Activité de Courtabœuf
BP 112
F-91944 Les Ulis Cedex A
France

Phone: +33 (0) 1 69 18 75 75
Fax: +33 (0) 1 69 28 84 91

contact-edps@webofconferences.org

Additional copies of this publication are available from:

Curran Associates, Inc.
57 Morehouse Lane
Red Hook, NY 12571 USA
Phone: 845-758-0400
Fax: 845-758-2633
Email: curran@proceedings.com
Web: www.proceedings.com

TABLE OF CONTENTS

 https://doi.org/10.1051/matecconf/202236200001

XXII International Conference on Computational Mechanics and Modern Applied Software Systems

CMMASS 2021

September 4-13, 2021

Preface

This issue comprises selected works presented on the 22nd International Conference on Computational Mechanics and Modern Applied Software Systems (CMMASS 2021) held from 4 September to 13 September 2021 in Alushta (The Republic of Crimea). The Conference was hosted by Moscow Aviation Institute (national research university) in cooperation with the Laboratory «Space flight mechanics» of the Moscow Aviation Institute, the Russian Academy of Sciences, the Ministry of Science and Higher Education of the Russian Federation, the Russian National Committee on Theoretical and Applied Mechanics.

The objective of the Conference is to raise and discuss up-to-date questions relating to computational mechanics problems, modern software systems, parallel computing problems, far-reaching numerical approaches and various aspects of application of computational tools to solve problems in mechanics of a deformable solid, fluid dynamics, aerospace mechanics, nanomechanics and other fields. Also during the Conference the problems of software and database support with analysis of instruments for data processing and data visualization and new achievements from researches are considered.

The Conference became a workspace for participants from Russia, the People's Republic of China, Japan, Australia, India, Poland, Serbia, France, Italy, Armenia, Republic of Kazakhstan.

Conference Committee:

Co-Chairs:

Prof. Dr. Irina G. Goryacheva (Academician of RAS)
Prof. Dr. Vladimir A. Levin (Academician of RAS)
Prof. Dr. Vladimir F. Tishkin (Corresponding member of RAS)
Prof. Dr. Ilia S. Nikitin (Institute of Computer Aided Design RAS)

Deputy Chairman:

Prof. Dr. Sergey S. Krylov (Moscow Aviation Institute, Moscow, Russia)

Program Committee:

Prof. Ara S. Avetisyan (Institute of Mechanics of AAS, Yerevan, Armenia),
Prof. Gevorg E. Bagdasaryan (Russian-Armenian University, Yerevan, Armenia),
Prof. Valentina Balas (Aurel Vlaicu University of Arad, Romania),
Prof. Boris V. Boshenyatov (Institute of Applied Mechanics of RAS, Moscow, Russia),
Prof. Maria B. Bulakina (Moscow Aviation Institute, Lomonosov Moscow State University, Moscow, Russia),
Prof. Irma A. Vardanyan (The National Academy of Sciences of the Republic of Armenia, Yerevan, Armenia),

Prof. Alexander N. Vasiliev (Peter the Great St. Petersburg Polytechnic University, Saint Petersburg, Russia),
Prof. Wang Ruoyu (Beihang University, Beijing, China),
Prof. Vladimir Yu. Gidaspov (Moscow Aviation Institute, Moscow, Russia),
Prof. Decheng Wan (Shanghai Jiao Tong University, Shanghai, China),
Prof. Yuriy I. Dimitrienko (Bauman Moscow State Technical University, Moscow, Russia),
Prof. Anna Esposito (Universita della Campania «Luigi Vanvitelli», Italy),
Prof. Igor E. Ivanov (Moscow Aviation Institute, Lomonosov Moscow State University, Moscow, Russia),
Prof. Valeriy O. Kaledin (Novokuznetsk Branch Kemerovo State University, Novokuznetsk, Russia),
Prof. Sergey S. Krylov (Moscow Aviation Institute, Moscow, Russia),
Prof. Evgenii B. Kuznetsov (Moscow Aviation Institute, Moscow, Russia),
Prof. Li Mingzhi (Beihang University, Beijing, China),
Prof. Vladimir N. Lukin (Moscow Aviation Institute, Moscow, Russia),
Prof. Andrzej Maciejewski (University of Zielona Gora, Zielona Gora, Poland),
Prof. Mayuri Mehta (Sarvajanik College of Engineering and Technology, India),
Prof. Josue Melguzo–Gavilanes (National Center for Scientific Research of France, Paris, France),
Prof. Nenad Mladenovic (SANU, Belgrade, Serbia),
Prof. Janmenjoy Nayak (Aditya Institute of Technology and Management, India),
Prof. Afanasiy V. Ostrik (Institute of Problems of Chemical Physics of RAS, Chernogolovka, Russia),
Dr. Mrutyunjaya Panda (Utkal University, India),
Prof. Yuriy N. Panov (Moscow Aviation Institute, Moscow, Russia),
Prof. Prasant Kumar Pattnaik (Kalinga Institute Of Industrial Technology, India),
Prof. Bernhard Peters (University of Luxembourg, Luxembourg, Germany),
Prof. Maria Przybylska (University of Zielona Gora, Zielona Gora, Poland),
Prof. Vijayalakshmi Ramasamy (University of Wisconsin, USA),
Prof. Dmitriy L. Reviznikov (Moscow Aviation Institute, Moscow, Russia),
Prof. Mika Sato-Ilic (University of Tsukuba, Japan),
Prof. Natal'ya S. Severina (Moscow Aviation Institute, Moscow, Russia),
Prof. Ordabay Z. Sembiyev (M. Auezov South Kazakhstan State University, Shymkent, Republic of Kazakhstan),
Dr. Milan Simic (RMIT University, School of Engineering, Melbourne, Australia),
Prof. Dmitriy Yu. Strelets (Moscow Aviation Institute, Moscow, Russia),
Prof. Sergey V. Strijhak (ISP RAS, Moscow, Russia),
Prof. Dmitriy A. Tarkhov (Peter the Great St. Petersburg Polytechnic University, Saint Petersburg, Russia),
Prof. Niu Han (Beihang University, Beijing, China),
Prof. Yuan Yuan (Shanghai Jiao Tong University, Shanghai, China)

Editor-in-Chief:

Dr. Natalia Severina, Moscow Aviation Institute, Russia
E-mail: severina@mai.ru

Guest Editor:

Dr. Marine Mikilyan, Institute of Mechanics, National Academy of Sciences of Armenia, Armenia
E-mail: marine.mimkilyan@rau.am

Prediction of Earth polar motion in short time interval

Natalia Anosova[1], *Alexandra* Filippova[2], *Sergej* Krylov[2], and *Vadim* Perepelkin[2,*]

[1]RUDN University, 6, Miklukho-Maklaya st., Moscow, 117198, Russia
[2]Moscow Aviation Institute (National Research University), 4, Volokolamskoe shosse, Moscow, 125993, Russia

Abstract. The accuracy characteristics of mathematical models of the Earth pole motion in a short time intervals have been investigated. The modeling was carried out using two approaches: a numerical-analytical model of the Earth's rotational motion relative to the center of mass, based on the weighted least squares method and a neural network model for predicting the Earth pole motion. A comparison is made between the standard deviations of short-term forecasts of the Earth pole motion for 1–2 days according to the developed models and the standard deviations of forecasts published by the International Earth Rotation Service in Bulletin A. The models under consideration did not use data smoothing. The model based on neural network approach showed a higher accuracy of short-term forecasting.

1 Introduction

A lot of scientific papers are devoted to studying the Earth axial rotation irregularities and the prediction of the Earth polar motion. Recently improved methods for predicting Earth orientation parameters (EOP) have been proposed, both using classical approaches based on the least squares method, Kalman filter, spectral and wavelet analyzes, and using widely used neural networks [1, 2].

In papers [3–5], the fundamental astrometric problem of modeling the Earth's motion relative to its center of mass and constructing high-precision forecasts of the Earth pole trajectory and irregularities in the Earth axial rotation was investigated. On the basis of the developed mathematical models of the rotational-oscillatory motion of the deformable Earth relative to the center of mass [3–5] and the neural network approach, in this work a comparative analysis is carried out and the accuracy characteristics of the models of short-term (for 1–2 days) forecast of the Earth polar motion are studied in order to use them in applied problems of astrometry and navigation systems such as GLONASS/GPS.

The importance of the problem is due to the significantly increased accuracy of measurements and the lack of a rational approach when constructing models of the polar motion forecast for the intervals of various durations with the correspondingly required accuracies.

The Earth orientation parameters have an important role in navigation and control of spacecraft motion. Knowledge of the current values of the EOP (precession and nutation angles, angular coordinates of the Earth pole, dUT1 difference between UT1 and UTC) is necessary for the exact mutual transformation of the j2000 inertial coordinate system into

*e-mail: vadimkin1@yandex.ru

the Greenwich coordinate systems WGS 84 and PZ 90-02, as well as accurate prediction of spacecraft orbits.

2 Forecast of a numerical-analytical model of the oscillatory process of the Earth pole

Numerical and analytical modeling of the Earth pole motion is based on the following structure of analytical expressions, that are convenient for numerical calculations [6, 7]:

$$x_p = c_x - a_x^c \cos N^* t + a_x^s \sin N^* t - d_x^c \cos \nu_h t - d_x^s \sin \nu_h t,$$
$$y_p = c_y + a_y^c \cos N^* t + a_y^s \sin N^* t - d_y^c \cos \nu_h t + d_y^s \sin \nu_h t.$$

Here c_x, c_y are coordinates of the mean Earth pole point; N^*, ν_h are the frequencies of the Chandler wobble and annual wobble, respectively. The values of the average Chandler and annual frequencies were selected on the basis of a spectral analysis of a number of observations of Earth pole coordinates in the time interval from 1970 to 2006. N^* is assumed to be 0.843 cycles per year, and the frequency ν_h is equal to one cycle per year. When determining the coefficients $a_x^{c,s}$, $a_y^{s,c}$, $d_x^{c,s}$, $d_y^{s,c}$ we should keep in mind the equalities $a_x^{c,s} \approx a_y^{s,c}$, $d_x^{c,s} \approx d_y^{s,c}$, that are the structural feature of the model [8]. This also means that the processes x_p and y_p are related.

For a situation that corresponds to the modern data of the International Earth Rotation Service (IERS), a short-term forecast can be given using a 5–6 parametric adaptive model, by reducing the number of processed daily measurements. This is due to a decrease in the dynamic error of approximation of the process and high measurement accuracy. The procedure for reducing the interpolation interval in the algorithmic implementation of the model (model adaptation procedure) and its impact on the forecast were evaluated in the algorithm cycle on the test interval immediately adjacent to the forecasting interval.

Table 1. Average standard deviations of the forecasts of the coordinates of the earth's pole of the developed model and the model of the IERS

Interval, day	σ_x, mas	σ_x^{IERS}, mas	σ_y, mas	σ_y^{IERS}, mas	σ_{xy}, mas	σ_{xy}^{IERS}, mas
1	0.331	0.308	0.239	0.242	0.408	0.392
2	0.793	0.488	0.396	0.362	0.886	0.608

Numerical modeling shows that the highest interpolation accuracy is usually achieved in the middle part of the interval. This property is one of the features of the least-squares method and is well known [9, 10]. In particular, theoretically it can be established in the case of polynomial filtration, i.e. a set of support functions in the form of polynomials (usually of a low degree) of the temporal smoothing parameter (for short-term forecasting). Therefore, an increase in the forecast accuracy for a short interval adjacent to the end of the interpolation interval can be achieved by introducing "weight" coefficients in the algorithm of the least-squares method and their relative increase towards the end of the interval. The filtering algorithm can contain a small number of parameters that allow to adjust the specified weighting factors and optimize the algorithm, depending on the previous results of interpolation and forecast.

To assess the accuracy of the developed model, the time interval from 53773 to 57060 of the MJD date was chosen. The filtering procedure and tuning of the algorithm by the "weighted" least-squares method for the first forecast for the time interval from 53773 MJD date was carried out on the adjacent interpolation interval. The standard deviations of the forecast for 1–2 days from the IERS observations were referred to the date of the forecast. Forecasts were calculated with a step equal to a day. The end date corresponding to the last constructed forecast was 57000 MJD date. For comparison, the forecasts of the IERS were taken, published in "Bulletin A", thc standard deviations of which were calculated in a similar way for 1–2 days. Table 1 shows the average deviations of 1–2 daily forecasts of the coordinates of the earth's pole and published forecasts of the IERS for the entire seven-year forecasting interval.

3 Neural Network approach in simulation of the Earth pole oscillations

Artificial neural network (ANN) is a mathematical model built on the principle of organization and functioning of biological neural networks—networks of nerve cells of a living organism. The ANN model is built on the basis of interconnected and interacting nodes—neurons, which form a structured network among themselves. Each of these neuron nodes can receive and process both input data and values generated by other network nodes. Based on the received input, each neuron, taking into account the weighting coefficients of the connections, forms a weighted average sum using it as an argument of the activation function. It allows to generate output values for each neural node. Such output values can serve as input data for other neural network nodes, or form the resulting values, the so-called network output.

A network based on a multilayer perceptron with end-to-end connections from the input layer to the hidden layers was used for modeling. This allows a more accurately simulation

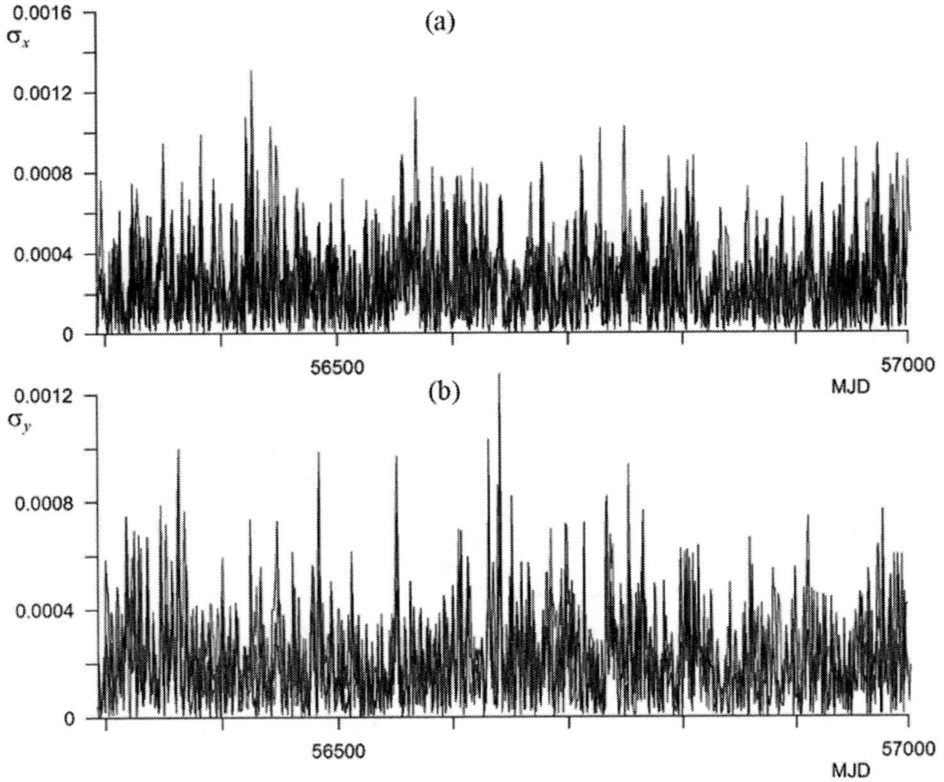

Figure 1. The absolute value of the deviation of the daily forecasts of the pole coordinates x_p (a), y_p (b) according to the developed neural network model (black), numerical-analytical model (red) and IERS forecasts (blue)

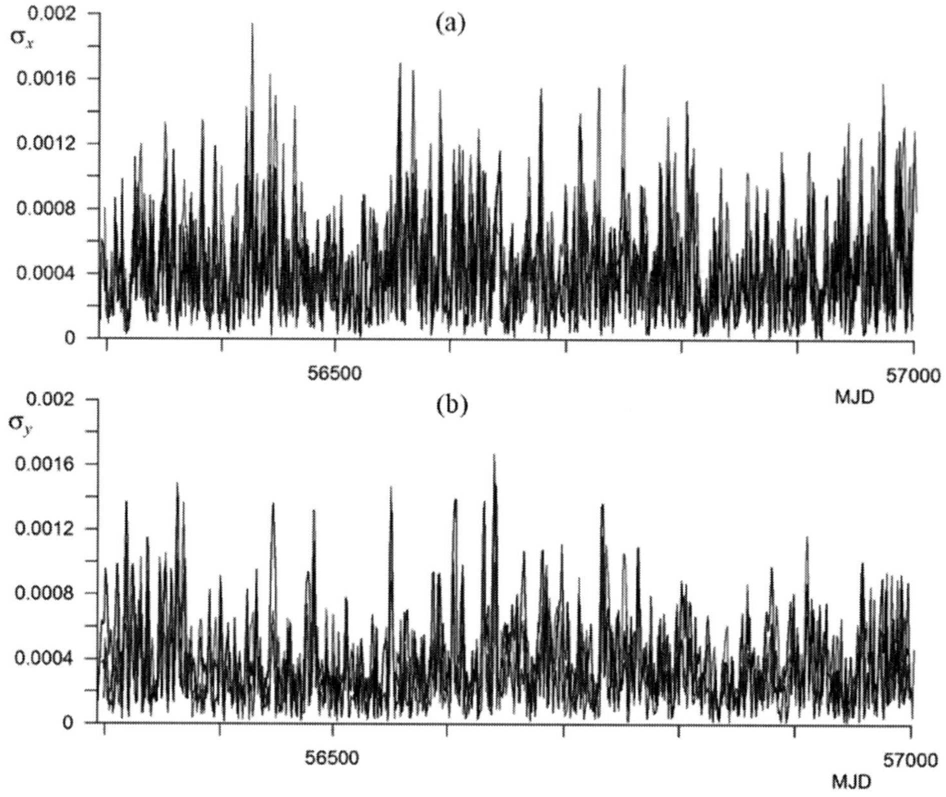

Figure 2. Standard deviations of 2-day forecasts of the pole coordinates x_p (a), y_p (b) according to the developed neural network model (black), numerical-analytical model (red) and IERS forecasts (blue)

Table 2. Average standard deviations of the forecasts of the coordinates of the earth's pole of the developed model and the model of the IERS

Interval, day	σ_x, mas	σ_x^{IERS}, mas	σ_y, mas	σ_y^{IERS}, mas	σ_{xy}, mas	σ_{xy}^{IERS}, mas
1	0.230	0.253	0.182	0.182	0.293	0.312
2	0.418	0.421	0.319	0.300	0.526	0.516

of the linear parts of functions calculated by hidden layers. In addition, a pyramidal structure of hidden layers was used. The first of which is slightly larger or comparable in the number of neurons with the number of inputs, and the subsequent ones decrease within 30–40% of the number of neurons in the previous layer.This makes it possible to achieve a more accurate and smoother result compared to the classic two-layer perceptron with one hidden layer, comparable in the number of neurons and neural connections.

As input data, the values of x_p and y_p were taken for 11 days immediately before the forecast and 31 days before this period with a step of 17 days to take into account the periodic nature of the input data (82 values in total). The result of the network operation is the forecast of the Earth pole coordinates for one and two days ahead from of the current value.

Since it is classical multilayer perceptrons with a sigmoidal activation function, the input data must be in the range from -1 to $+1$, so the data must be scaled before being processed by the network. In the case of direct use of the values of x_p and y_p, the scaling factor was about 600, which leads to the need to train the network with an accuracy of the order of 10^{-4} to achieve an accuracy comparable to polynomial modeling.

As one of the options for solving the problem of large input data scaling factors and, consequently, increased requirements for the accuracy of network training, we can use not the Earth pole coordinates themselves, but the differences between the current and previous values, which make it possible to reduce the scaling factor to 6. To restore the immediate values of the predicted coordinates, it is enough to add the results obtained to the current coordinates x_p and y_p. With such a data presentation, the features associated with high-frequency oscillations in the Earth pole coordinates become more noticeable.

The period from 1962 (the beginning of observations) to the end of 2012 (about 18 thousand values in total) was taken as a training sample. And the data for 2013 and 2014 were used as a test sample. Table 2 shows the average deviations of 1–2 daily forecasts of the coordinates of the Earth pole and published forecasts of the IERS for the test forecasting interval (2013–2014).

In figures 1 and 2 shows a comparison of the absolute values of the deviations of the first day of the forecast from the observed and standard deviations of 2-day and 3-day (respectively) forecasts of the developed neural network model (black), numerical-analytical model (red) and IERS model (blue).

4 Conclusion

The analysis of the numerical modeling results of the Earth pole oscillations based on the developed methods and algorithms, that was carried out in this work, shows the adequacy of the mathematical models of the Earth pole motion to the IERS observations and measurements data. The developed models, in terms of the approximation accuracy of the process, provide good agreement with the IERS data for short (1–2 days) time intervals. Comparative analysis with existing analogs shows that the obtained numerical results of the forecasting the Earth pole coordinates are fundamentally comparable with them.

References

[1] L. Chen, T. Geshi, J. Sun, S. Hu, W. Lu, *High Precision Determining and Predicting of Earth Orientation Parameters for Supporting Spacecraft Navigation*, in China Satellite Navigation Conference (CSNC) 2016 Proceedings: Vol. III (China, 2016)

[2] L. Yu, G. Min, H. Dan-Dan, C. Hong-Bing, Z. Dan-Ning, H. Zhao-Peng, G. Yu-Ping, Adv. Space Res., **59(2)**, 524 (2016)

[3] L.D. Akulenko, Yu.G. Markov, V.V. Perepelkin, L.V. Rykhlova, A.S. Filippova, Astron. Rep., **57**, 391 (2013)

[4] L.D. Akulenko, Yu.G. Markov, V.V. Perepelkin, Astronautics and rocket science, **2(55)**, 130 (2009)

[5] Yu.G. Markov, V.V. Perepelkin, L.V. Rykhlova, A.S. Filippova, Astron. Rep., **62(4)**, 299 (2018)

[6] International Earth Rotation and Reference Systems Service—IERS Annual Reports, URL `http://www.iers.org`

[7] S.A. Kumakshev, Mech. Solids, **53(2)**, 159 (2018)

[8] D.M. Klimov, L.D. Akulenko, S.A. Kumakshev, Dokl. Phys., **59(10)**, 472 (2014)

[9] S. Krylov, V. Perepelkin, A. Filippova, *Short-term prediction of universal time variations dUT1*, AIP Conf. Proc., **2181**, 020023 (2019)

[10] D.M. Klimov, L.D. Akulenko, S.A. Kumakshev, Dokl. Phys., **58(11)**, 505 (2013)

Computational method for solving boundary value problems of mechanics deformable body using non-orthogonal functions

Vladimir Bakulin[1,*] and *Victor* Revenko[2,**]

[1]Institute of Applied Mechanics of Russian Academy of Sciences, 7, Leningradsky prospekt, Moscow, 125040, Russia
[2]Pidstryhach Institute for Applied Problems of Mechanics and Mathematics NASU, 3-b, Naukova st., Lviv, 79060, Ukraine

Abstract. In this paper a complete system of non-orthogonal functions was built on the basis of orthogonal sines and cosines. It is shown that the known orthogonal systems of functions are a degenerate case of non-orthogonal systems of functions. It has been proven that the continuous function can be approximated non-orthogonal functions in such a way that one selected non-orthogonal function will not included in this amount. The boundary value problem of the elasticity theory has been considered for an inhomogeneous plate. A new method for solving the boundary value problem is developed for the fourth-order equation with variable coefficients. The proposed method is based on separating of the stress state of the plate, use of complete systems of non-orthogonal functions and a generalized quadratic form. Criteria have been established under which the constructed approximate solution coincides with the exact solution. This method has been adapted to solving a boundary value problem for high-order differential equations. The high accuracy of the method has been confirmed by numerical calculations.

1 Introduction

Currently, both continuous [1] and piecewise continuous [2, 3] systems of orthogonal functions are widely used. The use of modern computers allows us to develop new techniques for modeling functions and various processes in engineering. In [4], it is proposed to use non-orthogonal systems of functions to solve the boundary value problems of the elasticity theory. The development of computer techniques for non-orthogonal functions made it possible to propose a new analytical-numerical approach to solve the boundary value problems of the elasticity theory [5–7], as well as differential equations [8] using the least squares method [9, 10]. The computational method of solving boundary value problems of the mechanics of a deformable solids with variable elastic characteristics is considered in [11], where methods with use of non-orthogonal functions were proposed.

The non-orthogonal systems of functions naturally appear when constructing eigenfunctions of boundary value problems for differential equations [12]. They are also used to solve

*e-mail: vbak@yandex.ru
**e-mail: victor.rev.12@gmail.com

the boundary value problems of the elasticity theory [4]. However, up to date computational methods based on the non-orthogonal systems are not yet well developed for functions modelling, optimization of various processes, solving boundary value problems of mathematical physics, etc.

The purpose of research is to develop a new computer method to solve the boundary value problems involving both partial differential equations and ordinary differential equations and to devise analytical and computer tools for operating over the basis of the orthogonal sine-cosine families.

2 Approximation of continuous functions

Consider a complete non-orthogonal system of functions on the interval $[l_1, l_2]$. In this work, for its construction, we use the system of orthogonal functions (sinus and cosine), which will be considered on a smaller segment $[l_1, l_2]$ than the interval of their orthogonality $[A, B]$. An approximation of a continuous function $f(x)$ on the interval $[l_1, l_2]$ will be will be realized as a finite sum of the following form

$$f(x) \cong \sum_{k=0}^{M} c_k \varphi_k(x), \tag{1}$$

where c_k are unknown coefficients, $M = 2N$, N is a natural number, $\varphi_0(x) \equiv 1$, $\varphi_k(x) = \cos(k\omega x)$, $\varphi_{k+N}(x) = \sin(k\omega x)$ are the basic functions, $k = \overline{1, N}$; $\omega = 2\pi/(B - A)$, $[l_1, l_2] \subset [A, B]$, $l_2 - l_1 < 0.95(B - A)$. It is not difficult to verify that the selected system of functions $\{\varphi_k(x)\}$ will be complete and orthogonal in the interval $[A, B]$, and and complete but not orthogonal in the segment $x \in [l_1, l_2]$. Increasing the value N without bound in (1) leads to the conversion of the right hand side of equation (1) into a row.

The use of decomposition (1) by non-orthogonal functions allows us to satisfy the different values of the function at the edges of the segment $f(l_1) \neq f(l_2)$ without disrupting of the convergence conditions at points l_j, $j = \overline{1, 2}$. It requires a smaller quantity of number of the sum of the row (1) during approximation of functions with a given accuracy.

We formulate the important difference between non-orthogonal and orthogonal systems of functions.

Theorem 1. *A continuous function $f(x)$ is specified in the segment $[l_1, l_2]$ can be approximated by the sum of the row* (1) *in such a way that one arbitrarily selected function $\varphi_m(x)$ of the non-orthogonal basis* (1) *is not included in its representation.*

Proof. We have extended the function $f(x)$ over the segment $[A, B]$ and denoted it as $f_1(x)$. The function $f(x)$ can always be extended so that it remains continuous, and the integral over the segment $[A, B]$ of the function $f_1(x)\varphi_m(x)$ is equal to zero. According to [1], we construct the decomposition of a function $f_1(x)$ in the form of a series of orthogonal in segment $[A, B]$ of sinus and cosine, where equality have performed $c_m = 0$. Consider the function $f_1(x)$ on the segment $[l_1, l_2]$ and obtain its decomposition without a basic function $\varphi_m(x)$. The function $f_1(x)$ coincides with the function $f(x)$ on the segment $[l_1, l_2]$. So, we have received a decomposition of a function $f(x)$ in which there is no basic function $\varphi_m(x)$. End of the proof.

Corollary. *When satisfying the boundary conditions, depending on the physical nature of the problem being solved, we have an opportunity not to use one basic non-orthogonal function.*

The coefficients c_k are determined from the condition of the minimum functional, which characterizes the deviation between the approximation of the function (1) and its value [6, 7].

3 Solution of two-dimensional boundary value problems of the mechanics of the deformable solid body

Consider a two-dimensional boundary problem of the elasticity theory for a plate with variables Young's modulus and Poisson's ratio [13]. The median surface of the plate occupies a rectangular region $\Pi = \{(x, y) \in ([0, a] \times [-b, b])\}$, with a contour L on which the load is given. Stresses in the region Π are expressed through stresses function $F(x, y)$

$$\sigma_x = \frac{\partial^2 F}{\partial y^2}, \quad \tau_{xy} = \tau = -\frac{\partial^2 F}{\partial x\, \partial y}, \quad \sigma_y = \frac{\partial^2 F}{\partial x^2}. \tag{2}$$

The stresses function (2) satisfies the partial differential equation [13]

$$TF(x, y) \equiv \left\{ \Delta(\gamma\Delta) - q_y'' \frac{\partial^2}{\partial x^2} + 2q_{xy}'' \frac{\partial^2}{\partial y\, \partial x} - q_x'' \frac{\partial^2}{\partial y^2} \right\} F = 0, \tag{3}$$

where $\Delta = \frac{\partial^2}{\partial x^2} + \frac{\partial^2}{\partial y^2}$ is two-dimensional Laplace operator; T is fourth-order two-dimensional operator defined by the relation (3), $\gamma(x, y) = \frac{1}{E(x, y)}$, $E(x, y)$ is the Young's modulus, $q = \frac{1}{2G} = \frac{1 + \nu}{E}$, $\nu(x, y)$ is the Poisson's ratio. We assume that the elastic constant of material E, ν are even with respect to the coordinate y and twice continuously differentiable functions in the region Π.

Let us consider in detail when $a \gg b$ and normal loads are given only on the transverse sides, so that the tangent loads in the angular points of the plate are zero. Consider even normal stresses, odd tangent stresses

$$\sigma_x(a_j, y) = \sigma_j(y), \quad \tau_{xy}(a_j, y) = \tau_j(y), \quad \tau_{xy}(a_j, \pm b) = 0, \quad j = \overline{1, 2}, \tag{4}$$

where $\sigma_j(y)$ are even normal, $\tau_j(y)$ are odd tangent loads, $a_1 = 0$, $a_2 = a$. For problem (4) we will find the basic stress state of plate Π [5]

$$\sigma_0 = \frac{1}{b} \int_0^b \sigma_1(y) = \frac{1}{b} \int_0^b \sigma_2(y),$$

so only one stress component $\sigma_x^0(x, y) = \sigma_0$ is not zero. Thus, we separate of the stress-strain state (SSS) of the plate by the sum of the basic SSS ($\sigma_0 = \sigma_x^0$) with one component of the forces, and the perturbed self-balanced state with zero basic vectors of forces and moments [5].

Given the condition $a \gg b$ the symmetric self-balanced problem will be divided into two tasks. Consider the first of these tasks in which only one side of the rectangle is loaded. We have wrote boundary conditions

$$\sigma_x(0, y) = \left. \frac{\partial^2 F}{\partial y^2} \right|_x = 0 = \sigma^1(y), \quad \tau_{xy}(0, y) = -\left. \frac{\partial^2 F}{\partial x\, \partial y} \right|_x = 0 = \tau_1(y), \tag{5}$$

$$\sigma_x(a, y) = 0, \quad \tau_{xy}(a, y) = 0, \quad y \in [0, b], \tag{6}$$

$$\sigma_y(x, b) = 0 \quad \tau_{yx}(x, b) = 0, \quad x \in [0, a], \tag{7}$$

where $\sigma^1(y) = \sigma_1(y) - \sigma_0$ is self-balanced load. On two longitudinal sides of the plate, zero boundary conditions (7) are presented. The Saint-Venant's principle implies that the self-balanced stress state of the plate will be rapidly decreased by removing from the transverse side $x = 0$. Therefore, the function of stresses that satisfies equations (3), (5)–(7) we can search with the symmetry of the stress state in such form:

$$F = E_0 \sum_{k=1}^{N} [(a_k + a_{k+N} y^2)\varphi_k + (d_k y + d_{k+N} y^3)\varphi_{k+N}] e^{-k\omega x}, \tag{8}$$

where a_k, d_k are unknown coefficients; N is natural number; $\varphi_k(x) = \cos(k\omega y)$, $\varphi_{k+N}(x) = \sin(k\omega y)$, $k = \overline{1, N}$, $\omega = 0.9\pi/b$.

We substitute the function of stresses (8) in relations (2) and we obtain an obvious type of components stresses

$$\sigma_x = E_0 \sum_{k=1}^{N} \Big\{ -k^2\omega^2 a_k \varphi_k + a_{k+N}[(2 - k^2\omega^2 y^2)\varphi_k - 4k\omega y \varphi_{k+N}] +$$
$$+ (2k\omega\varphi_k - k^2\omega^2 y \varphi_{k+N}) d_k + d_{k+N}[6y\varphi_{k+N} + 6y^2 k\omega \cos(k\omega y) - k^2\omega^2 \varphi_{k+N}] \Big\} e^{-k\omega x},$$
$$\sigma_y = E_0 \sum_{k=0}^{N} k^2\omega^2 [(a_k + a_{k+N} y^2)\varphi_k + (d_k y + d_{k+N} y^3)\varphi_{k+N}] e^{-k\omega x}, \tag{9}$$
$$\tau = \sum_{k=0}^{N} k\omega \Big\{ -k\omega a_k \varphi_{k+N} + [2y\varphi_k - k\omega y^2 \varphi_{k+N}] a_{k+N} + d_k (k\omega y \varphi_k + \varphi_{k+N}) +$$
$$+ d_{k+N}[3y^2 \varphi_{k+N} + k\omega y^3 \varphi_k] \Big\} e^{-k\omega x}.$$

The expression of the stress function (8) has been chosen so that it had a part that coincides with the expression of the stress function for a homogeneous isotropic material [5].

We introduce the norm [14] in the space of two-dimensional of two-dimensional functions $F(x, y)$specified in the rectangular region Π

$$\|F(x, y)\| = \sqrt{\int_0^a \int_0^b [F(x, y)]^2 \, dx \, dy.} \tag{10}$$

It is known [14] that if the function $F(x, y)$ is a continuous function and $\|F(x, y)\| = 0$, then $F(x, y) = 0$.

Note that unknown coefficients a_k, d_k of stress function (8) must satisfy the equation (3) in the region Π, which will be present in the following form:

$$|TF(x, y)| = \left| \sum_{k=1}^{N} \sum_{n=0}^{1} \{ T_{k+nN}(x, y) a_{k+nN} + S_{k+nN}(x, y) d_{k+nN} \} \right| \to 0, \tag{11}$$

where

$$T_k(x, y) = 2\chi_{k,0}^{0,1} - \chi_{k,0}^{0}, \quad T_{k+N}(x, y) = 2\psi_{k,1}^{0} - 4k\omega\psi_{k,2}^{1} + 2\chi_{k,0}^{2,1} - \chi_{k,0}^{2},$$
$$S_k = 2k\omega\psi_{k,1}^{0} + 2\chi_{k,1}^{1,1} - \chi_{k,1}^{1}, \quad S_{k+N} = 6k\omega\psi_{k,1}^{2} + 6\psi_{k,2}^{1} + 2\chi_{k,1}^{3,1} - \chi_{k,1}^{3},$$
$$\psi_{k,1}^{m}(x, y) = (\Delta\gamma) y^m \cos(k\omega y) e^{-k\omega x}, \quad \psi_{k,2}^{m}(x, y) = (\Delta\gamma) y^m \sin(k\omega y) e^{-k\omega x},$$
$$\chi_{k,n}^{m} = \left(\frac{\partial^2 g}{\partial y^2}\frac{\partial^2}{\partial x^2} + \frac{\partial^2 g}{\partial x^2}\frac{\partial^2}{\partial y^2} \right) y^m \varphi_{k+nN} e^{-k\omega x}, \tag{12}$$
$$\chi_{k,n}^{m,1} = \frac{\partial^2 g}{\partial x \partial y}\frac{\partial^2}{\partial x \partial y} y^m \varphi_{k+nN} e^{-k\omega x}, \quad n = \overline{0, 1}, \quad k = \overline{1, N}.$$

Equation (11) defines the continuous function in the region Π. We minimize the deviations of the function (11) from zero. To do this, we substitute expressions (12) and after using the norm (10) in the region Π we get a non-negative quadratic form

$$||TF(x,y)||^2 = \int_0^a \int_0^b \left[\sum_{k=1}^{M} c_k T_k(x,y) \right]^2 dx\,dy = \sum_{k,j=1}^{M} c_k c_j B_{kj} \to 0, \tag{13}$$

where $M = 4N$, $T_{k+2N} = S_k$; $c_k = a_k$, $c_{k+2N} = d_k$, $k = \overline{0,2N}$ are variables and $B_{kj} = = \int_0^a \int_0^b T_k T_j \, dx\,dy$, $k = \overline{1,M}$ are the coefficients of the quadratic form. The given method allows us to reduce of numerical solution of the equation (3) to an estimation of the minimum of the quadratic form (13).

By means of selecting the exponentially reduced presentation of the stress state, the conditions (6) will be approximately satisfied. It remains to satisfy the boundary conditions (5), (7). After substitution the stresses (9) in them they are reduced to a compact form

$$\sum_{k=1}^{M} c_k A_{m,k}(\gamma_m) = P_m(\gamma_m), \quad \gamma_m \in [0, \alpha_m], \quad m = \overline{1,K}, \tag{14}$$

where

$$K = 4, \quad P_1 = \sigma^1(y), \quad P_2 = \tau_1(y), \quad P_j = 0, \quad j = \overline{3,4}, \quad \alpha_1 = \alpha_2 = b, \quad \alpha_3 = \alpha_4 = a,$$
$$A_{1,k} = (2k\omega \cos(k\omega y) - k^2\omega^2 y \sin(k\omega y)), \quad A_{2,k} = -k^2\omega^2 \sin(k\omega y),$$
$$A_{3,k} = k^2\omega^2 \cos(k\omega b) e^{-k\omega x}, \quad A_{4,k} = -k^2\omega^2 \sin(k\omega b) e^{-k\omega x}, \quad k = \overline{1,N};$$
$$A_{1,k} = [(2 - k^2\omega^2 y^2)\cos(k\omega y) - 4k\omega y \sin(k\omega y)], \quad A_{2,k} = k\omega[2y\cos(k\omega y) - k\omega y^2 \sin(k\omega y)],$$
$$A_{3,k} = b^2k^2\omega^2 \cos(k\omega b) e^{-k\omega x}, \quad A_{4,k} = k\omega[2b\cos(k\omega b) - k\omega b^2 \sin(k\omega b)] e^{-k\omega x},$$
$$k = \overline{N+1, 2N}; \quad A_{2,k} = -k^2\omega^2 \sin(k\omega y), \quad A_{1,k} = (2k\omega\cos(k\omega y) - k^2\omega^2 y \sin(k\omega y)),$$
$$A_{3,k} = k^2\omega^2 b \sin(k\omega b) e^{-k\omega x}, \quad A_{4,k} = -k^2\omega^2 \sin(k\omega b) e^{-k\omega x}, \quad k = \overline{2N+1, 3N};$$
$$A_{1,k} = [6y\sin(k\omega y) + 6y^2 k\omega \cos(k\omega y) - k^2\omega^2 \sin(k\omega y)],$$
$$A_{2,k} = k\omega[3y^2 \sin(k\omega y) + k\omega y^3 \cos(k\omega y)], \quad A_{3,k} = k^2\omega^2 b^3 \sin(k\omega b) e^{-k\omega x},$$
$$A_{4,k} = k\omega[3b^2 \sin(k\omega b) + k\omega b^3 \cos(k\omega b)] e^{-k\omega x}, \quad k = \overline{3N+1, 4N}.$$

We denote the left-hand sides of equation (14) by

$$f_{m,N}(\gamma_m) = \sum_{k=1}^{M} c_k A_{m,k}(\gamma_m), \quad m = \overline{1,K}. \tag{15}$$

To solve equations (14), we had used the analytical-numerical method developed in [5, 6]. The method is based on the convergence of all expressions in the left parts of equations (14), (15) to the loads $P_m(\gamma_m)$. We have written these expressions

$$\left| \sum_{k=1}^{M} c_k A_{m,k}(\gamma_m) - P_m(\gamma_m) \right| \to 0, \quad \gamma_m \in [0, \alpha_m], \quad m = \overline{1,K}. \tag{16}$$

An effective analytic and numerical methodology has been developed, which allowed us to simultaneously minimize all K expressions (16) in the norms $L_2[0, \alpha_m]$ and quadratic form (13) [11]. The given method allows us to have been reduced numerical solution of equations (13), (16) to the definition of a minimum of the generalized quadratic form

$$\|TF(x,y)\|^2 + \sum_{m=1}^{4} \|f_{m,N}(\gamma) - P_m(\gamma)\|_m^2 = \sum_{k,j=1}^{M} c_k c_j W_{kj} - 2\sum_{k=1}^{M} c_k V_k + P^2, \qquad (17)$$

where $\|f(\gamma)\|_m = \sqrt{\int_0^{\alpha_m} f^2(\gamma)\, d\gamma}$ are norms $L_2[0,\alpha_m]$, $P^2 = \sum_{m=1}^{2} \|P_m\|_m^2$, $m = \overline{1,K}$,

$$W_{k,j} = B_{k,j} + \int_0^{\alpha_m} \sum_{m=1}^{4} A_{m,k}(\gamma_m) A_{m,j}(\gamma_m)\, d\gamma_m, \quad V_k = \int_0^{\alpha_m} \sum_{m=1}^{2} A_{1,k}(\gamma_m) P_m(\gamma_m)\, d\gamma_m,$$
$$W_{k,j} = W_{j,k}, \quad k,j = \overline{1,M}.$$

The minimum of a generalized quadratic form (17) for the specified N is denoted $\Lambda(N)$, and unknown variables c_k on which it is achieved, we denote c_k^N. Variables c_k^N are defined from the condition of its minimum. The function of the stresses is defined by the found coefficients c_k^N is denoted $F_N(x,y)$.

4 Solutions of the differential equations with continuous coefficients

Consider the boundary value problem for the equation 2K-order:

$$T_K y \equiv \sum_{k=0}^{2K} f_k(x) y^{(k)}(x) = 0; \quad y^{(k)}(l_j) = u_k^j, \quad k = \overline{0,K-1}, \quad j = \overline{1,2}, \quad x \in [l_1, l_2], \quad (18)$$

here T_K is differential operator. Let us represent an approximate solution of problem (18) as the amount of finite series (1)

$$y_N(x) = \sum_{n=1}^{K} \sum_{k=1}^{2N} c_{k+2N(n-1)} x^{n-1} \varphi_k(x). \qquad (19)$$

The coefficients c_k are determined after the solution substitution (19) in conditions (18). In the one-dimensional case the relation (17) is simplified to the following generalized quadratic form:

$$\int_{l_1}^{l_2} [T_1 y_N(x)]^2 dx + \sum_{k=0}^{K-1} \sum_{j=1}^{2} (y^{(k)}(l_j) - u_k^j)^2 = \sum_{k,j=1}^{M} c_k c_j W_{kj} - 2\sum_{k=1}^{M} c_k V_k + P^2, \qquad (20)$$

where $M = 2NK$.

Two-point edge task has been solved for the method testing

$$T_1 y \equiv xy'' + y' + xy = 0, \quad x \in [l_1, l_2], \quad y(l_1) = y_1, \quad y(l_2) = y_2, \qquad (21)$$

where $0 < l_1 < l_2$, T_1 is the operator for Bessel's equation of order zero. The coefficients of form (20) for equation (21) are found in the analytical form [8] and calculated by the computer. A numerical experiment was carried out: $l_1 = 1$, $l_2 = 12$, $y_1 = 3$, $y_2 = 5$.

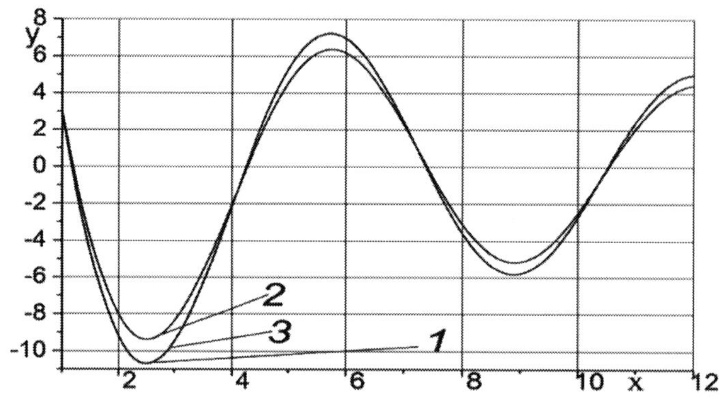

Figure 1. Graphs of solutions: 1 is accurate solution; 2, 3 are approximate solutions: $N = 10$, $N = 16$

The simulation results are shown in figure 1. The approximate solution of the equation (21) at $N = 10$ is not essentially different from the exact solution, and at $N = 16$ coincides with the exact solution with the error of less than 10^{-3}.

5 Convergence criteria for the constructed solution

Lemma [7]. *The function* $\Lambda(N)$ *is nonnegative and not increasing.*

If we assume that $\Lambda(N) = 0$, then all separate terms that are included in the left part of the expression (17) will be zero. Consequently, all equations (13), (16) will be satisfied and the initial task is solved.

We assume that the proposed computational process is stable [6]. The following theorem is faithful.

Theorem 2. *If there exists N such that* $\Lambda(N) < \varepsilon^2/4$ *for any* $\varepsilon > 0$, *then the stresses written through function* $F(x, y) = \lim_{N\to\infty} F_N(x, y)$ *exactly satisfy the conditions* (5), (7), *as well as equation* (3).

Proof. Consider the sequence of the small positive numbers ε_N convergent to zero. This sequence corresponds to the natural number sequence $N \to \infty$. Now due to Lemma and theorem 2 assumption there exist N and $\varepsilon_N \leq \varepsilon$ for any $\varepsilon > 0$ so that the relation

$$\Lambda(N + k) \leq \Lambda(N) < \frac{\varepsilon_N^2}{4}$$

are satisfied for any natural number k. Hence if we tend, then the limit $\Lambda(N)$ turns into

$$\lim_{N\to\infty} \Lambda(N) = 0.$$

Let us show that decrease of ε_N and relative increase of N lead to the following: The stresses constructed with the coefficients c_k^N and functions (9) meet conditions (14) with the given error ε in the metrics of the spaces $L_2[0, \alpha_m]$. Indeed, condition (17) and the triangle inequality ([14, p. 378]) imply for functions $f_{m,N}(\gamma_m)$ since

$$||f_{m,j} - f_{m,k}||_m \leq ||f_{m,j} - P_m||_m + ||f_{m,k} - P_m||_m \leq 2\sqrt{\Lambda(N)} < \varepsilon_N \leq \varepsilon, \quad m = \overline{1, K}$$

for arbitrary $k, j \geq N$, the function sequences $f_{m,N}(\gamma_m)$ are Cauchy sequences in the metrics of $L_2[0, \alpha_m]$. So, there are limits of functions sequences $f_{m,N}(\gamma_m)$ that we denote by

$$f_m(\gamma_m) = \lim_{N\to\infty} f_{m,N}(\gamma_m), \quad \gamma_m \in [0, \alpha_m], \quad m = \overline{1, K}. \tag{22}$$

According to (17), we write down an assessment of satisfaction of conditions (13) and boundary conditions (14) in the metric $L_2[0, \alpha_m]$

$$||TF_N(x,y)|| \leq \sqrt{\Lambda(N)} < \varepsilon_N/2,$$

$$||f_{m,N}(\gamma_m) - P_m||_m \leq \sqrt{\Lambda(N)} < \varepsilon_N/2, \quad m = \overline{1,K}. \tag{23}$$

We tend to the limit $\varepsilon_N \to 0$ in inequalities (23) and we obtain that stress function $F(x, y)$ satisfies the conditions (3), and the functions $f_m(\gamma_m)$ accurately satisfy the boundary conditions (14) in the metric of spaces L_2

$$||TF(x,y)|| = 0, \quad ||f_m(\gamma_m) - P_m||_m = 0, \quad m = \overline{1,K}. \tag{24}$$

Since all functions in conditions (24) are continuous, then according to [14], we will have

$$TF(x,y) = 0, \quad f_m(\gamma_m) = P_m(\gamma_m), \quad m = \overline{1,K}. \tag{25}$$

End of the proof of Theorem 2.

6 Conclusion

It has been proven that the boundary conditions can be satisfied without using one basic non-orthogonal function. In numerical solutions, the stress state of the plate has been separated by the main state and the perturbed state, which decreases with distance from the loaded edges. An effective way to solve boundary value problems in a rectangular plate with variable elastic characteristics is proposed. The proposed approach replaces the satisfaction of the boundary conditions and the solution of the partial differential equation, by the search a minimum of a generalized quadratic form. The generalized quadratic form is introduced, which is not linear, but its minimum tends to zero and determines the convergence and accuracy of the solution. It has been proven that the accuracy of the approximate solution of the partial differential equations with variable coefficients is estimated by one number, namely, a minimum of a generalized quadratic form. It has been established that a slight amount of non-orthogonal functions very well approximates the exact solution of the differential equation. Calculations confirmed the high accuracy of the proposed method.

References

[1] G.P. Tolstov, *Ryady Fur'e* (Nauka, Moscow, 1980)
[2] B.M. Mohan, S.K. Kar, *Continuous Time Dynamical Systems: State Estimation and Optimal Control With Orthogonal Functions* (CRC Press, Boca Raton, 2018)
[3] A. Patra, G.P. Rao, *General Hybrid Orthogonal Functions and Their Applications in Systems and Control* (Springer, London, 1996)
[5] M.A. Aleksidze, *Solving boundary tasks by non-orthogonal function decomposition* (Nauka, Moscow, 1991)
[5] V.P. Revenko, Int. Appl. Mech., **44**, 89 (2008), DOI: 10.1007/s10778-008-0021-6
[6] V.P. Revenko, Int. Appl. Mech., **45**, 730 (2009), DOI: 10.1007/s10778-008-0021-6
[7] V.N. Bakulin, V.P. Revenko, Russ. Math., **60**, 1 (2016), DOI: 10.3103/S1066369X16060013
[8] V.P. Revenko, Lvov University Visnyk, Applied Mathematics and Computer Science, **16**, 54 (2010)

[9] R.S. Varga, *Functional Analysis and Approximation Theory in Numerical Analysis* 76 (V 3. Regional Conference Series in applied mathematics, Soc. for Industrial and Applied Math., 1979)

[10] J.H. Bramble, A.H. Schatz, *Least squares method for 2th order elliptic boundary value problems*, Math. Compout., **25**, 1 (1971)

[11] V.N. Bakulin, V.P. Revenko, in Proceedings of the Twenty-second International Conference on Computational Mechanics and Modern Applied Software Systems, CMMASS'2021, 4--13 September 2021, Moscow (2021), URL `http://www.cmmass.ru/files/cmmass2021_web.pdf`

[12] E.C. Titchmarsh, *Eigenfunction Expansions Associated with Second-order Differential Equations* (Oxford University Press, Part I, 1946, Part 2, 1958)

[13] V.A. Lomakin, *Theory of elasticity of inhomogeneous bodies* (Publishing House of Moscow University, Moscow, 1976)

[14] G. Korn, T. Korn, *Mathematical Handbook for Scientists and Engineers. Definitions, Theorems, Formulas* (McGraw-Hill, New York, 1961; Nauka, Moscow, 1974)

Analysis of the orbital stability of periodic pendulum motions of a heavy rigid body with a fixed point under the Goryachev–Chaplygin condition

Boris Bardin[1,2,3,*], *Badma* Maksimov[1], and *Vladimir* Zarodov[1]

[1]Department of Mechatronics and Theoretical Mechanics, Moscow Aviation Institute (National Research University), 4, Volokolamskoe shosse, Moscow, 125993, Russia
[2]Department of Higher Mathematics, Moscow Automobile and Road Construction State Technical University, 64, Leningradsky prospect, Moscow, 125319, Russia
[3]Mechanical Engineering Research Institute of the Russian Academy of Sciences, 4, M. Kharitonyevskiy pereulok, Moscow, 101990, Russia

Abstract. In this paper the motion of a rigid body with a fixed point in a uniform gravity field is considered. It is assumed that the main moments of inertia of the body correspond to the case of Goryachev–Chaplygin, i.e., they are in the ratio of 1:1:4. In contrast to the integrable case of Goryachev–Chaplygin, there are no restrictions imposed on the position of the center of mass of the body. The problem of the orbital stability of periodic pendulum motions of a body (oscillations and rotations) is investigated. The equations of perturbed motion were obtained and the problem of orbital stability was reduced to the stability problem of the equilibrium position of a second-order linear system with 2π-periodic coefficients, the right-hand sides of which depend on two parameters. Based on the analysis of the linearized system, it has been established that the rotations are orbitally unstable for all possible values of the parameters. Moreover, a diagram of stability of pendulum oscillations has been constructed, on which the regions of orbital instability (parametric resonance) and regions of orbital stability in the linear approximation are indicated. At small values of oscillations amplitudes, a nonlinear analysis was performed. Based on the analysis of the coefficients of the normalized Hamilton function using theorems of the KAM theory, rigorous conclusions on the orbital stability of pendulum oscillations with small amplitudes were obtained.

1 Introduction

In the dynamics of a rigid body periodic motions play a special role. Their study often allows one to draw important conclusions about the nature of the body's motion and qualitatively describe the structure of the phase space of the system. Both for theoretical mechanics and for its applications the problem of orbital stability of periodic pendulum motions of a heavy rigid body with a fixed point is of considerable interest. Modern methods of dynamical systems: the method of normal forms, the methods of KAM theory and the general theory of stability make it possible to obtain rigorous conclusions about the orbital stability of periodic solutions of this type.

*e-mail: bsbardin@yandex.ru

In the general case, the problem of stability of pendulum motions of a heavy rigid body with one fixed point contains four parameters. Therefore, it is rather difficult to obtain a complete solution to this problem. That is why a various number of special cases were investigated. In particular, the cases of Kovalevskaya, Goryachev–Chaplygin, Bobylev–Steklov and some cases of a dynamically symmetric body were considered. In the case of Kovalevskaya, the pendulum rotations of a rigid body are orbitally unstable, and the stability of pendulum oscillations depends on their amplitudes [1–5].

The case of Goryachev–Chaplygin was studied in [6–8]. In the case of Bobylev–Steklov, the problem has two parameters and, depending on the values of the parameters, both stability and instability might take place [9, 10]. For a dynamically symmetric rigid body, the case was investigated when body's center of mass lies in the equatorial plane of the ellipsoid of inertia [11]. A special case of a dynamically symmetric body was considered in [12] when the center of mass does not lie in the equatorial plane and certain conditions are imposed on the moments of inertia of the body ($A = C = 2B$). Moreover, the problem of orbital stability of periodic motions in linear approximation for Bobylev–Steklov case was considered in [13]. Also, pendulum motions of a rigid body carrying rotor were studied in [14].

2 Problem statement

Let us consider the motion of a rigid body of mass m around a fixed-point O in a uniform gravity field. To describe the motion of a body, we introduce a fixed coordinate system $OXYZ$, the OZ axis of which is vertically upward, and a moving coordinate system $Oxyz$, rigidly connected to the body, the axes of which are along the main axes of inertia of the body for point O. Furthermore, it is assumed that for the main moments of inertia A, B, C of the body relative to the axes Ox, Oy, Oz respectively equality $A = C = 4B$ holds.

We do not impose any additional conditions on the position of the center of mass. Let l be the distance from the center of mass of the body to the origin, α—the angle between the position vector of the center of mass and the equatorial plane of the ellipsoid of inertia. We assume without loss of generality that $0 \le \alpha \le \pi/2$. In this case, pendulum motions of the body relative to the axis of inertia lying in the equatorial plane of the ellipsoid of inertia are possible. It is worth noting that for $\alpha = 0$, the classical case of Goryachev–Chaplygin takes place. As we have mentioned, this case was carried out in [6, 15]. The aim of this work is to solve the problem of the orbital stability of the periodic pendulum motion of the body at $0 < \alpha \le \pi/2$.

The position of the rigid body we define by using the Euler angles ψ, θ, φ. The equations of motion of a body can be written in form of canonical equations with the Hamiltonian

$$H = \frac{1}{2}\left(\frac{(p_\theta \cos\varphi - p_\varphi \cot\theta \sin\varphi)^2}{A} + \frac{(p_\theta \sin\varphi + p_\varphi \cot\theta \cos\varphi)^2}{B} + \frac{p_\varphi^2}{C}\right) + mgl \sin\theta \sin(\varphi + \alpha), \tag{1}$$

where p_ψ, p_θ, p_φ are momenta corresponding to the Euler angles. The angle ψ is a cyclic coordinate, that is why $p_\psi = \text{const}$. In what follows, we put $p_\psi = 0$.

The equations of motion have a particular solution describing the motion such that the Oz axis keeps a constant horizontal position, and the body performs pendulum motions relative to this axis. Depending on the initial conditions, pendulum motions can be periodic motions (oscillations or rotations about this axis), or they can be asymptomatic motions approaching an unstable equilibrium position. Since pendulum periodic motions are unstable in the sense of Lyapunov, the problem of the orbital stability of these motions is of interest.

To begin with let us introduce the dimensionless time $\tau = \mu t$, $\mu^2 = mgl/C$. In addition to that, to describe the behavior of a body in the neighborhood of its periodic pendulum motions, it is convenient to carry out the following dimensionless coordinates

$$\varphi = q_1 - \alpha + \frac{3\pi}{2}, \quad \theta = q_2 + \frac{\pi}{2}, \quad p_1 = \frac{p_\varphi}{C\mu}, \quad p_2 = \frac{p_\theta}{C\mu}. \tag{2}$$

In the new variables the Hamiltonian of the problem takes the form

$$H = \frac{1}{2}\Big[(p_2 \sin(q_1 - \alpha) - p_1 \tan q_2 \cos(q_1 - \alpha))^2 + \\ + 4(p_2 \cos(q_1 - \alpha) + p_1 \tan q_2 \sin(q_1 - \alpha))^2 + p_1^2\Big] - \cos q_2 \cos q_1. \tag{3}$$

On the pendulum motions the equality $q_2 = p_2 = 0$ holds and evolution of variables q_1, p_1 is determined by the canonical system with the Hamiltonian

$$H_0 = \frac{1}{2}p_1^2 - \cos q_1. \tag{4}$$

The type of pendulum motions depends on the value of the constant of the energy integral $H_0 = h$: if $|h| < 1$, then pendulum oscillations take place, if $|h| > 1$, then pendulum rotations take place.

3 Local variables and isoenergetic reduction

In order to study the orbital stability of pendulum periodic motions we follow the method described in [7]. In accordance of this method, we introduce local variables in a neighborhood of an unperturbed periodic motion. To this end we perform the following canonical change of variables

$$q_1 = f + \frac{\sin f}{V^2}\eta - \frac{\sin f}{2V^4}\eta^2 + O(\eta^3), \quad p_1 = g + \frac{g}{V^2}\eta - \frac{g\cos f}{2V^4}\eta^2 + O(\eta^3), \tag{5}$$

where $V^2 = g^2 + \sin^2 f$.

Depending on the type of periodic motion, the functions f and g are introduced in different ways: in case of oscillations

$$f = \arcsin[k_1 sn(\xi, k_1)], \quad g = 2k_1 cn(\xi, k_1), \quad k_1^2 = \frac{h+1}{2}, \tag{6}$$

and in case of rotations

$$f = 2am(\xi, k_2), \quad g = 2k_2^{-1} dn(\xi, k_2), \quad k_2^2 = \frac{2}{h+1}, \tag{7}$$

In these cases, the periods are equal to $2\pi/\omega$, where in the case of oscillations $\omega = \pi/2K(k_1)$, in the case of rotations $\omega = \pi/k_2 K(k_2)$. Here we use the generally accepted notation for elliptic functions and integrals.

Let us carry out one more canonical change of variables by the formulas

$$\xi = \frac{1}{\omega}w, \quad \eta = \omega r. \tag{8}$$

The expansion of the Hamilton function in a power series in a neighborhood $q_2 = p_2 = \eta = 0$ has the form

$$\Gamma = \Gamma_2 + \Gamma_4 + \dots \tag{9}$$

where

$$\Gamma_2 = \omega r + \sum_{i+j=2} \varphi_{ij} q_2^i p_2^j, \quad \Gamma_4 = \omega^2 \chi(w) r^2 + \omega r \sum_{i+j=2} \psi_{ij} q_2^i p_2^j + \sum_{i+j=4} \varphi_{ij} q_2^i p_2^j, \tag{10}$$

The coefficients in (10) 2π-periodically depend on w. They can be calculated by means following formulas

$$\chi(w) = \frac{(\cos f - 1)(\sin^2 f - g^2)}{2V^4},$$
$$\psi_{20} = \frac{6g^2\cos(2f-2\alpha) - 3g^2\cos(f-2\alpha) + 3g^2\cos(3f-2\alpha) - 10g^2 - \cos 2f - 1}{4V^2},$$
$$\psi_{11} = \frac{-3g\sin(2f-2\alpha) - 3g\sin(3f-2\alpha) + 3g\sin(f-2\alpha)}{2V^2},$$
$$\psi_{02} = \frac{3\cos(f-2\alpha) - 3\cos(3f-2\alpha)}{4V^2}, \quad \varphi_{20} = \frac{-3g^2\cos(2f-2\alpha) + 2\cos f + 5g^2}{4}, \tag{11}$$
$$\varphi_{11} = \frac{3g\sin(2f-2\alpha)}{2}, \quad \varphi_{02} = \frac{3\cos(2f-2\alpha)+5}{4},$$
$$\varphi_{40} = \frac{-12g^2\cos(2f-2\alpha) - \cos f + 20g^2}{24},$$
$$\varphi_{31} = \frac{g\sin(2f-2\alpha)}{2}, \quad \varphi_{13} = \varphi_{22} = \varphi_{04} = 0,$$

where f, g are 2π-periodic functions of the variable w. Their explicit and analytic form is obtained by substituting (8) into (6) and (7), respectively. Note that the moduli of the elliptic integral, k_1 and k_2, depend on h (see (6) and (7)). Thus, coefficients (11) are 2π-periodic functions of the variable w and analytically depend on parameters h, α.

Let us consider the motion on zero isoenergetic level $\Gamma = 0$. The evolution of the variables q_2, p_2 at the level $\Gamma = 0$ can be described by means of a reduced canonical system (Whittaker equations)

$$\frac{dq_2}{dw} = \frac{\partial K}{\partial p_2}, \quad \frac{dp_2}{dw} = -\frac{\partial K}{\partial q_2}, \tag{12}$$

where w plays the role of a new independent variable. The Hamiltonian K is obtained as a solution to the equation $\Gamma = 0$ with the respect to r. For small r Hamiltonian K can be represented as a power series expansion of q_2, p_2. Taking formulas (9)–(10) into account, we have the following expansion of the Hamilton function K in a power series of q_2, p_2

$$K = K_2 + K_4 + \ldots, \tag{13}$$
$$K_2 = \frac{1}{\omega} \sum_{i+j=2} \varphi_{ij} q_2^i p_2^j, \tag{13.1}$$
$$K_4 = \frac{1}{\omega}\left[\chi(w)\left(\sum_{i+j=2} \varphi_{ij} q_2^i p_2^j\right)^2 + \left(\sum_{i+j=2} \psi_{ij} q_2^i p_2^j\right)\left(\sum_{i+j=2} \varphi_{ij} q_2^i p_2^j\right) + \sum_{i+j=4} \varphi_{ij} q_2^i p_2^j.\right], \tag{13.2}$$

The problem of orbital stability of periodic motions of the rigid body is equivalent to the problem of stability of the trivial equilibrium position of system (12).

4 Linear stability analysis

First, we study the stability problem in linear approximation. To this end we consider linear canonical system with the Hamiltonian K_2

$$\frac{dq_2}{dw} = \varphi_{11}q_2 + 2\varphi_{02}p_2, \quad \frac{dp_2}{dw} = -2\varphi_{20}q_2 - \varphi_{11}p_2. \tag{14}$$

The conclusion about the stability of the linear system (14) can be made on the basis of the analysis of the roots of its characteristic equation

$$\rho^2 - 2\kappa\rho + 1 = 0, \tag{15}$$

where κ is the sum of the diagonal elements of the monodromy matrix of the system (14).

Let us recall that coefficients of the linear system (14) depend on parameters h and α. Hence, the coefficient κ is a function of the parameters h and α. Based on the general theory of stability of A.M. Lyapunov, linear system is stable if all roots of characteristic equation are simple and have modulus equal to one. If there is at least one root with modulus not equal to one, then linear system is unstable. The instability of a linear system implies instability in a complete nonlinear system with Hamiltonian (13). The latter means the orbital instability of the unperturbed periodic motion. If $|\kappa| > 1$, then the characteristic equation has a root with modulus greater than one; therefore, linear system (14) is unstable. If $|\kappa| < 1$, then the trivial equilibrium position of the linear system is stable, i.e., unperturbed periodic motion is orbitally stable in the linear approximation.

The question on orbital stability of pendulum rotations ($h > 1$) was solved on the basis of linear analysis. To this end, by integrating the system (14) from 0 to 2π the coefficient κ was calculated. The calculations were performed for all parameters values chosen from the intervals $1 < h < 100$, $0 < \alpha \leq \pi/2$ with tolerance 0.01. It appeared that in the above-mentioned range of parameter values the inequality $|\kappa| > 1$ holds, hence, the linear system is unstable. It yields orbital instability of pendulum rotations of the rigid body.

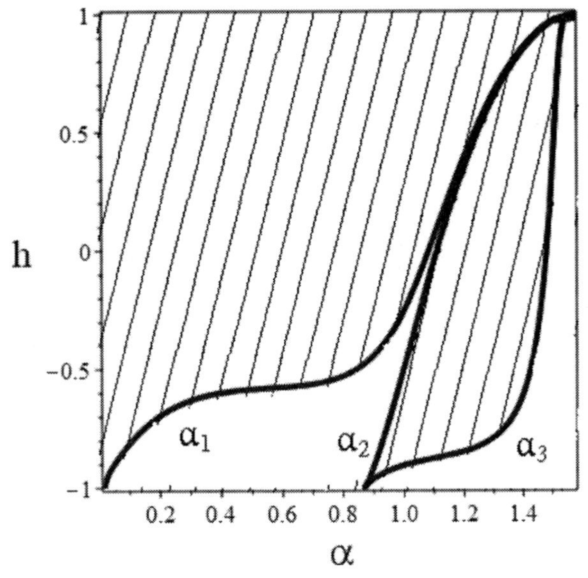

Figure 1. The diagram of orbital stability of pendulum oscillations

In the case of pendulum oscillations ($|h| < 1$) linear analysis showed that either inequality $|\kappa| > 1$ or the inequality $|\kappa| < 1$ holds, that is, depending on parameter values, the oscillations of a rigid body can be either orbital stable or orbital unstable. Based on numerical calculations in the plane of the parameters, a diagram of the orbital stability of pendulum oscillations was obtained (see figure 1). The regions of instability are shaded. In the unshaded areas, the pendulum oscillations are stable in the linear approximation. By α_i ($i = 1, 2, 3$) the boundaries of instability regions are denoted. At $\alpha = 0$ the equation (15) has multiple root $\rho = 1$. In [15] it was shown that in this case the so-called identical resonance takes place. The rigorous study of orbital stability in this case has performed in [6].

5 Investigation of stability for the case of small amplitude oscillations

In this section we suppose that amplitudes of oscillations are sufficiently small. In this case we can introduce a small parameter and perform rigorous analytical study by using small parameter method and the KAM theory. As a small parameter, it is convenient to choose the modulus of the elliptic integral k. At the first stage we study the linear system with the Hamiltonian K_2 whose series expansion in powers of k has the form

$$K_2 = \frac{1}{2}q_2^2 + \frac{1}{2}p_2^2(1 + 3\cos^2\alpha) + (-3\cos w \sin 2\alpha p_2 q_2 + 3p_2^2 \sin w \sin 2\alpha \cos\alpha)k + O(k^2). \quad (16)$$

For $k = 0$, the linear system is autonomous and describes harmonic oscillations with a frequency $\Omega_0 = \sqrt{1 + 3\cos^2\alpha}$. If $\Omega_0 \neq N/2$, $N \in Z$, then for sufficiently small k stability takes place in the linear approximation. If $\Omega_0 \neq N/2$, $N \in Z$, then at $k \ll 1$ the phenomenon of parametric resonance is possible, leading to instability. In our problem, parametric resonance takes place in two cases: $\Omega_0 \approx 2$ ($N = 4$) and $\Omega_0 \approx 3/2$ ($N = 3$). The regions of parametric resonance in the plane of parameters (α, h) emanate at $k = 0$ from the points $\alpha = 0$ and $\alpha = \arccos(5/\sqrt{12})$.

At $k \ll 1$ the boundaries α_i ($i = 1, 2, 3$) of parametric resonance regions can be obtained in a form of convergent power series in small parameter k

$$\alpha_i = \alpha_{i0} + k\alpha_{i1} + \ldots \quad (17)$$

In order to obtain the coefficients of the above series we use the technique developed in [16]. In accordance with this technique, we have to perform a linear canonical change of variables $(q_2, p_2) \to (X, Y)$, which brings the Hamiltonian K_2 into the simplest form (the so-called normal form). In new variables the normalized Hamiltonian K_2 does not depends on independent variable w and takes the following form

$$K_2 = k_{20}X^2 + k_{11}XY + k_{02}Y^2, \quad (18)$$

where k_{02}, k_{20}, k_{11} analytically depend on k and α. To construct the above-mentioned linear change of variables the Depri–Hori method [6] can be used. It also allows to calculate the expressions for coefficients k_{02}, k_{20}, k_{11} in a form of series in powers k.

Simple analysis of the canonical system with the Hamiltonian (16) shows that it is stable if $k_{11}^2 < 4k_{20}k_{02}$, otherwise if $k_{11}^2 > 4k_{20}k_{02}$ it is unstable. Thus, the boundaries of the parametric resonance region are determined by following equation

$$k_{11}^2 = 4k_{20}k_{02}. \quad (19)$$

Now we recall that k_{02}, k_{20}, k_{11} depend on k and α. By substituting (17) into (18) and equating the terms of at equal power of k in right- and left-hand sides of (18) we will get equations for coefficients of series (17).

Our calculations based on this approach have shown that in the case of resonance $\Omega_0 \approx 2$, the boundary of the parametric resonance region is determined by the following asymptotic formula

$$\alpha_1 = \frac{\sqrt{3}}{2}k_1^2 + O(k_1^3), \tag{20}$$

In the case of resonance $\Omega_0 \approx 3/2$ we have

$$\alpha_2 = \arccos\sqrt{\frac{5}{12}+\frac{27\sqrt{35}}{160}k_1^2+\frac{315}{128}k_1^3+O(k_1^4)},\ \ \alpha_3 = \arccos\sqrt{\frac{5}{12}+\frac{27\sqrt{35}}{160}k_1^2-\frac{315}{128}k_1^3+O(k_1^4)}, \tag{21}$$

Outside the regions of parametric resonance, the linear system is stable. However, it does not mean the stability of the original nonlinear system. Hence, to obtain rigorous conclusions about orbital stability for parameter values outside the regions of parametric resonance a nonlinear analysis is required.

We start nonlinear study of stability from the simplification of the quadratic part of the Hamilton function. Outside the regions of parametric resonance, we can construct a canonical change of variables

$$q_2 = a_{11}(w)x + a_{12}y, \quad p_2 = a_{21}(w)x + a_{22}(w)y, \tag{22}$$

to bring the Hamiltonian (16) into the following normal form, which does not depend on w

$$K_2 = \frac{1}{2}\Omega(x^2 + y^2), \tag{23}$$

The coefficients $a_{11}(w)$, $a_{12}(w)$, $a_{21}(w)$, $a_{22}(w)$ are 2π-periodic functions of the variable w and are convergent series in the small parameter k. Ω is an analytic function of k and α. As above we used the Depri–Hori method to obtain the power series expansion of Ω with respect to k. Our calculations have shown that $\Omega = \Omega_0 + \Omega_1 k^2 + O(k^3)$, where

$$\Omega_1 = \frac{(144\Omega_0^2 + 288)\cos^4\alpha + (-144\Omega_0^2 - 288)\cos^2\alpha - 16\Omega_0^6 + 48\Omega_0^4 + 69\Omega_0^2 - 20}{16\Omega_0^3 - 4\Omega_0}. \tag{24}$$

Now it is convenient to pass to the canonical polar coordinates by the formulas

$$x = \sqrt{2\rho}\sin\vartheta, \quad y = \sqrt{2\rho}\cos\vartheta, \tag{25}$$

The Hamiltonian K reads

$$K = \Omega\rho + G_4(\vartheta, w)\rho^2 + O(\rho^3), \tag{26}$$

The expression for $G_4(\vartheta, w)$ is omitted here due to the cumbersomeness. Rigorous conclusions on stability of the trivial solution of the system with Hamiltonian (26) can be obtained on basis of the KAM theory. For this purpose, let us perform further simplification of the Hamiltonian $K$and reduce it to a normal form up to terms of the order R^2.

If there are no third and fourth order resonances in the system, that is, $\Omega \neq n/3$, $n \in Z$ and $\Omega \neq n/4$, $n \in Z$, respectively, then the Hamilton function (26) can be reduced to a simpler form using a close to identity, analytic in k, canonical change of variables $(\vartheta, \rho) \to (\varphi, R)$

$$\Phi = \Omega R + c_2R^2 + O(R^3), \quad c_2 = \frac{1}{4\pi}\int_0^{2\pi}\int_0^{2\pi} G_4(\vartheta, w)\, dw\, d\vartheta. \tag{27}$$

Calculations have shown that

$$c_2 = -\frac{27\sin^2\alpha(2\cos^2\alpha + 1)(\cos^2\alpha + 1)}{8(4\cos^2\alpha + 1)(3\cos^2\alpha + 1)} + O(k_1). \tag{28}$$

The equation $c_2 = 0$ has no solutions in the interval $0 < \alpha \leq \pi/2$. Hence $c_2 \neq 0$, then by the Arnold–Moser theorem [17], the equilibrium point of system (12) is Lyapunov stable. It yields the orbital stability of pendulum oscillations with small amplitudes in non-resonant case.

To complete the study of orbital stability of the pendulum oscillations with small amplitudes we have to consider the cases of the third- and fourth-order resonances. Since the original Hamilton function does not contain third-order terms with respect to variables q_2, p_2, third-order resonances do not appear in the system. The fourth-order resonances are possible for $\Omega = n/4$, $n \in Z$. In our problem they occur only for two values of n: $n = 5$, $n = 7$. The resonance values of the parameters corresponding to the resonances $\Omega = 5/4$ and $\Omega = 7/4$ are respectively determined by such relations

$$\alpha = \arccos \frac{\sqrt{3}}{4} + \frac{3\sqrt{39}}{7}k_1^2 + O(k_1^3), \quad \alpha = \arccos \frac{\sqrt{11}}{4} + \frac{3\sqrt{55}}{55}k_1^2 + O(k_1^3). \tag{29}$$

In the case of fourth-order resonances the Hamiltonian (26) can be reduced to the form

$$\Phi = \Omega R + (c_2 + a_4 \cos(4\vartheta - nw) + b_4 \sin(4\vartheta - nw))\, R^2 + O(R^3), \tag{30}$$

where

$$a_4 = \frac{1}{2\pi^2}\int_0^{2\pi}\int_0^{2\pi} G_4(\vartheta, w)\cos(4\vartheta - nw)\, dw\, d\vartheta,$$

$$b_4 = \frac{1}{2\pi^2}\int_0^{2\pi}\int_0^{2\pi} G_4(\vartheta, w)\sin(4\vartheta - nw)\, dw\, d\vartheta.$$

Our calculations have shown that in the case of $\Omega = 7/4$ the following coefficients are

$$c_2 = -\frac{4617}{3136}k_1 + O(k_1^3) \quad a_4 = O(k_1^3), \quad b_4 = O(k_1^3). \tag{31}$$

In the case of $\Omega = 5/4$, the calculated coefficients are

$$c_2 = -\frac{73359}{11200}k_1 + O(k_1^3), \quad a_4 = O(k_1^3), \quad b_4 = O(k_1^3). \tag{32}$$

In accordance with Markeev's theorem [17] the stability condition has a form

$$|c_2| > \sqrt{a_4^2 + b_4^2}. \tag{33}$$

It can be seen that in both cases, this condition is fulfilled, that is, there is Lyapunov stability for system (12), which implies the conclusion about the orbital stability of the original nonlinear system. Therefore, outside the regions of parametric resonance, the oscillations are orbitally stable.

6 Conclusion

In this paper, a complete study of the orbital stability of pendulum periodic motions of a heavy rigid body with a fixed point was carried out under the assumption when the main moments of inertia of the body satisfy the equality $A = C = 4B$ (Goryachev–Chaplygin condition). On the basis of linear analysis, it was proved that pendulum rotations are orbitally unstable. Pendulum oscillations can be either orbitally stable or orbitally unstable depending on parameters values. The result of orbital stability study in linear approximation have been presented in stability diagram for $0 < \alpha \leq \pi/2$. The domains of parametric resonances leading to instability were shown. Outside these domains the pendulum oscillations are orbitally stable in linear approximation. This result is in a good agreement with the results of [15] where the case $\alpha = 0$ was considered. At small amplitudes of oscillations, the study was carried out analytically and rigorous conclusions on orbital stability have been obtained. In particular, it was shown that outside the regions of parametric resonance, the oscillations are orbitally stable.

This work was financially supported by the Russian Foundation for Basic Research, project No. 20-01-00637.

References

[1] A.P. Markeev, *The stability of the plane motions of a rigid body in the Kovalevskaya case* J. Appl. Math. Mech., **65(1)**, 51 (2001)

[2] A.P. Markeev, S.V. Medvedev, T.N. Chekhovskaya, *To the problem of stability of pendulum-like vibrations of a rigid body in Kovalevskaya's case*, Mech. Solids, **38(1)**, 1 (2003)

[3] V.D. Irtegov, *The Stability of the Pendulum-like Oscillations of a Kovalevskaya Gyroscope*, Trudy Kazan. Aviats. Inst. Matematika i Mekhanika, **97**, 38 (1968)

[4] A.Z. Bryum, *A Study of Orbital Stability by Means of First Integrals*, J. Appl. Math. Mech., **53(6)**, 689 (1989)

[5] A.Z. Bryum, A.Ya. Savchenko, *On the orbital stability of a periodic solution of the equations of motion of a Kovalevskaya gyroscope*, J. Appl. Math. Mech., **50(6)**, 748 (1986)

[6] B.S. Bardin, *Stability problem for pendulum-type motions of a rigid body in the Goryachev–Chaplygin case*, Mech. Solids, **42(2)**, 177 (2007)

[7] B.S. Bardin, *On a Method of Introducing Local Coordinates in the Problem of the Orbital Stability of Planar Periodic Motions of a Rigid Body*, Rus. J. Nonlin. Dyn., **16(4)**, 581 (2020)

[8] A.P. Markeev, *The pendulum-like motions of a rigid body in the Goryachev–Chaplygin case*, J. Appl. Math. Mech., **68(2)**, 249 (2004)

[9] B.S. Bardin, T.V. Rudenko, A.A Savin, *On the Orbital Stability of Planar Periodic Motions of a Rigid Body in the Bobylev–Steklov Case*, Regul. Chaotic Dyn., **17(6)**, 533 (2012)

[10] B.S. Bardin, *Local coordinates in problem of the orbital stability of pendulum-like oscillations of a heavy rigid body in the Bobylev–Steklov case*, J. Phys.: Conf. Ser., **1925**, 012016 (2021)

[11] B.S. Bardin, A.A. Savin, *On the Orbital Stability of Pendulum-like Oscillations and Rotations of a Symmetric Rigid Body with a Fixed Point*, Regul. Chaotic Dyn., **17(3–4)**, 243 (2012)

[12] B.S. Bardin, A.A. Savin, *The stability of the plane periodic motions of a symmetrical rigid body with a fixed point*, J. Appl. Math. Mech., **77(6)**, 806 (2014)

[13] H.M. Yehia, S.Z. Hassan, M.E. Shaheen, *On the orbital stability of the motion of a rigid body in the case of Bobylev–Steklov*, Nonlinear Dyn., **80(3)**, 1173 (2015)

[14] H.M. Yehia, E.G. El-Hadidy, *On the Orbital Stability of Pendulum-like Vibrations of a Rigid Body Carrying a Rotor*, Regul. Chaotic Dyn., **18(5)**, 539 (2013)

[15] A.P. Markeev, *On the identity resonance in a particular case of the problem of stability of periodic motions of a rigid body*, Mech. Solids, **38(3)**, 22 (2013)

[16] A.P. Markeev, *Linear Hamiltonian Systems and some problems of stability of the satellite center of mass* (Research Center "Regular and Chaotic Dynamics", Institute of Computer Research, Izhevsk, 2009)

[17] A.P. Markeev, *Libration Points in Celestial Mechanics and Space Dynamics* (Nauka, Moscow, 1978)

Mechanism of heterogeneous flow—solid substrate interaction on the formation of coatings of different thicknesses using different types of spray accelerators

Sergey Borisov[1], *Julia* Gloukhovskaya[1], *Sergey* Dobrovolskiy[1,*], *Alexander* Myakochin[1], and *Igor* Podporin[1]

[1]Moscow Aviation Institute (National Research University), 4, Volokolamskoe shosse, Moscow, 125993, Russia

Abstract. Physical processes of the inleakage of a supersonic heterogeneous flow of air-powder mixture to a solid substrate, tested on the experimental setup designed for applying protective coatings by a cold spray method, are considered. The interaction of a particle with a flat plane and cylindrical solid substrates, the multilayer coatings and the application of coatings using spray accelerators of different configurations are considered. A mathematical model is demonstrated based on the equation of energy balance in the impact area intended to estimate the interaction between a particle and a solid substrate during impact, as well as examples of results calculated using this model. Methods for calculation of the thickness of the coating applied to flat plane and cylindrical substrates are described. The features of one-step and multi-step coating applications with and without additional exposures are described. As an example, the results of testing the coating for porosity are given. A list of factors and additional exposures affecting the strength of the coating is given.

1 Introduction

The cold spray method of applying coatings is a process in which powder particles are introduced into a high-speed heated gas jet creating a heterogeneous flow, and then accelerating to a supersonic speed and collide with a substrate surface, which leads to the forming a coating on that surface [1, 2]. The coating is formed in the process of severe plastic deformation of particles during the impact on a surface.

Before the impact, particles have a temperature well below the melting point of the powder material. When a cold (not molten) particle hits the surface, either the erosion of a substrate material or the adhesion of a particle into the substrate occurs, i.e. the formation of a coating occurs, which depends on the concentration and flow rate of particles in the air-powder mixture, the particle velocity and the flow temperature. In one case, bonds are formed between a surface of the particle and the substrate, i.e. the adhesion occurs. In another case, elastic deformation energy accumulates in the particle volume, which is released in the form of the rebound kinetic energy. When the elastic deformation energy becomes greater than the total energy of adhesion, the particle bounces off the substrate. Otherwise, the particle is adhere to a substrate surface [3].

*e-mail: dobrovolskiy_s@mail.ru

After forming one layer of particles on a substrate surface, the materials of the inleakage particles and the formed coating come into the contact, i.e. a mutual adhesion of the powder particles occurs. Thus, a coating of the desired thickness is created. The surface profile of a single coating layer is determined by the material properties, the distance from the accelerator nozzle exit to a substrate, the angle between heterogeneous flow jet and a substrate surface, a substrate properties, a substrate deformation, a local heat transfer, etc. If to sum a large number of surfaces profiles of individual coating layers, providing a given thickness, the distribution of the total coating surface profile can be considered approximately as a continuous curve [4].

When a movable supersonic spray accelerator with a circular cross-section of small area is used for the cold spray coatings, the coating is created in one pass of accelerator along a substrate (scan step). Scan step of such accelerator is usually quite narrow compared to other types of accelerators, and the gap between two sequential scan steps on the same coating layer is usually small in order to ensure the uniform thickness of a coating on the layer. In some publications, for example, in [4], the angle between a heterogeneous flow jet and a substrate surface has been analyzed, and how it influences on the coating characteristics. It was determined that the relative efficiency of the coating formation is maximum at spraying angles from 80 to 90°. When the angle decreases approximately to 40°, almost no deposited particles adhere to a substrate, and the relative efficiency of the coating formation tends to zero.

In our experimental setup [5], designed for the application of protective coatings on pipes with a diameter of 150 and 300 mm, spray accelerators with the circular slotted nozzles are used. They apply the air-powder mixture over the entire circumference of the pipe cross-section at once. There is no need to optimize a path of accelerator in this case, and the uniformity and strength of the coating are influenced mainly by accelerator displacement velocity, mass flow rate of the powder, and thermal parameters of processes.

At present, the cold spray method of coatings is being intensively studied and developed in the world of science and technology and adopts a high profile among protective coatings technologies, hardening technologies, and additive technologies.

2 The equation of energy balance for the particle impact on a substrate

To simulate physical processes in the area of particle impact on a substrate, we can use the equation of energy balance, which is written in symbolic form [3]:

$$E_{kin.\,p} = E_{def.\,p} + E_{def.\,wall} + E_{heat.\,p} + E_{heat.\,wall}. \quad (1)$$

The equation (1) takes into account only the main stationary physical factors affecting the energy balance. In the equation (1), for example, there no terms for the heating energy of particles in the carrier-gas flow and the energy of substrate heating by the gas flow. These energies are implicitly taken into account by parameters characterizing the mechanical properties of the particle and substrate materials.

The kinetic energy of the particle during impact is defined by particle mass m_p and particle velocity V_p and is represented in the form:

$$E_{kin.\,p} = \frac{m_p \cdot V_p^2}{2}. \quad (2)$$

Upon impact, the particle penetrates the substrate to a small depth, and it is assumed that the deformation of the particle is a compression and free spreading over the surface [6].

Particle deformation energy is spent on transforming the particle shape from spherical to approximately ellipsoidal (spheroidal) during impact on the substrate:

$$E_{def.p} = \pi d_p \sigma_T \left[2a^2 + \frac{b^2}{\alpha} \ln \frac{1+\alpha}{1-\alpha} - d_p^2 \right] (1-\varepsilon), \quad (3)$$

where d_p is the particle diameter, σ_T is the limit of the particle material yield, $\varepsilon = 2b/d_p$ is the degree of particle deformation during the impact, a and b are the large and small semi-axes of the deformed particle spheroid (figure 1), and $\alpha = \sqrt{1-(b/a)^2}$.

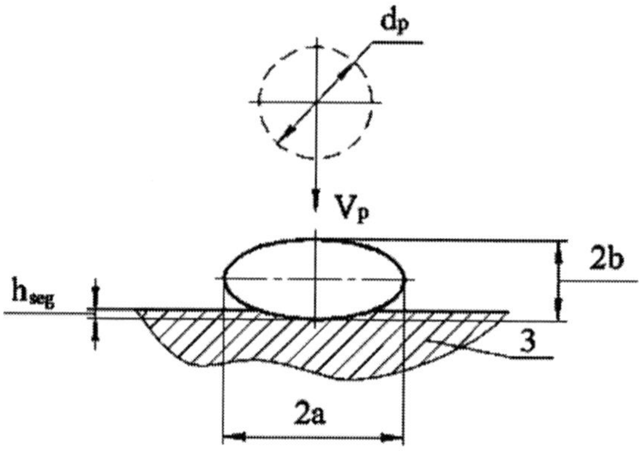

Figure 1. Deformation of the metal particle upon impact with the substrate

Substrate deformation energy is spent on the plastic compression of a substrate material. This energy is determined by the depth of particle penetration into the substrate bulk and by the compression resistance force of the substrate material. To estimate the depth of spherical particle penetration into the substrate material, a geometric model of the particle interaction with a substrate is used (figure 1), and the final expression has the following form:

$$E_{def.wall} = 2\pi h_{seg}^2 d_p H_\beta, \quad (4)$$

where H_β is the Brinell hardness of the substrate material.

The expression for the energy spent on particle heating has the form:

$$E_{heat.p} = m_{\tau,p} c_p (\bar{T}_{\tau,p} - T_{st.p}), \quad (5)$$

where $m_{\tau,p}$ is the fraction of the particle mass heated during the impact time period τ; $\bar{T}_{\tau,p}$ is the weight-averaged temperature of the heated particle segment; c_p is the specific heat of the particle material at a temperature of $\bar{T}_{\tau,p}$; $T_{st.p}$ is the average particle temperature before the impact.

The expression for the energy spent on substrate heating has the form:

$$E_{heat.wall} = m_{\tau,wall} c_{wall} (\bar{T}_{\tau,wall} - T_{st.wall}), \quad (6)$$

where $m_{\tau,wall}$ is the segment of the substrate mass heated during the impact time period τ; $\bar{T}_{\tau,wall}$ is the weight-averaged temperature of the heated substrate segment; c_{wall} is the specific heat of the substrate material at the temperature $\bar{T}_{\tau,wall}$; $T_{st.wall}$ is the substrate temperature before the impact.

Table 1. Results for the interaction of an Al particles and a Cu substrate

Particle diameter d_p, [m]	Particle velocity V_p, [m/sec]	Degree of deformation ε	Particle mechanical properties σ_T, [Pa]	Impact time τ, [sec]	Particle specific heat c_p, [J/kg K]	Substrate specific heat c_{wall}, [J/kg K]	Brinell hardness of the substrate H_β, [MPa]	Change in the temperature in the impact area $\bar{T}_\tau - T_{st}$, [°C]
$5\cdot10^{-5}$	410	0.701	$6.37\cdot10^{-7}$	$1.49\cdot10^{-7}$	929	400	400	230
$5\cdot10^{-5}$	508	0.598	$6.37\cdot10^{-7}$	$1.27\cdot10^{-7}$	929	400	400	323
$5\cdot10^{-5}$	615	0.503	$6.37\cdot10^{-7}$	$1.10\cdot10^{-7}$	929	400	400	470
$5\cdot10^{-5}$	1000	0.600	$6.37\cdot10^{-7}$	$1.00\cdot10^{-7}$	929	400	400	1770

By substituting the dependences (2)–(6) into equation (1), we obtain the final form of the energy equation:

$$\frac{m_p \cdot V_p^2}{2} = \pi d_p \sigma_T \left[2a^2 + \frac{b^2}{\alpha} \ln \frac{1+\alpha}{1-\alpha} - d_p^2\right](1-\varepsilon) + 2\pi h_{seg}^2 d_p H_\beta +$$
$$+ m_{\tau,p} c_p (\bar{T}_{\tau,p} - T_{st.p}) m_{\tau,wall} c_{wall} (\bar{T}_{\tau,wall} - T_{st.wall}). \quad (7)$$

Equation (7) lets calculate weight-averaged temperatures of the heated particle segment and also the average temperature of the heated substrate segment in the contact area during impact. In this case, the following must be known: the particle diameter, the physical and mechanical properties of powder and substrate materials, the velocities of particles at the instant of impact with the substrate, and the degree of their deformation on impact. For example, table 1 shows results for the interaction of an Al particles and a Cu substrate [3].

3 Calculation of the coating thickness in case of movable spray accelerator with a circular cross-section of small area

For such a case, the kinematic scheme of the coating process is shown in figure 2. When applying the coating (3), its quality essentially depends on the following kinematic parameters of the accelerator displacement (2) relative to the substrate (1): spray distance L, spray angle α, accelerator velocity V relative to the substrate, scan step S, and the number of passes. The influence of these parameters is studied in detail in a number of publications, for example, in [7, 8].

For that kinematic scheme, a coating band in one scan step is quite narrow, and the gap between two sequential scan steps is small in order to ensure the uniform thickness of the coating.

It was shown in [4] that the Gaussian distribution curve can be used to calculate the height of a coating surface profile on a flat plane substrate (figure 3).

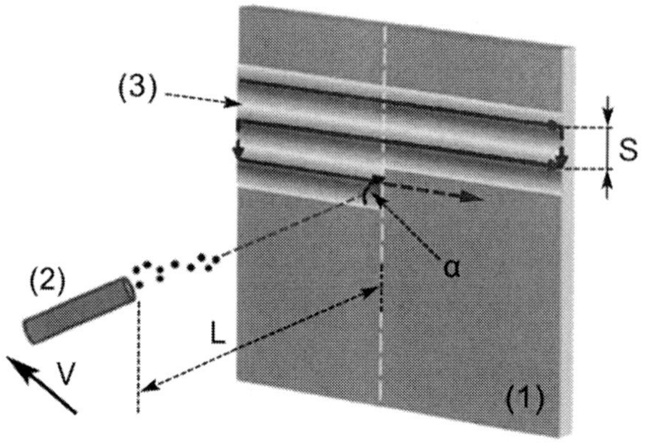

Figure 2. Kinematic scheme of the coating process

When spray parameters such as accelerator displacement velocity and spray distance are changed, the Gaussian curve will also change, but this does not affect the estimation of the coating thickness using the following equation:

$$y = \frac{K}{\sigma\sqrt{2\pi}} \cdot e^{-\frac{x^2}{2\sigma^2}}, \tag{8}$$

where coefficient K and standard deviation σ can be fit based on experimental results to determine desired shape of the Gaussian curve. Coefficient K is a crest factor that affects the thickness of individual coating profile.

To calculate the coating height on the cylindrical substrate, a polar coordinate system is used due to a specific shape of the coating surface profile, and the Gaussian curve only cannot be used in this case [9].

The coating surface profile (2) on the cylindrical substrate (1) with a radius R is shown in figure 4.

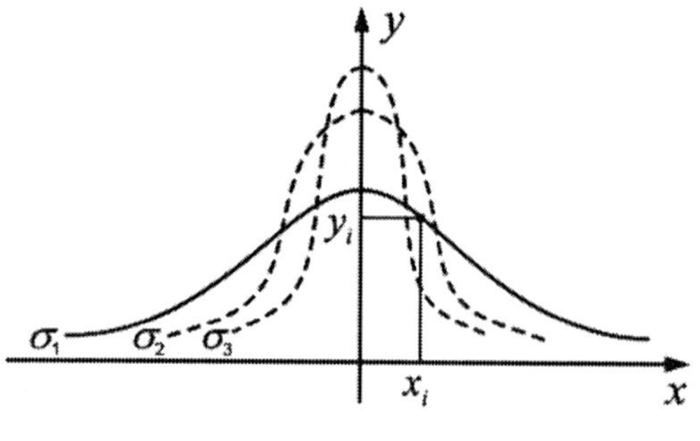

Figure 3. Coating surface profiles on the flat plane substrate

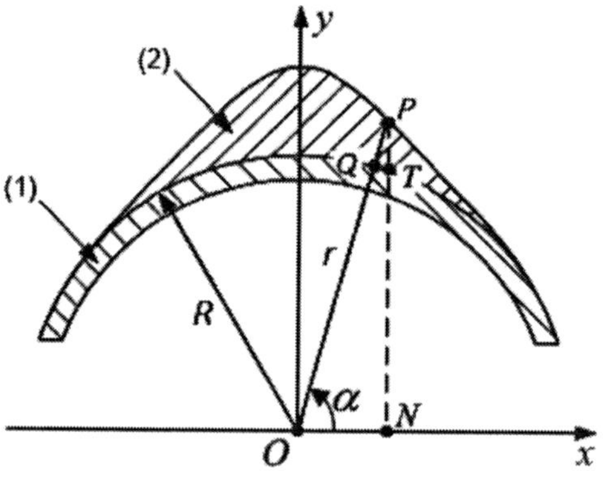

Figure 4. Coating surface profile on the cylindrical substrate

Polar coordinates r and α of the point P laying on the coating profile can be transformed to rectangular coordinates: $x_p = r \cdot \cos\alpha$, $y_p = r \cdot \sin\alpha$. Next, we need to determine the coating thickness $|PQ|$ at point P. In accordance with figure 5 and the equation (8), the following can be written:

$$r \cdot \sin\alpha = |PT| + |TN|, \tag{9}$$

$$|PT| = \frac{K}{\sigma\sqrt{2\pi}} \cdot e^{-\frac{(r\cdot\cos\alpha)^2}{2\sigma^2}}, \tag{10}$$

$$|TN| = \sqrt{R^2 - (r\cdot\cos\alpha)^2}. \tag{11}$$

Substituting (10) and (11) in (9) gives the following:

$$r \cdot \sin\alpha = \frac{K}{\sigma\sqrt{2\pi}} \cdot e^{-\frac{(r\cdot\cos\alpha)^2}{2\sigma^2}} + \sqrt{R^2 - (r\cdot\cos\alpha)^2}. \tag{12}$$

Having solved equation (12) for r, we can determine the coating thickness on the substrate surface at point Q:

$$|PQ| = r - R.$$

Finally, summing the number of coatings profiles for all scan steps S, we can consider the resulting profile with the averaged height H as a continuous curve for definition of the resulting coating surface (figure 5).

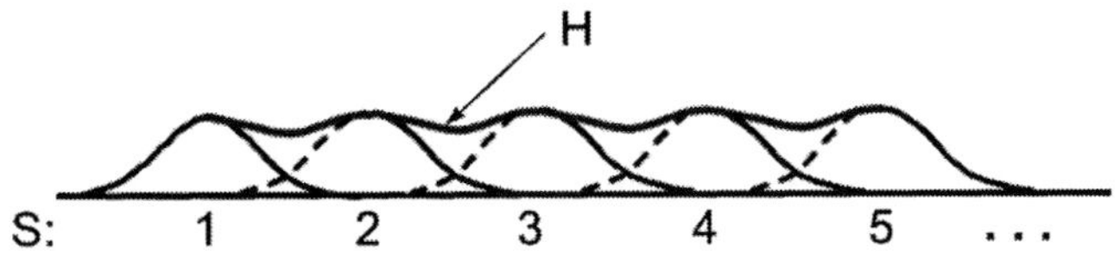

Figure 5. Resulting coating surface profile

4 Applying a coating layer to the entire cylindrical surface at once

To apply coatings on cylindrical surfaces, we have created the experimental setup with special prefabricated accelerators of heterogeneous flow, which make it possible to apply a continuous coating layer on the inner and outer sides of pipes of diameter of 150 and 300 mm [5]. The principle of operation of accelerators is shown in figures 6 and 7. Dashed lines with arrows on the figures indicate a heterogeneous flow. The flow enters the accelerator and evenly distributes within its cavities, and then leaves the circular slotted nozzle of accelerator. Thus, a continuous coating layer is applied to the entire cylindrical surface at once.

One of the key characteristics of a thick coating, including ones applied in a continuous layer, are its strength and porosity [10]. In order to increase the strength of a coating with a thickness of more than 2.5 mm, special techniques are used. For example, using step-by-step and layer-by-layer spray with additional exposure intended to recover residual stresses in the coating, it is possible to create coatings that are significantly thicker than obtained in a one-step process. After applying a coating layer with a thickness of 0.15–0.2 mm, the powder supply stops. Further, the created coating layer is blown with heated air until the recovery temperature of the coating material structure (0.2–0.3 of melting point) is reached, and then the powder begins to feed and the next coating layer is applied. The cycle is repeated until the specified coating thickness is obtained. For example, table 2 shows the porosity test results for a stainless steel coating applied to a flat plane substrate of a Steel 20.

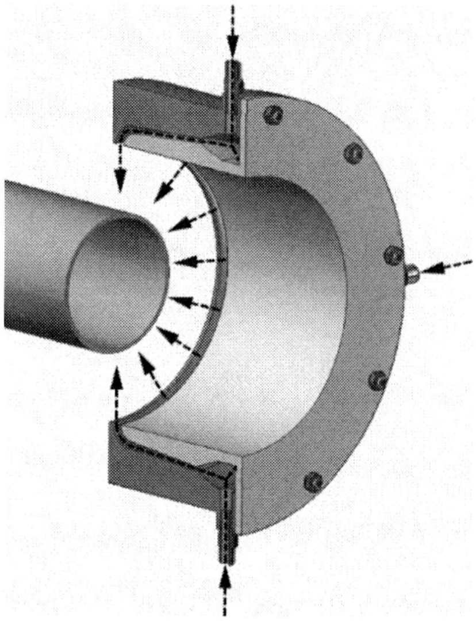

Figure 6. Circular slotted accelerator for outside coating (a cross section shown)

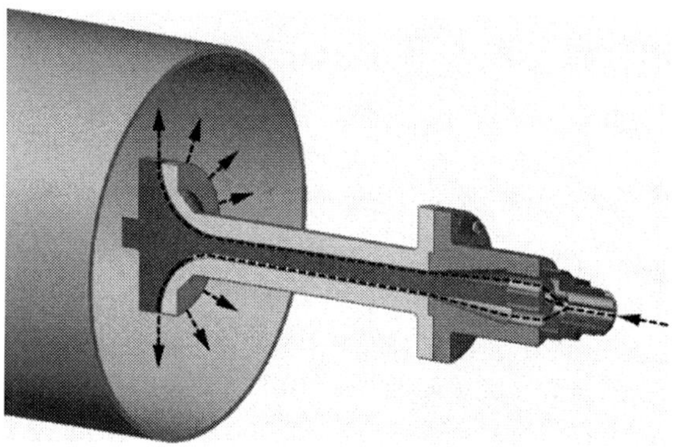

Figure 7. Circular slotted accelerator for inside coating (a cross section shown)

Table 2. Porosity test results for a stainless steel coating

Parameter	Measured values	
	Coating without additional exposure	Coating with additional exposure
Number of measurements	160	160
Total pore area	28,057 μm^2	27,839 μm^2
Pore size	172 μm^2	167 μm^2
Porosity percentage	3.21%	3.09%

5 Conclusion

Using considered calculation methods, it is possible to estimate the energy balance in the area of particle impact on a substrate and the coating thickness for the given process parameters and material properties. Regarding the strength of thick coatings, well-established engineering estimation methods have not been developed yet, despite the known basic physical processes that cause residual stresses in cold spray coatings. This can be explained not only by the complexity of processes, but also by the thermal, physical and chemical phenomena taken into account.

Experiments performed on the experimental setup let to highlight the following factors and additional exposures, which make it possible to increase the coating strength:

- matching the material properties of a coating and a substrate and, first of all, their thermal-expansion coefficients;
- regulation of the thermal effect of the gas-dynamic flow on the particles and substrate by changing the distribution of its thermal power over the heating spot as well as adjusting the distance from the edge of accelerator nozzle to a substrate or changing the accelerator displacement velocity;

- decreasing the elastic modulus of the coating material, for example, by introduction of additives of plastic material into the heterogeneous flow;
- using intermediate layers between a substrate and a coating, that provide a smooth transition of properties from a substrate material to a coating material;
- changing the coating thickness as well as applying multilayer coatings with alternate layers of different materials.

References

[1] A. Papyrin, V. Kosarev, S. Klinkov, A. Alkhimov, V. Fomin, *Cold spray technology*, 1st ed. (Elsivier, Amsterdam, 2006)

[2] A.P. Alkhimov, S.V. Klinkov, V.F. Kosarev, V.M. Fomin, *Cold spray. Theory and practice* (Fizmatlit, Moscow, 2010)

[3] S.V. Dobrovolskiy, P.V. Nikitin, S.A. Borisov, Yu.I. Glukhovskaya, *Energy exchange in the area of particle—substrate interaction on inleakage of a supersonic heterogeneous flow to a solid substrate*, in Proceedings of the XI International Conference on Non-Equilibrium Processes in Nozzles and Jets, NPNJ'2016, 26 June—1 July 2006, Saint-Petersburg, Russia (2016)

[4] Z. Cai, S. Deng, H. Liao, C. Zeng, G. Montavon, *The effect of spray distance and scanning step on the coating thickness uniformity in cold spray process*, J. Therm. Spray Technol., **23(3)**, 362 (2014)

[5] S.A. Borisov, Yu.I. Glukhovskaya, S.V. Dobrovolskiy, P.V. Nikitin, I.V. Podporin, *Concept of an experimental setup for testing the technology for the formation of new anti-corrosion coating materials using low-temperature supersonic heterogeneous flows*, TEM J., **9(2)**, 566 (2020)

[6] R.V. Alyakretskiy, D.V. Ravodina, T.V. Trushkina, E.V., Vakhteev, E.G. Alekseeva, *Development of anti-erosion coatings for elements of spacecraft power systems*, Electronic journal "Trudy MAI", **74** (2014)

[7] C.-J. Li, W.-Y. Li, Y.-Y. Wang, H. Fukanuma, *Effect of Spray Angle on Deposition Characteristics in Cold Spraying*, in Thermal Spray 2003: Advancing the Science and Applying the Technology, 5–8 May, 2003, Orlando, FL (2003)

[8] F.I. Trifa, G. Montavon, C. Coddet, *On the relationships between the geometric processing parameters of aps and the Al_2O_3–TiO_2 deposit shapes*, Surf. Coat. Technol., **195**, 54 (2005)

[9] Zhang Yanjun, Wenbo Li, Li Dayu, Xiao Jinkun, Chao Zhang, *Modeling of thickness and profile uniformity of thermally sprayed coatings deposited on cylinders*, J. Therm. Spray Technol., **27(3)**, 288 (2018)

[10] Yu.S. Korobov, *Analysis of Properties of Thermal Coatings. Part 2. Evaluation of Coating Parameters* (Publishing House of Ural University, Ekaterinburg, 2016)

Mechanism of the powder material particle in different phase states—solid substrate interaction

Sergey Borisov[1], *Julia* Gloukhovskaya[1], *Sergey* Dobrovolskiy[1,*], *Alexander* Myakochin[1], and *Igor* Podporin[1]

[1]Moscow Aviation Institute (National Research University), 4, Volokolamskoe shosse, Moscow, 125993, Russia

Abstract. The paper discusses impact of solid and molten particles of a powder material in heterogeneous flow acting on solid surface and its effect on characteristics of coating applied by a gas-dynamic (cold spray) method. An equation of energy balance in impact zone of the particle on the substrate is given. Obtained equation accounts the particle size, mass average temperatures in heated portions of the particle and the substrate, before impact temperatures of the particle and the substrate, fraction of the heated mass of the particle and the substrate during impact, specific heat capacities of the particle and the substrate material, the Brinell characteristic, and deformation of a solid particle. Results of the bibliographic study of the molten metal droplet formation, their kicks in the substrate depending on its velocity, substrate temperature and surrounding gas pressure are presented. The differences in the impact mechanism of solid and molten particles on the substrate and their influence on the coating formation and its properties are determined. Recommendations for operating parameters of high-quality coating are given.

1 Introduction

The protective coating application by a cold spray method is widely used in many industries (aerospace, chemical, fuel, military, mining, medicine, electronics and mechanical engineering) [1, 2].

The quality of applied coating is generally determined by conditions of interaction of the powder material particles with the substrate. These conditions include the speed of the particles as well as their size, temperature and phase state. In addition, pretreatment of the contact surface is of great importance [3–10]. Therefore, the study of fundamental physical processes occurring during the interaction of the particle with the substrate is of great practical importance.

Each particle attached to the substrate surface is a part of the coating. Theoretical and experimental studies of recent years are devoted to the study of the characteristics of such parts and their interaction with each other and the substrate.

Interaction of the particle with the substrate surface upon impact can be considered as several simultaneously occurring processes due to separate or joint influence of the components of particles flux and properties of the surface.

*e-mail: dobrovolskiy_s@mail.ru

Experimental study of such processes are carried out on gas-dynamic installations generating high-speed heterogeneous flows [1–4].

The kinetic energy changes depending on the type of impact (elastic or plastic), and limiting values correspond to absolutely elastic and absolutely plastic impact. The type of impact is determined by the particle velocity, which directly depends on the velocity of the carrier gas.

According to the physics laws, there is no energy transformation from one type to another during the elastic impact of the particle on the substrate. The particle rebounds from the substrate with a velocity equal to the rebound velocity. In this case, the coating is not applied.

Let's call the particle velocity at which the elastic impact occurs, the first critical one. Then we can assume the existence of a second critical velocity, which is the particle velocity at which the solid surface begins to collapse (due to erosion) when particles bombard the surface.

Indicated critical velocities can be calculated using continuum mechanics laws, which make it possible to determine the motion parameters of the high-speed (supersonic) heterogeneous flow [2].

When applying coatings by a cold spray method, it is of practical interest to establish the range of particle velocities at which the coatings are formed. This range should be the gap between the first and second critical velocities, i.e. between the velocity at which the elastic impact of the particle against the surface occurs, and the speed at which the material of the bombarded surface begins to collapse (undergo erosion) [2–5].

Limiting values of these velocities depend on properties of the particles and the surface. The particle velocity at the moment of impact depends on characteristics of the heterogeneous flow at the exit from the gas-dynamic accelerator. It also depends on the flow nature on the final interval of the particle flight when it overcomes the shock wave and compressed layer, as well as on the mechanical properties of the particle and substrate materials.

Factors influencing the interaction of the particle with the surface are the particle shape, the particle orientation at the contact with the surface, the particle velocity, the particle incidence angle, and properties of the particle and surface materials [3–7].

In context of a cold spray technology application for the coating deposition, we consider the impingement of a supersonic heterogeneous jet of finite dimensions onto an obstacle (figure 1).

Depending on properties of the particle material and characteristics and temperature of heterogeneous flow, the particle can have several phase states. We will consider a cold (not molten) and partially molten state.

Figure 1. Leakage of a supersonic heterogeneous jet of finite dimensions onto an obstacle

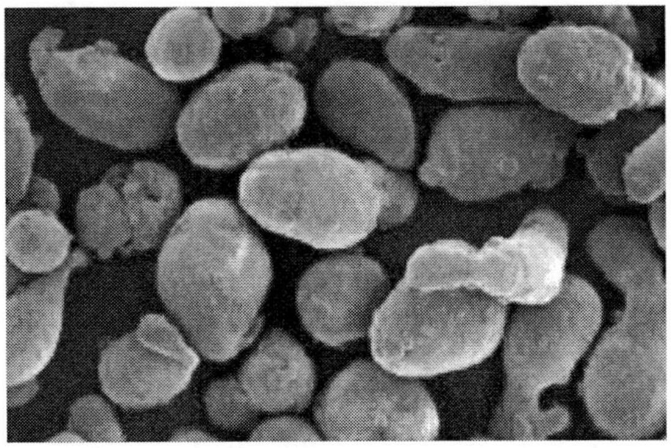

Figure 2. Shape of deformed particles

The particle shape at the moment of impact on the substrate is close to ellipsoidal, which was established by visual examination of metal coatings at multiple magnifications. And only at particle velocities of ~ 1000 m/s their deformed shape is close to the disk shape (figure 2).

2 Deformation of solid particles

Upon impact, the solid particle penetrates the substrate to a shallow depth, and deformation of the particle is compression and spreading over the surface. The defining process is spreading. It means that the particle deformation energy is spent on overcoming the forces of resistance to yield strength during deformation of its surface from the initial spherical shape to an ellipsoid. The metal particle deformation upon impact on the substrate is shown in figure 3.

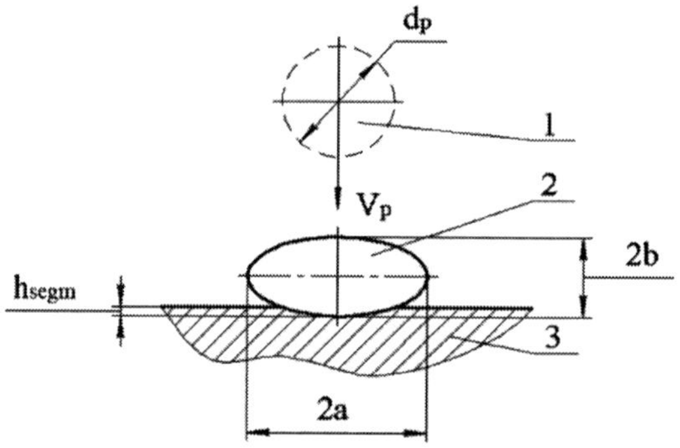

Figure 3. Deformation of a solid metal particle upon impact with a substrate: 1—particle shape before impact; d_p—particle diameter; V_p—particle velocity at the moment of impact on the substrate; 2—particle segmental shape after impact, $2b$—diameter of the segment; h_{segm}—penetration depth of the particle 1 into the substrate 3

In symbols, the energy balance equation in the impact zone of the particle on the substrate can be written as follows:

$$E_{kin.\,p} = E_{def.\,p} + E_{def.\,wall} + E_{heat.\,p} + E_{heat.\,wall}, \tag{1}$$

where $E_{kin.\,p}$—the particle kinetic energy; $E_{def.\,p}$—the particle deformation energy; $E_{def.\,wall}$—the substrate deformation energy; $E_{heat.\,p}$—the particle heating energy; $E_{heat.\,wall}$—the substrate heating energy.

The particle kinetic energy at the moment of impact is determined by the particle mass m_p and velocity V_p.

Particles strain energy is spent for the change of particle shape from spherical to close to ellipsoid (spheroid) at the moment of impact on the substrate, where d_p—the particle diameter; σ_T—the tensile yield strength of the particle material; a and b—the large and small semi-axes of the of the spheroid, respectively, and $\alpha = \sqrt{1 - \left(\frac{b}{a}\right)^2}$.

Degree of solid particles deformation upon impact on the substrate is calculated by the formula:

$$\varepsilon = \frac{2b}{d_p}. \tag{2}$$

The substrate deformation energy is spent to a plastic compression of the substrate material, the value of which is determined by penetration depth of the particle into the substrate body h_{segm} and the substrate compression resistance strength (H_β—the substrate Brinell hardness index).

The energy spent on heating the particle $E_{heat.p}$ depends on $m_{\tau,p}$—the mass of the particle heated during the time of the impact τ; $\bar{T}_{\tau,p}$—the average mass temperature of the heated portion of the particle; c_p—the specific heat capacity of the particle material at temperature $\bar{T}_{\tau,p}$; $T_{st.p}$—the average temperature of the particle before the impact.

The energy spent on heating the substrate $E_{heat.wall}$ depends on $m_{\tau,wall}$—the mass of the substrate heated during the time of the impact τ; $\bar{T}_{\tau,wall}$—the average mass temperature of the heated portion of the substrate; c_{wall}—the specific heat capacity of the substrate material at temperature $\bar{T}_{\tau,wall}$; $T_{st.wall}$—the substrate temperature before the impact.

The heating scheme of the particle and the substrate during their interaction is shown in figure 4.

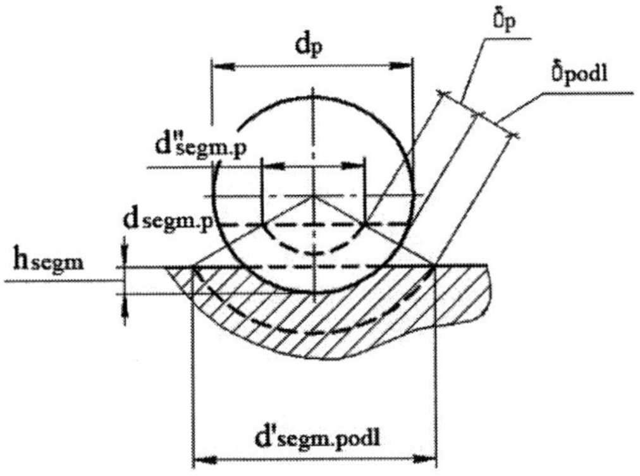

Figure 4. Heating scheme of the particle and the substrate on interaction: d_p—diameter of non-deformed particle; h_{segm}—particle penetration depth into the substrate; $d'_{segm.podl}$—diameter of the heated segment deformation of the substrate; $d_{segm.p}$—diameter of the heated segment of the particle; $d''_{segm.p}$—diameter of the cold segment of the particle; δ_p—particle heating depth; δ_{podl}—substrate heating depth

Substituting the corresponding expressions in (1), we obtain the final form of the energy balance in the impact zone of the particle on the substrate [3–5]:

$$\frac{m_p \cdot V_p^2}{2} = \pi d_p \sigma_T \left[2a^2 + \frac{b^2}{\alpha} \ln \frac{1+\alpha}{1-\alpha} - d_p^2\right](1-\varepsilon) + 2\pi h_{segm}^2 d_p H_\beta + \\ + m_{\tau,p} c_p (\bar{T}_{\tau,p} - T_{st.p}) + m_{\tau,wall} c_{wall} (\bar{T}_{\tau,wall} - T_{st.wall}). \quad (3)$$

The equation accounts the particle size, mass average temperatures in heated portions of the particle and the substrate, before impact temperatures of the particle and the substrate, fraction of the heated mass of the particle and the substrate during impact, specific heat capacities of the particle and the substrate material, the Brinell characteristic, and deformation of a solid particle, etc.

3 Deformation of melted particles

During application, the molten particles kick the colder surface of the substrate with subsequent spreading and simultaneous deformation and solidification.

At the initial moment of impact, a spherical particle, flattening under the action of inertial forces, spreads over the surface from the place of primary contact. These processes take place simultaneously. Under the action of inertial forces, the particle spreading velocity over the surface will differ from the particle velocity. Simultaneously with the particle spreading, the boundary of its solidification will also move, and the movement direction is perpendicular to the substrate surface. The final solidification and spreading of the particle occurs simultaneously, thus its shape is retained in the coating [6]. When the particle kicks the substrate at the angle of 90 degrees, the particle shape is close to a disk (see figure 2) with a thickness of 1/10–1/20 of the initial diameter of a spherical particle. However, in a real coating, the shape resulting from spreading can be completely different, since particles fall on the substrate at different angles [7–10].

Together with the circular surface spreading, heat transfer from the particle material occurs in the direction perpendicular to the substrate surface, and the solidification front propagates in the same direction.

We assume that the solidified particle, when kicks at the right angle on the flat substrate surface, acquires the disk shape from the initial spherical one.

Deformation of the molten particle upon impact on the substrate is calculated by the formula:

$$\varepsilon = 1 - \frac{h}{d_p}, \tag{4}$$

where h—the height of the fixed particle; d_p—the initial diameter of the particle.

Deformation of the molten metal particle upon impact with the substrate is shown in figure 5 [10–17].

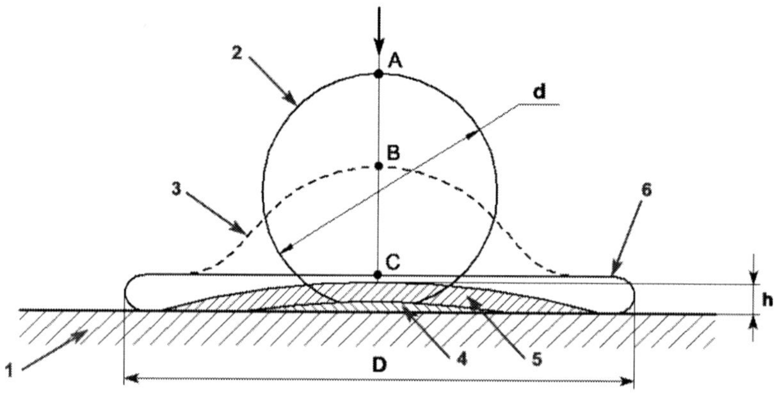

Figure 5. Deformation of the molten metal particle upon impact with the substrate: 1—substrate; 2—particle with the diameter d; 3—intermediate form of the particle; 4—solid phase of the particle; 5—crystallization phase of the particle; 6—final form of the crystallized particle (disk with the diameter D and the height h)

Substituting expressions (2) and (4) into the energy balance equation, we obtain the final form of the energy balance in the impact zone of solid (5) and molten (6) particles on the substrate [3–5]:

$$\frac{m_p \cdot V_p^2}{2} = \pi d_p \sigma_T \left[2a^2 + \frac{b^2}{\alpha}\ln\frac{1+\alpha}{1-\alpha} - d_p^2\right]\left(1 - \frac{2b}{d_p}\right) + 2\pi h_{segm}^2 d_p H_\beta + \\ + m_{T,p} c_p(\bar{T}_{T,p} - T_{st.p}) + m_{T,wall} c_{wall}(\bar{T}_{T,wall} - T_{st.wall}), \tag{5}$$

$$\frac{m_p \cdot V_p^2}{2} = \pi d_p \sigma_T \left[2a^2 + \frac{b^2}{\alpha}\ln\frac{1+\alpha}{1-\alpha} - d_p^2\right]\frac{h}{d_p} + 2\pi h_{segm}^2 d_p H_\beta + \\ + m_{T,p} c_p(\bar{T}_{T,p} - T_{st.p}) + m_{T,wall} c_{wall}(\bar{T}_{T,wall} - T_{st.wall}). \tag{6}$$

Characteristics of the coating microstructure—porosity and adhesion—depend on the solidified particle shape [15].

The solidified particle shape can have a different morphology: from splashes to a disc of regular shape. The disk shape is one of the main factors influencing the production of a high-quality coating with low porosity for most materials [13–17]. In [12], the dependence of the shape of the formed molten metal particle on the rate of its deposition on the substrate was investigated. With an increase in the rate of deposition of a molten tin particle of the same diameter on the steel substrate, a flattened drop is formed; then a donut, the central part of which is filled with a thin layer of metal; then a disc with weakly expressed edges; then a disk with star-shaped edges; and then the molten particle breaks up into small splashes.

As the particle diameter decreases, the shape change pattern shifts towards higher velocities, because there is an increase in the total force of the surface tension of the resulting drop in relation to its volume. An increase in the substrate temperature to a value sufficient for its penetration by a molten particle leads to an increase in the values of the velocity ranges.

Experimental studies have shown that the shape of a solidified drop depends on both the substrate temperature and the pressure of the surrounding gas [12–16]. It was also shown that there is a critical substrate temperature, exceeding which leads to the formation of disks in 50% of their total amount. Further increase in the substrate temperature increases the fraction of solidification of droplets in the form of disks [17].

Thus, in order to increase the adhesion and decrease the porosity of the applied coating, it is advisable to select such application modes in which the heated powder particles do not splatter upon contact with the substrate but form a coating consisting of molten particles having the disk shape upon solidification.

Operational application modes will depend on the particles and substrate material, carrier gas, etc. The selection of such modes is a separate scientific and practical task.

4 Conclusion

Improvement of characteristics of coatings obtained by a cold spray method, namely: an increase in the adhesion of the coating, a decrease in its porosity, an increase in the coefficient of use of the powder due to the greater proportion of the powder forming the coating, as well as an increase in the rate of formation of the coating—all of them depends on many factors that must be taken into account when setting the parameters heterogeneous flow. One of the important factors is the metal particles temperature and velocity in heterogeneous flow that forms the coating. The temperature range of particles for coating formation is from unheated to molten. The particle velocity range for coating formation is from the elastic impact of the particle on the substrate to the velocity at which the molten particle breaks up into small splashes. Certain combinations of these parameters make it possible to obtain coatings with specified and desired characteristics.

References

[1] A.P. Alkhimov, S.V. Klinkov, V.F. Kosarev, V.M. Fominm, *Cold gas dynamic spraying. Theory and practice* (Fizmatlit, Moscow, 2010)

[2] P.V. Nikitin, *Heterogeneous flows in innovative technologies* (Yanus-K, Moscow, 2010)

[3] S.A. Borisov, Yu.I. Glukhovskaya, S.V. Dobrovolskiy, P.V. Nikitin, I.V. Podporin, *Concept of an Experimental Setup for Testing the Technology for the Formation of New Anti-Corrosion Coating Materials Using Low-Temperature Supersonic Heterogeneous Flows*, TEM J., **9(2)**, 566 (2020)

[4] S.V. Dobrovolsky, P.V. Nikitin, S.A Borisov, Yu.I. Glukhovskaya, *Energy exchange in the interaction of a particle with a substrate when a supersonic heterogeneous flow flows around a flat solid wall*, in Proceedings of the XI International Conference on Nonequilibrium Processes in Nozzles and Jets, 26 June—1 July 2006, Saint-Petersburg, Russia (2016)

[5] S.A. Borisov, S.V. Dobrovolsky, Yu.I. Glukhovskaya, P.V. Nikitin, *Investigation of erosive wear of coatings using a mobile jet of heterogeneous flow of a limited area*, Surface. X-ray, Synchrotron and Neutron Techniques, **3**, 106 (2019)

[6] A.Yu. Pavlov, V.V. Ovchinnikov, A.D. Shlyapin, *The Basics of thermal spraying of protective coatings* (Infra-Engineering, Moscow, Vologda, 2020)

[7] L.H. Baldaev, V.N. Borisov, V.A. Vakhalin, G.I. Gannochenko, A.E. Zatoka, B.M. Zakharov, A.V. Ivanov, V.M. Ivanov, V.I. Kalita, V.V. Kudinov, A.F. Puzryakov, Yu.P. Collectors, B.G. Khamitsev, E.Ya. Shkolnikov, V.M. Yaroslavtsev, *Textbook under the general editorship of L.H. Baldaev* (Market DS, Moscow, 2007)

[8] D.N. Garkunov, *Tribotechnics (wear and tear): Textbook*, 4th ed. (MSHA, Moscow, 2001)

[9] F.I. Panteleenko, V.P. Lyalyakin, V.P. Ivanov, V.M. Konstantinov, *Restoration of machine parts handbook*, ed. by V.P. Ivanov (Mechanical Engineering, Moscow, 2003)

[10] T.N. Nemova, G.V. Kuznetsov, G.Ya. Mamontov, E.A. Kolesnikova, *Experimental study of the effect of the temperature of the melt particles and the substrate on the formation of protective coatings*, Physics. Proceedings of Higher Educational Institutions, **12/2**, 200 (2010)

[11] V.A. Klimenov, Zh.G. Kovalevskaya, K.V. Zaitsev, A.I. Tolmachev, *Investigation of the adhesion of coatings obtained by high-speed gas-plasma spraying*, Bulletin of the Tomsk Polytechnic University, **3**, 57 (2007)

[12] T.N. Nemova, A.A. Kolesnikov, *Dynamics of changes in the shape and size of metal droplets when they are deposited on the substrate surface*, Physics. Proceedings of higher educational institutions, **53(12/2)**, 161 (2010)

[13] I.A. Gainova, E.I. Sagaidak, V.N. Popov, *Modeling of droplet deposition on a metal substrate*, Sibirskii Zhurnal Industrial'noi Matematiki, **7(4)**, 47 (2004)

[14] M.R. Predtechensky, A.H. Cherepanov, V.H. Popov, Yu.D. Varlamov, *Investigation of the dynamics of collision and crystallization of a liquid metal drop with a multilayer substrate*, Applied mechanics and technical physics, **43(1)**, 112 (2002)

[15] M. Fukumoto, H. Hayashi, T. Yokoyama, *Relationship between particle's splat pattern and coating adhesive strength of HVOF sprayed Cu-alloy*, Journal of Japan Thermal Spraying Society, **2(3)**, 149 (1995)

[16] M. Fukumoto, S. Katoh, I. Okane, *Splat behavior of plasma sprayed particles on flat substrate surface*, in Proceedings of the 14th International Thermal Spray Conference, 22–26 May 1995, Kobe, Japan (1995)

[17] M. Fukumoto, H. Nagai, T. Yasui, *Influence of surface character change of substrate due to heating on flattening behavior of thermal sprayed particles*, Journal of Thermal Spray Technology, **15(4)**, 759 (2006)

Momentary finite element for elasticity 3D problems

Dmitry Chekmarev[1,*] and *Yasser* Abu Dawwas[1]

[1]Lobachevsky State University of Nizhny Novgorod, 23, Gagarina prospekt, Nizhny Novgorod, 603022, Russia

Abstract. A description of a new 8-node finite element in the form of a hexahedron is given for solving elasticity 3D problems. This finite element has the following features. This is a linear approximation of functions in the element, one point of integration and taking into account the moments of forces in the element. The finite element is based on "rare mesh" FEM schemes—finite element schemes in the form of n-dimensional cubes (square, cube, etc.) with templates in the form of inscribed simplexes (triangle, tetrahedron, etc.). Among the rare mesh schemes, schemes in 3-dimensional and 7-dimensional spaces are successful, in which the simplex can be arranged symmetrically with respect to the center of the n-dimensional cube. The rare mesh FEM schemes have not the hourglass instability due to the fact that the template of the finite element operator has the form of a simplex. Compared to traditional linear finite elements in the form of a simplex, rare mesh schemes are more economical and converge better, since they do not have the effect of overestimated shear stiffness. Moment FEM schemes are constructed by rare mesh schemes higher dimensional projection, respectively, on a two-dimensional or three-dimensional finite element mesh. The resulting finite elements are close to the known polylinear elements and surpass them in efficiency. The schemes contain parameters that allow you to control the convergence of numerical solutions. The possibility of applying this approach to the construction of numerical schemes for solving other problems of mathematical physics is discussed.

1 Introduction

When solving solid dynamic problems in combination with an explicit "cross" scheme, 4-node two-dimensional and 8-node three-dimensional finite elements with one integration point are widely used. In this case, as a rule, the problem of "hourglass instability" arises [1–3], associated with the incompleteness of the system of basic operators and leading to strong nonphysical distortions of the computational grid. To combat it, special means are used that require additional computational costs. Traditional schemes based on 2-dimensional triangular elements and 3-dimensional tetrahedral elements have poorer stability and slower convergence. One of the possible approaches to increasing the efficiency is the use of finite element schemes on rare meshes [4–7]. In rare mesh FEM schemes, the computational elements cover the problem area with regular intervals, which makes it possible to noticeably reduce the number of elements on grids of the same size and thereby reduce the amount of computations.

*e-mail: 4ekm@mm.unn.ru

The most successful rare mesh FEM scheme for solving 3D elasticity and plasticity problems is a scheme based on a hexahedral grid, in each cell (hexahedron) of which there is one linear element in the form of a central tetrahedron (figure 1), and the remaining 4 tetrahedrons are not involved in the calculations. In this case, in comparison with the traditional scheme of linear finite element on the same grids, the number of elements is reduced by a factor of 5, and the number of nodes is reduced by a factor of 2. With better efficiency, the rare mesh scheme has the second order of accuracy (no worse than that of schemes based on linear and bilinear finite elements). On a number of test problems, it also showed better convergence in comparison with the indicated schemes. As noted in [4], successful rare mesh schemes are obtained with a symmetric arrangement of a linear element (simplex) inside an n-dimensional cube. So, for $n = 2$ (a triangle in a square), the arrangement is asymmetric (figure 2), and for $n = 3$ (a tetrahedron in a cube), it is symmetric (the center of the tetrahedron coincides with the center of the cube (figure 1)). The next case of a lucky dimension is $n = 7$ (a regular 8-node simplex can be inscribed symmetrically into a 7-dimensional cube). When solving problems of "unsuccessful" dimension, effective FEM circuits can be obtained by projection a rare mesh circuit of a higher "successful" dimension. At the same time, it is natural to assume that the high quality of the circuit is preserved during projection. An approach to constructing new numerical FEM schemes by projection rare mesh schemes on a mesh of a lower dimension is described in [8].

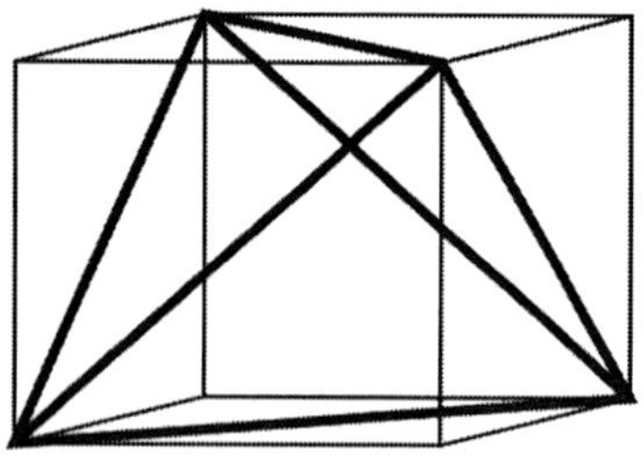

Figure 1. The cell of 3D rare mesh

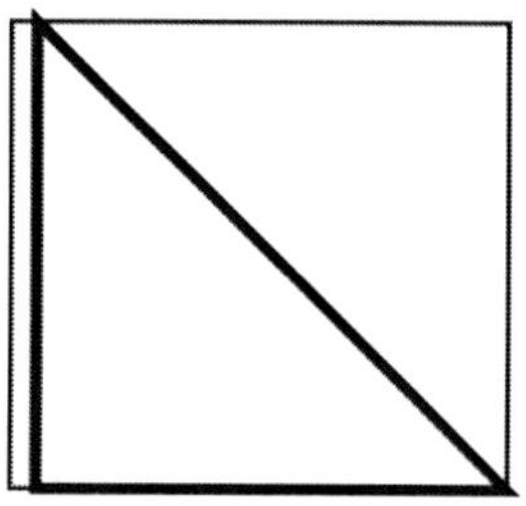

Figure 2. The cell of 2D rare mesh

This paper describes the implementation of the three-dimensional moment scheme proposed in [8]. Further presentation is based on the material [8].

2 2D FEM scheme for elasticity problems based on a 3D rare mesh scheme

Rare mesh FEM scheme for solution 3D dynamic elasticity problem on a uniform orthogonal mesh $x_i^1 = x_0^1 + h_1 i$, $x_j^2 = x_0^2 + h_2 j$, $x_k^3 = x_0^3 + h_3 k$; can be written in the following finite-difference form (see [4]):

$$(\lambda + \mu)\begin{vmatrix} D_{11}u_1 + D_{12}u_2 + D_{13}u_3 \\ D_{21}u_1 + D_{22}u_2 + D_{23}u_3 \\ D_{31}u_1 + D_{32}u_2 + D_{33}u_3 \end{vmatrix} + \mu D_\Delta \begin{vmatrix} u_1 \\ u_2 \\ u_3 \end{vmatrix} + \rho \begin{vmatrix} F_1 \\ F_2 \\ F_3 \end{vmatrix} = \rho D_{tt} \begin{vmatrix} u_1 \\ u_2 \\ u_3 \end{vmatrix}. \quad (1)$$

This form is similar to the Lamé system of equations form

$$(\lambda + \mu)\,\mathrm{grad}\,\mathrm{div}\,u + \mu\Delta u + \rho F = \rho\frac{\partial^2 u}{\partial t^2}. \quad (2)$$

Here λ, μ are the Lame parameters, ρ—is the density of the medium, F—is the mass force, u—is the displacement field, D_{ij}, D_Δ—are the grid operators approximating the second derivatives and the Laplace operator. Operators are constructed as follows. Basic operators approximating the first partial derivatives in an element ($d_m^+ \approx \partial/\partial x^m$) are written as

$$\begin{aligned}
(d_1^+ f)_{ijk} &= \frac{1}{2h_1}\left(f_{i+1,j+1,k+1} + f_{i+1,j,k} - f_{i,j+1,k} - f_{i,j,k+1}\right),\\
(d_2^+ f)_{ijk} &= \frac{1}{2h_2}\left(f_{i+1,j+1,k+1} + f_{i,j+1,k} - f_{i+1,j,k} - f_{i,j,k+1}\right),\\
(d_3^+ f)_{ijk} &= \frac{1}{2h_3}\left(f_{i+1,j+1,k+1} + f_{i,j,k+1} - f_{i,j+1,k} - f_{i+1,j,k}\right).
\end{aligned} \tag{3}$$

Operators

$$\begin{aligned}
(d_1^- f)_{ijk} &= \frac{1}{2h_1}\left(-f_{i-1,j-1,k-1} - f_{i-1,j,k} + f_{i,j-1,k} + f_{i,j,k-1}\right),\\
(d_2^- f)_{ijk} &= \frac{1}{2h_2}\left(-f_{i-1,j-1,k-1} - f_{i,j-1,k} + f_{i-1,j,k} + f_{i,j,k-1}\right),\\
(d_3^- f)_{ijk} &= \frac{1}{2h_3}\left(-f_{i-1,j-1,k-1} - f_{i,j,k-1} + f_{i,j-1,k} + f_{i-1,j,k}\right)
\end{aligned} \tag{4}$$

are equal to the conjugate to (3), taken with the sign "–". We define the operators D_{ij} in terms of superpositions of operators (3) and (4): $D_{ij} = \frac{1}{2}\left(d_i^+ d_j^- + d_j^+ d_i^-\right)$, $D_\Delta = D_{11} + D_{22} + D_{33}$.

A numerical scheme for solving a two-dimensional elasticity problem (plane deformation) is obtained by projecting a three-dimensional one on a plane x_1Ox_2. Assuming that the three-dimensional computational domain has the form $\Omega \times [0, h_3]$, where Ω—area in R^2, take one layer of cells along the x^3 coordinate and impose the constraint on the solution $u_3 = 0$. In this case, the difference scheme (1) takes the form

$$\begin{aligned}
(\lambda + \mu)(D_{11}u_1 + D_{12}u_2) + \mu(D_{11}u_1 + D_{22}u_1 + D_{33}u_1) + \rho F_1 &= \rho D_{tt}u_1,\\
(\lambda + \mu)(D_{21}u_1 + D_{22}u_2) + \mu(D_{11}u_2 + D_{22}u_2 + D_{33}u_2) + \rho F_2 &= \rho D_{tt}u_2.
\end{aligned} \tag{5}$$

In (5), the operators D_{ij} are obtained by projecting the operators considered above onto a two-dimensional grid, while the operators

$$\begin{aligned}
(d_3^+ f)_{ij} &= \frac{1}{2h_3}\left(f_{i+1,j+1} + f_{i,j} - f_{i,j+1} - f_{i+1,j}\right),\\
(d_3^- f)_{ij} &= \frac{1}{2h_3}\left(f_{i-1,j-1} + f_{i,j} - f_{i,j-1} - f_{i-1,j}\right)
\end{aligned} \tag{6}$$

approximate, up to a factor, the operator of the second mixed derivative $\dfrac{\partial^2}{\partial x^1 \partial x^2}$.

The element size h_3 along the coordinate x^3 in the constructed numerical scheme turns into an adjustable parameter of the scheme. Thus, a one-parameter family of numerical FEM schemes for solving the plane elasticity problem was obtained. By adjusting the parameter h_3, the influence of the moment component in the element and thereby the convergence of numerical solutions can be regulated.

3 3D FEM scheme for elasticity problems based on a 7D rare mesh scheme

Generalizing this approach, we construct the 8-node finite element for solving 3D problems of elasticity theory, similar to the 8-node multilinear finite element.

3.1 Seven-dimensional rare mesh scheme

Among the vertices of a unit 7-dimensional cube, you can choose 8 vertices that form a regular simplex, the center of which coincides with the center of the cube. An example is a set of vertices

$$\begin{bmatrix}0\\0\\0\\0\\0\\0\\0\end{bmatrix}, \begin{bmatrix}0\\0\\1\\0\\1\\1\\1\end{bmatrix}, \begin{bmatrix}0\\1\\0\\1\\0\\1\\1\end{bmatrix}, \begin{bmatrix}0\\1\\1\\1\\1\\0\\0\end{bmatrix}, \begin{bmatrix}1\\0\\0\\1\\1\\0\\1\end{bmatrix}, \begin{bmatrix}1\\0\\1\\1\\0\\1\\0\end{bmatrix}, \begin{bmatrix}1\\1\\0\\0\\1\\1\\0\end{bmatrix}, \begin{bmatrix}1\\1\\1\\0\\0\\0\\1\end{bmatrix}. \quad (7)$$

All edges of a given simplex have the same length. Note also that the sets of the first three coordinates of the given vector system form the set of vertices of the unit cube in R^3.

Consider a hypothetical "theory of elasticity" in 7-dimensional space. We define the tensor of deformations in R^7 as a generalization of the Cauchy relations:

$$\varepsilon_{ij} = \frac{1}{2}(u_{i,j} + u_{j,i}), \quad i, j = 1, \dots, 7.$$

The relationship between stresses and strains is established on the basis of "Hooke's law":

$$\sigma_{ij} = \lambda\delta_{ij}\varepsilon_{kk} + 2\mu\varepsilon_{ij}, \quad i, j = 1, \dots, 7.$$

As a result, the 7-dimensional equilibrium equations of the "elasticity theory" will be written in the form (2), where the gradient, divergence and the Laplace operator are defined, respectively, in R^7. Rare mesh FEM scheme based on a linear finite element on a uniform orthogonal mesh $x^1_{i_1} = x^1_0 + h_1 i_1, \dots, x^7_{i_7} = x^7_0 + h_7 i_7$ will take a form similar to (1):

$$(\lambda+\mu)\begin{vmatrix}D_{11}u_1 + \dots + D_{17}u_7\\ \dots \\ D_{71}u_1 + \dots + D_{77}u_7\end{vmatrix} + \mu D_\Delta \begin{vmatrix}u_1\\ \dots\\ u_7\end{vmatrix} + \rho\begin{vmatrix}F_1\\ \dots\\ F_7\end{vmatrix} = 0. \quad (8)$$

To construct a three-dimensional scheme, we will proceed similarly to the case of a two-dimensional scheme considered above. Assuming at all nodes $u_4 = u_5 = u_6 = u_7 = 0$, we get the FEM scheme in the form

$$(\lambda+\mu)\begin{vmatrix}D_{11}u_1 + D_{12}u_2 + D_{13}u_3\\ D_{21}u_1 + D_{22}u_2 + D_{23}u_3\\ D_{31}u_1 + D_{32}u_2 + D_{33}u_3\end{vmatrix} + \mu D_\Delta\begin{vmatrix}u_1\\u_2\\u_3\end{vmatrix} + \mu(D_{44} + D_{55} + D_{66} + D_{77})\begin{vmatrix}u_1\\u_2\\u_3\end{vmatrix} + \rho\begin{vmatrix}F_1\\F_2\\F_3\end{vmatrix} = 0 \quad (9)$$

after projection into R^3. In (9), the operators $D_{ij} = \frac{1}{2}\left(d_i^+ d_j^- + d_j^+ d_i^-\right)$, $D_\Delta = D_{11} + D_{22} + D_{33}$ expressed in terms of basic operators $d_1^+, d_2^+, d_3^+, d_4^+, d_5^+, d_6^+, d_7^+$ and their conjugates (with a— sign) $d_1^-, d_2^-, d_3^-, d_4^-, d_5^-, d_6^-, d_7^-$ defined on 8-node templates. Moreover, the operators d_1^+, d_2^+, d_3^+ and d_1^-, d_2^-, d_3^- approximate the first derivatives with respect to the corresponding coordinates, and the operators $d_4^+, d_5^+, d_6^+, d_7^+$ and $d_4^-, d_5^-, d_6^-, d_7^-$ approximate, up to a positive factor, the differentiation operators $\frac{\partial^2}{\partial x^1 \partial x^2}, \frac{\partial^2}{\partial x^1 \partial x^3}, \frac{\partial^2}{\partial x^2 \partial x^3}, \frac{\partial^3}{\partial x^1 \partial x^2 \partial x^3}$. Similar approximations of the moment components take place in the traditional scheme of a multilinear 8-node finite element (see, for example, [9]).

Thus, a 4-parameter family of numerical FEM schemes for solving a 3D elasticity problem was obtained. The parameters h_4, h_5, h_6, h_7 can be adjusted for changing the influence of the moments in the element.

4 Implementation of a 3D moment FEM scheme

The implementation of a moment element can be performed in the traditional FEM technique using vector and matrix notation, building basis functions, mapping finite elements to a standard cubic element, etc. [9–12]. It is more convenient and economical to make the implementation in a form similar to Wilkins' scheme [13], because stresses, strains and Jacobian of the mapping are constants in the element. The derivatives in an arbitrary hexahedral element are calculated by the formula

$$\frac{\partial f}{\partial x^i} \approx d_i^+ f = \frac{\det V_i}{\det V},$$

where the matrix V has the form, $V = \begin{bmatrix} 1 & x_1^1 & x_1^2 & x_1^3 & 0 & 0 & 0 & 0 \\ 1 & x_2^1 & x_2^2 & x_2^3 & 0 & h_5 & h_6 & h_7 \\ 1 & x_3^1 & x_3^2 & x_3^3 & h_4 & 0 & h_6 & h_7 \\ 1 & x_4^1 & x_4^2 & x_4^3 & h_4 & h_5 & 0 & 0 \\ 1 & x_5^1 & x_5^2 & x_5^3 & h_4 & h_5 & 0 & h_7 \\ 1 & x_6^1 & x_6^2 & x_6^3 & h_4 & 0 & h_6 & 0 \\ 1 & x_7^1 & x_7^2 & x_7^3 & 0 & h_5 & h_6 & 0 \\ 1 & x_8^1 & x_8^2 & x_8^3 & 0 & 0 & 0 & h_7 \end{bmatrix}$, and matrices V_i are obtained from the matrix V by replacing the column of the corresponding coordinate with the column of function values, for example $V_1 = \begin{bmatrix} 1 & f_1 & x_1^2 & x_1^3 & 0 & 0 & 0 & 0 \\ 1 & f_2 & x_2^2 & x_2^3 & 0 & h_5 & h_6 & h_7 \\ 1 & f_3 & x_3^2 & x_3^3 & h_4 & 0 & h_6 & h_7 \\ 1 & f_4 & x_4^2 & x_4^3 & h_4 & h_5 & 0 & 0 \\ 1 & f_5 & x_5^2 & x_5^3 & h_4 & h_5 & 0 & h_7 \\ 1 & f_6 & x_6^2 & x_6^3 & h_4 & 0 & h_6 & 0 \\ 1 & f_7 & x_7^2 & x_7^3 & 0 & h_5 & h_6 & 0 \\ 1 & f_8 & x_8^2 & x_8^3 & 0 & 0 & 0 & h_7 \end{bmatrix}$ etc.

In this case, the strain tensor takes the form:

$$\varepsilon_{ij} = \begin{bmatrix} \varepsilon_{11} & \varepsilon_{12} & \varepsilon_{13} & \varepsilon_{14} & \varepsilon_{15} & \varepsilon_{16} & \varepsilon_{17} \\ \varepsilon_{21} & \varepsilon_{22} & \varepsilon_{23} & \varepsilon_{24} & \varepsilon_{25} & \varepsilon_{26} & \varepsilon_{27} \\ \varepsilon_{31} & \varepsilon_{32} & \varepsilon_{33} & \varepsilon_{34} & \varepsilon_{35} & \varepsilon_{36} & \varepsilon_{37} \end{bmatrix}, \quad \varepsilon_{ij} = \begin{cases} \frac{1}{2}\left(d_i^+ u_j + d_j^+ u_i\right), & j \le 3, \\ \frac{1}{2} d_j^+ u_i, & j > 3 \end{cases}$$

and the stress tensor—respectively

$$\sigma_{ij} = \begin{bmatrix} \sigma_{11} & \sigma_{12} & \sigma_{13} & \sigma_{14} & \sigma_{15} & \sigma_{16} & \sigma_{17} \\ \sigma_{21} & \sigma_{22} & \sigma_{23} & \sigma_{24} & \sigma_{25} & \sigma_{26} & \sigma_{27} \\ \sigma_{31} & \sigma_{32} & \sigma_{33} & \sigma_{34} & \sigma_{35} & \sigma_{36} & \sigma_{37} \end{bmatrix}.$$

The last 4 columns correspond to three bending moments and one torque moment in the finite element.

Further, the integrals of the components of the stress tensor determine the forces applied to the nodes of the element, as a result of which the equations of motion of the nodes are formed.

5 Numerical Results

Let us consider two problems of vibrations of a square bar under the action of an instantaneously applied pressure. A beam with dimensions $L = 10$ cm, $H_1 = 1$ cm, $H_2 = 1$ cm with rigidly embedded ends, loaded with instantly applied pressure 0,17 GPa, uniformly distributed over the site $4 < x^1 < 6$ (figure 3). Properties of the material: $\rho = 7.8$ g/cm^3, $E = 210$ GPa, $\nu = 0.3$.

The problem was solved on two grids: $40 \times 4 \times 4$ elements and $80 \times 8 \times 8$ elements. The second (asymmetric) problem was solved in a similar setting, but the loaded area was half of that considered in the first problem (figure 4).

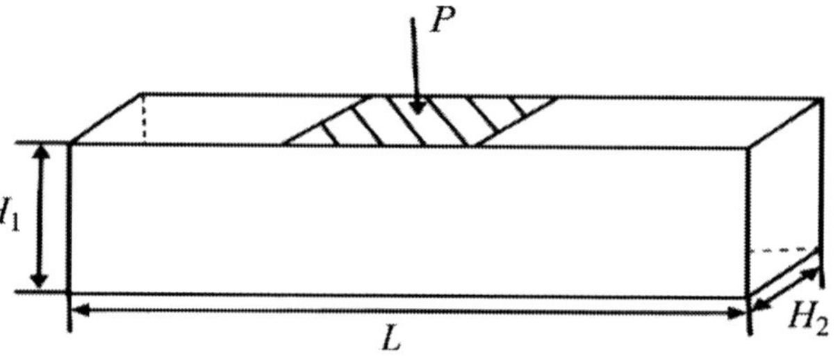

Figure 3. Symmetric problem

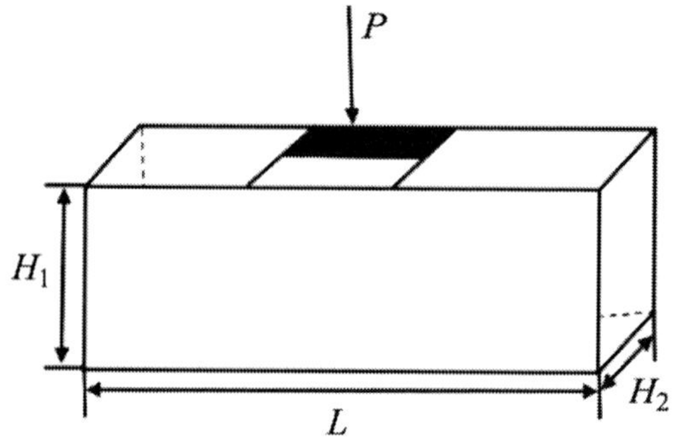

Figure 4. Asymmetric problem

For comparison, the same problem were solved using the rare mesh scheme and Wilkins' scheme.

The symmetrical problem was previously solved using the Wilkins scheme and the rare mesh scheme. The solution results are given in [6]. The solution based on the moment scheme is shown in figure 5. It show the graphs of the velocity v_3 in the center of the face opposite to the loaded one, depending on time. The graphs show the results of calculations up to 747 μs. Solid lines correspond to calculations on a grid of $40 \times 4 \times 4$ elements, dashed lines correspond to calculations on a grid of $80 \times 8 \times 8$ elements. Note that all three schemes showed approximately the same results. The slowest convergence was observed for the Wilkins scheme.

Figures 6–8 show similar results of solving the asymmetric problem.

Numerical results show good qualities of the rare mesh and moment schemes for both problems and their better convergence in comparison with the Wilkins scheme. Note the difference between the numerical solutions of the symmetric and asymmetric problems. The hourglass instability of the Wilkins scheme can be realized to varying degrees depending on the grid size, boundary conditions, and other factors. Therefore, it is not observed in the symmetric problem. Wilkins scheme on an asymmetric problem shows completely unsatisfactory

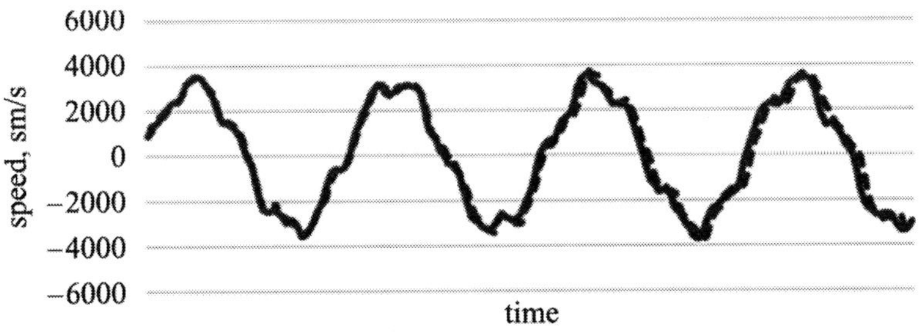

Figure 5. Symmetric problem. Momentary scheme

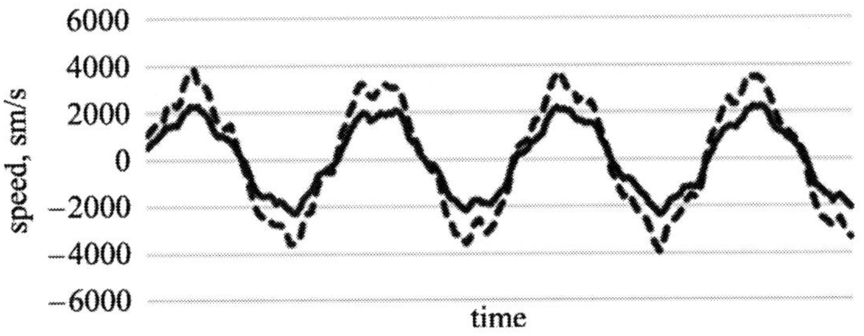

Figure 6. Asymmetric problem. Rare mesh scheme

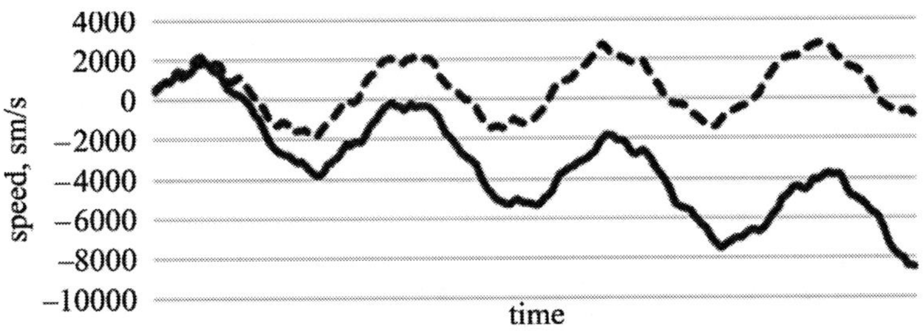

Figure 7. Asymmetric problem. Wilkins scheme

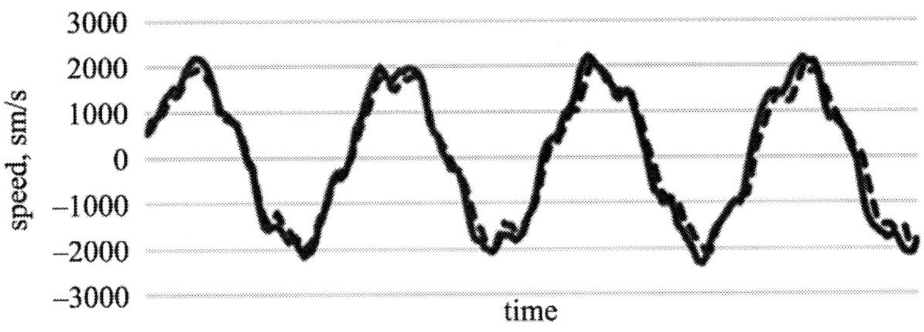

Figure 8. Asymmetric problem. Momentary scheme

results, which is due to the hourglass instability. Note that in the Wilkins scheme, no means were used to combat this type of instability. According to the results of solving both test problems, the moment scheme showed the best convergence.

6 Conclusion

A new moment FEM scheme is implemented for solving 3D dynamic elasticity problems. It is free from hourglass instability effect. This scheme is very economical since it has one integration point. The circuit has adjustable parameters.

The creation of new effective numerical methods for solving problems of mathematical physics is relevant both for the present and for the future due to the obvious trend of an ever wider and more detailed application of mathematical modeling in various fields of science. In particular, the discontinuous Galerkin method is currently being actively developed [14]. The finite element considered in the paper is a development of the concept of rare mesh finite element schemes. This concept appears to be very promising. Note that rare mesh schemes are similar in some aspects to the discontinuous Galerkin method, since they ensure the continuity of physical fields only on the edges of finite elements. Rare mesh schemes for solving 3D elasticity and plasticity problems have been theoretically studied in detail. They were verified and validated on model problems, compared with numerical solutions obtained using traditional FEM schemes and experimental data. The conducted studies have shown that rare mesh schemes are not inferior to traditional schemes in terms of accuracy and exceed them in efficiency, in particular, in speed - several times. Note that the concept of rare mesh schemes, as well as the moment schemes built on their basis, are based on purely geometric ideas that are in no way connected with the physical features of the simulated processes. Therefore, there is every reason to believe that schemes built within the framework of this concept, including moment schemes, can be used not only in the mechanics of a deformable solid body, but also in fluid and gas mechanics, electromagnetic theory and other areas of mathematical physics. Schemes based on this approach can be applied to solving problems of dimension more than three (if any).

References

[1] O.P. Jacquotte, J.T. Oden, E.B. Becker, *Numerical control of the hourglass instability*, Int. J. Numer. Methods Eng., **22(1)**, 219 (1986)

[2] V.G. Bazhenov, A.I. Kibets, O.V. Tulintsev, *Application of momentary FEM scheme to analysis of nonlinear dynamic 3D problems of solid and shell structural elements*, Applied problems of strength and plasticity. Solution methods, **47**, 46 (1991)

[3] R.D. Cook et al., *Concepts and Applications of Finite Element Analysis* (John Willey and Sons, New York, 1989)

[4] D.T. Chekmarev, *Finite element schemes on rare meshes*, Probl. At. Sci. Techn. Ser. Math. Model. Phys. Proc., **2**, 49 (2009)

[5] S.V. Spirin, D.T. Chekmarev, A.V. Zhidkov, *Solving the 3D Elasticity Problems by Rare Mesh FEM Scheme*, Finite Difference Methods, Theory and Applications, **9045**, 379 (2015)

[6] K.A. Krutova, S.V. Spirin, D.T. Chekmarev D.T., *On the influence of the mutual arrangement of finite elements on the accuracy of the numerical solution of elasticity problems*, Problems of strength and plasticity, **75(4)**, 312 (2013)

[7] A.V. Zhidkov, K.A. Krutova, A.A. Mironov, D.T. Chekmarev, *Numerical solution of 3D dynamic elastic-plastic problems using the rare mesh FEM scheme*, Problems of strength and plasticity, **79(3)**, 327 (2017)

[8] D.T. Chekmarev, *On one method of constructing 2D 4-node and 3D 8-node finite elements for solving elasticity problems*, Uchenye Zapiski Kazanskogo Universiteta. Seriya Fiziko-Matematicheskie Nauki, **155(3)**, 150 (2013)

[9] A.I. Golovanov, D.V. Berezhnoy, *The finite element method in solid mechanics* (DAS Publishing, Kazan, 2001)

[10] O. Zienkiewicz, *The Finite Element Method in Engineering Science* (McGraw-Hill, London, New York, 1971)

[11] O. Zienkiewicz, K. Morgan, *Finite elements and approximation* (Dover Publications, Inc., Mineola, New York, 2006)

[12] R.H. Gallager, *Finite element analysis. Fundamentals* (Prentice-hall, Inc., Englewood Cliffs, New Jersey, 1975)

[13] M. Wilkins, *Calculation of Elastic-Plastic Flow*, Methods Comput. Phys., **3**, 211 (1964)

[14] Bernardo Cockburn, *An Introduction to the Discontinuous Galerkin Method for Convection—Dominated Problems*, Advanced Numerical Approximation of Nonlinear Hyperbolic Equations. Lecture Notes in Mathematics, **1697**, 151 (1998)

The experience in teaching mathematical disciplines using information technologies

Olga Demidova[1] and *Nataliya* Malinina[1,*]

[1]Moscow Aviation Institute (National Research University), 4, Volokolamskoe shosse, Moscow, 125993, Russia

Abstract. The presented material is devoted to the long-term experiment on the introduction of computer technologies in the process of teaching students of technical (and not only technical) specialties and lectors in various courses and disciplines at MAI. The experiment was carried out in teaching completely different courses, such as: combinatorial analysis, operations research, computational mathematics, as well as in the course of information technology, designed to improve the qualifications of teachers.

1 Introductions

The development, search and implementation of innovative ways in organizing of the educational process, which that meets the trends of modern education, are mentioned in the Federal Law "On Education in the Russian Federation" and has become a standard practice at the Moscow Aviation Institute.

The direction, which began to develop actively in the 20th century under the names "computational mathematics", "numerical methods", "operations research" and etc. is aimed at introducing the theoretical possibilities of higher mathematics, both in modern engineering and in the field of production management.

The pandemic has accelerated the process of introducing computer technologies. Two years of the pandemic have shown that it is quite possible to give lectures in a remote format, for example, using the LMS.MAI platform, although conducting practical classes and laboratory work in a remote format is still not available for all disciplines. This is reflected in the current training programs. Such tactic should continue in the future.

Classical mathematical disciplines are certainly important for the formation of an engineer's worldview. They open up the possibilities of interaction between objects and explain the principles of their existence in the real world. But the solution of practical problems is impossible without the use of computer technology.

We will share the experience of conducting a number of courses that were taught at the aerospace institute and other technical institutes both before the pandemic and during the pandemic. This approach was certainly liked by students when lectures were presented in the form of presentations.

2 Experience first. Combinatorial analysis

The course "Combinatorial Analysis" is taught to students if two faculties: "Aerospace institute" and "Aviation technology institute". Despite the fact that this course is devoted to the

*e-mail: malinina806@gmail.ru

study of the basics of modern mathematics, namely: set theory and combinatorics, topology and graph theory, as well as the theory of algorithms and logic, there still is the possibility of using and implementing the basics of computer technologies even in such a course [1]. Computer technologies, which are used in this course:

- MS Word is taught in order to write term papers and essays correctly. Students need to gain basic knowledge and working skills in MS Word (document structuring, lists, through numbering, insertion of literature sources, etc.) regardless of what and how they were taught at school.
- MS Excel must be taught in order to get skills in possibilities to perform complex calculations (autocomplete, working with formulas, etc.). Students use spreadsheets to solve all combinatorial problems, also they use spreadsheets to order graphs for the purpose to reduce an arbitrary graph to the form of Koenig graph, which obtains the recursive property.
- MS Visio is taught in order to get skills in for the competent creation of almost any illustrations for present and future term papers. Students learn to draw graphs, matrices and various diagrams, which the will need in their future study.

All these computer technologies are integrated into the practice of the course along with gaining theoretical knowledge, and all the tasks of the Combinatorial Analysis course are solved exclusively with the help of information technologies. Students write essays, solve combinatorial problems and problems of graph theory using tools of MS Office (Word, Excel, Visio).

3 Experience second. Computational mathematics

Unfortunately, not much time is given to combine the solution of applied mathematics problems with the use of programming capabilities in high-level languages, in the engineering faculties of many universities, for almost any specialty, in accordance with the curricula.

The objective of the course "Numerical Methods" is to acquaint future masters, specialists and bachelors with the basic rules for solving engineering problems: interpolation, approximation, optimization and solution of differential equations, without which modern technical developments are impossible [2–4].

For the numerical solution of ordinary differential equations, a sufficient number of marching explicit and implicit, one-step and multi-step methods of various orders have been developed, which allow you to effectively build calculation schemes in the spreadsheet environment. A large number of numerical schemes make it possible to organize calculations according to individual options for each listener.

Thus, the course covers 24 explicit methods of different orders from 1 to 6 for solving first-order ordinary differential equations based on Runge–Kutta computational schemes, including eight nested schemes that correct the choice of the integration step. Students can solve even stiff differential equations by reviewing the collection of 18 implicit circuits.

In addition, for the final stages of calculations, it is proposed to use sets of multi-step methods. Why are there so many methods offered for study? Only so that future engineers have a choice of methods, and there is confidence that you can always build your own computational scheme that will be able to solve a new separate problem.

Problems leading to the solution of linear partial differential equations have always been included only in term papers or computational papers due to certain difficulties in modeling physical processes, choosing the equations themselves, and also in selecting computational procedures.

It makes sense to abstract from the semantic and physical loads that are assigned to equations built according to the laws of classical physics. It is quite possible to work only with well-classified [5] model types of equations. Then the decision problem passes into the plane of choosing a specific computational method and into algorithms for calculating the accompanying parameters.

At the same time, plane and even axisymmetric problems of mathematical physics can be modeled and analyzed within the framework of practical or laboratory work, where it is convenient to illustrate the algorithms in spreadsheets, which can later serve as a source of control calculations in cases where it is necessary to debug the developed software products.

So, for example, for equations of parabolic and hyperbolic types with linear coefficients and with any initial and boundary conditions, the finite element method is most often used, which leads to a marching integration scheme. Due to the certain complexity of such tasks, programs should be written in arbitrary programming languages, when implementing the selected algorithms.

At the same time, for explicit finite-difference schemes it is necessary to simulate the calculations in Excel tables, and tabular calculations can be used to debug the corresponding programs. For the implicit difference schemes, which are usually solved using methods for solving systems of linear algebraic equations, computing also can be done with the help of tabular calculation, using simplified methods for solving systems.

So, the study of partial differential equations in the spreadsheet environment can be included in the plans of practical and laboratory classes in order to develop skills in performing numerical calculations and parametric studies, as well as for graphical interpretation of the results.

The solution of linear differential equations in partial derivatives has always been included only in term papers or computational-graphic works. But it is impossible to conduct classes without mass calculations. Undoubtedly, computing systems have been created and are currently actively used, such as Matlab or Mathcad. They are positioned as universal applications for researchers.

Such complexes, which run on modern operating systems, including Linux, MacOS, Solaris and Windows, belong to the class of computer-aided design systems and are intended only for calculations using algorithms selected for each specific task. However, these systems do not have a training link. Most of these systems are based on the "black box" principle, which indicates its answer (calculation result) for any question (source data option).

Such systems are good for mass standard parametric calculations or for multiple statistical processing, but are not suitable for initial acquaintance with numerous engineering problems that can be solved by methods of applied computational mathematics.

In these cases, spreadsheets come to the rescue as a learning tool. It is possible to train students in almost all most common methods in a fairly short time, with the help of such an application. In the future, students easily will switch to creating programs for such methods in high-level languages, which will be available to them in other applications.

Thus, we can offer a spreadsheet environment (Excel) as a modern universal computing platform, because it has a large mathematical content and is easy to use. The main advantage of spreadsheets is the availability of this application, since it is included with MS Office by default.

Thanks to the widespread use of Excel, as well as learning the basic skills of working with MS Office in high school, most students already used spreadsheets for simple calculations, and therefore had some experience, although they considered spreadsheets like a large calculator. It should be noted that there is a great opportunity to use spreadsheets for other, more complex, engineering tasks, for example, for processing statistical data obtained in natural experiments.

The input and output of data is clear and simple in spreadsheets, and the possibility of a simple and high-quality graphical representation of the obtained results, allows students quickly to acquire the necessary computing skills. In addition, iterative processes are well modeled in spreadsheets.

The visibility of calculations in spreadsheets makes it possible not to hide solution algorithms in subprograms, but to show its step-by-step execution, while graphic support in tables illustrates even a small change in parameters. In addition, within the framework of one calculation, it is possible to perform parametric studies, which, for example, may be required when discussing the problem of launching a payload by a multistage vehicle into a given orbit.

The disadvantages of spreadsheets include the following situations: solving problems of small dimensions and working with relatively simple functions. Naturally, these shortcomings narrow the scope of spreadsheets for scientific research, but this software product is quite suitable for teaching students and completing term papers. Subsequently, even partial calculations performed using Excel can be used to test the performance of programs written in high-level languages.

4 Experience third. Operational research

Spreadsheet programming is also quite possible to apply in some other engineering courses. The courses, under general name of "Operational Research" are studied by students not only of two faculties: "Aerospace institute" and "Aviation technology institute", but also by some other technical institutes. In particular, there is experience in using the option "Search for a solution" in order to work with problems related to linear and non-linear programming, as well as extreme programming and solving probabilistic and game problems in courses of disciplines called "Operations Research".

Of course, outside the programming environment in high-level languages, and using only spreadsheets, we can only solve small tasks.

In such disciplines, the essence of the use of spreadsheets is as follows. The student creates problem models, using the spreadsheets and solves them without programming [6, 7]. On the one hand, this seems to be a disadvantage, but there are also advantages: it becomes possible to show the student a wide class of operations research problems [8–10].

The course contains the following tasks:

1. Linear tasks of operational research:

   a) distributive tasks (this section includes various tasks of production planning, subject to the variation of the model range of similar tasks);
   b) tasks, which apply transport models (mainly logistical tasks);
   c) assignment tasks (efficient organization of transport routes, distribution employees by vacancies);
   d) tasks, which are based on the guillotine cutting.

2. Non-linear tasks of the operational research:

   a) tasks, which apply inventory management models;
   b) financial tasks (Markowitz and Sharp models with different variations, and many others);
   c) the tasks of queuing systems;
   d) game tasks, including games with nature;
   e) the statistically distributed (stochastic) tasks with application of Monte-Carlo method.

3. Extremal problems:

    a) various tasks based on the salesman problem;

    b) scheduling tasks;

    c) tasks, based on the knapsack problem, etc.;

4. Tasks of searching and monitoring.

In addition we can say that since by now there are a huge number of proposals on the Internet for programming simple tasks of operations research (not the fact that they are created without errors). And presented approach with applying MS Excel (subject to the possibility of checking the student's work by the teacher for plagiarism) creates good conditions for fixing the material of practical classes and laboratory work.

The course also contains term papers, where model tasks can be solved not only applying spreadsheets, but also using high-level languages. The use of graphic technologies is also required to create illustrations accompanying term papers.

In the future, the transition to the programming of individual tasks is carried out as part of the work on course projects. However, spreadsheets allow program verification for course projects created in high-level algorithmic languages.

Three semesters of studying different tasks of operational research are combined by a common and single ideology of solving practical problems.

5 Experience fourth. Training in MS Office products for teachers of various disciplines MAI

For many teachers, the transition to distance education was accompanied by the need to create presentation lectures. And this is a kind of art. Some of the teachers needed to introduce spreadsheet calculations into the discipline. Almost everyone needed to transform their lectures into electronic format and create tutorials and textbooks also into electronic form. Many of them needed to provide lectures and textbooks with graphic illustrations. It requires the ability to use the MS Visio (graphic vector editor).

For these purposes, a two-semester advanced training course for MAI teachers was developed, which has been in existence for three years [11, 12]. This course, which is intended for teacher's training, combines the publishing functions of MS Word, the drawing tools of MS Visio, the computing power of MS Excel, and the demonstration tools of MS Power Point, which corresponds to the rules of a seminar or colloquium. This allows teachers not only quickly create presentations of their lectures for LMS.MAI, but also practice the layout of both monographs and study guides, in conditions of strict regulation of the number of printed sheets.

Transition from paper technology of knowledge transfer to paperless is currently important and opens up additional opportunities for knowledge acquisition. A possibility to create textbooks in the form of an original layout, which is ready to run on a printing press, in addition, such technology allows to do without a printing machine at once creating electronic textbooks.

6 Conclusions

Based on the methods already developed, digital technologies should be introduced not only to transfer information, as was done during the pandemic and, apparently, will remain with us, but also in face-to-face communications with students.

In the future, it is advisable that practical and laboratory work should be carried out not so much in a notebook using a calculator, but on a computer using appropriate software applications.

Various forms should be used for calculations: from spreadsheets to building programs in any available programming languages. It is best to integrate one computing environment into another and/or combine different technologies with each other.

The variety of existing technologies makes it possible to expand the scope of existing courses of disciplines, adding not only the formulation of new topical problems, but also offering various modern methods for solving them.

References

[1] N.L. Malinina, O.L. Demidova, *Combinatorial Analysis* (MAI Publishing House, Moscow, 2021)

[2] I.B. Petrov, A.I. Lobanov, *Lectures on computational mathematics* (Internet University of Information Technologies, Moscow, 2006)

[3] O.L. Demidova, N.L. Malinina, *Solution of systems of linear equations, algebraic and transcendental equations in spreadsheets* (MAI Publishing House, Moscow, 2017)

[4] O.L. Demidova, N.L. Malinina, *Nonlinear systems. The theory of approximation of functions. Practice of calculations in spreadsheets* (MAI Publishing House, Moscow, 2021)

[5] A.D. Polyanin, *Handbook of linear differential equations* (FIZMATLIT, Moscow, 2001)

[6] N.L. Malinina, O.L. Demidova, *Solution of extremal problems of operations research in spreadsheets* (MAI Publishing House, Moscow, 2021)

[7] N.L. Malinina, O.L. Demidova, *Solution of non-linear problems of operations research in spreadsheets* (MAI Publishing House, Moscow, 2017)

[8] H.M. Markowitz, *Mean-variance analysis in portfolio choice and capital markets* (Blackwell, Oxford–Cambridge 1992)

[9] J. Sharpe, F. William, *Investments* (Prentice Hall, Upper Saddler River, NJ, 1998)

[10] M. Pattnaik, *Inventory Models: A Management Perspective: Methods & Application of EOQ and Fuzzy EOQ Models* (LAP LAMBERT Academic Publishing, 2014)

[11] O.L. Demidova, N.L. Malinina, *Programming in spreadsheets. Excel 2010* (Buki Vedi, Moscow, 2013)

[12] O.L. Demidova, N.L. Malinina, *Programming of text documents. Word 2010* (Buki Vedi, Moscow, 2013)

Mathematical methods of predicting determining factors characteristics of composites based on generalized durability models

Evgeny Gusev[1,2,*] and *Vladimir* Bakulin[3]

[1]Institute of the problems to Oils and Gas of Research Scientific Center "Yakut Scientific Center of the Siberian Branch RAS", 2, Petrovskogo st., Yakutsk, 677980, Russia
[2]Institute of Mathematics and Informatics, North-East Federal University, 48, Kulakovskogo st., Yakutsk, 677000, Russia
[3]Institute of Applied Mechanics of Russian Academy of Sciences, 7, Leningradsky prospekt, Moscow, 125040, Russia

Abstract. The possibility of creating effective methods for predicting the defining characteristics of composites in refined variational formulations has been investigated. The methodology being developed is based on methods related to the theoretical justification of the admissibility of schemes for narrowing the set of acceptable options, in which the desired globally optimal solution continues to be found on the resulting set of smaller dimensions. Comparative computational experiments were carried out on the basis of the developed methods based on the application of necessary and sufficient extremum conditions. The conducted computational experiments have shown the effectiveness of the developed approach associated with the gradual complication of forecasting models and the corresponding sequential refinement of the optimal parameters of the model.

1 Introduction

Currently, polymer composite materials, composite structures, composite coatings are becoming increasingly common in modern fields of physics, engineering, and instrumentation, which determine scientific and technological progress [1–4]. Such areas that determine modern scientific and technological progress, and are characterized by the ever-increasing use of polymer composites, include aircraft and space technology, shipbuilding, oil and gas industry, modern technology operating in the extreme conditions of the Arctic and Subarctic zones. In connection with the ever-increasing use of polymer composites in technology, the problem of predicting the reliability, durability, and residual life of composite materials and structures made of them is acquiring significant relevance [5–8].

A constructive analysis of promising ways of increasing the potentialities associated with solving the problems of predicting the residual resource, reliability, durability of polymer composites under the influence of extreme environmental factors and operational loads is carried out. A comparative analysis of existing approaches and their possible modifications made it possible to conclude that the most promising opportunities are associated with the

*e-mail: elgusev@mail.ru

development of a methodology for effectively narrowing search areas containing globally optimal solutions.

2 Formulation of the problem

As a defining characteristic of a polymer composite, its residual resource can be considered [9–11]. Models describing the time dependence of the residual resource R under the influence of extreme climatic factors in general can be presented in the form of functional dependencies of the following form [12–14]:

$$\begin{gathered} R(u_1, u_2, \ldots, u_n; t) = R_0 + \Phi(u_1, u_2, \ldots, u_n; t), \\ \Delta R(u, t) = \Phi(u, t), \quad \Delta R = R - R_0. \end{gathered} \tag{1}$$

The parameters of the optimal forecasting model $u^* = (u_1^*, u_2^*, \ldots, u_m^*)$ deliver an extreme value to the special efficiency criterion J, which is a complex multi-parameter function of the parameters of the forecasting problem. As an efficiency criterion as a complex function of the parameters of a forecasting problem of the form (1), it is advisable to choose the standard deviation of the approximation of experimental data $\tilde{R}_i$ from the theoretical values $R(u, t)$:

$$J(u_1, u_2, \ldots, u_n) = \frac{1}{m} \sum_{i=1}^{m} \left[R(u_1, u_2, \ldots, u_n; t_i) - \tilde{R}_i \right]^2. \tag{2}$$

We will assume that the experimental data adequately reflect the structure of the dependence of the change in the residual resource of the composite on the effect of extreme environmental factors, and the experimental data were obtained with insignificant errors that slightly distort the regularities of the behavior of real dependencies. Then the problem of reconstructing the parameters of models of composite materials from the influence of extreme environmental factors can be reduced to finding the absolute minimum of the quality function (2) in the extreme problem:

$$J(u^*) = \min_u J(u). \tag{3}$$

The vector of parameters $u^* = (u_1^*, u_2^*, \ldots, u_n^*)$ that deliver the absolute minimum to the efficiency indicator $J(u)$ (3) determines the optimal predicted dependence of the change in the residual life of the polymer composite under the influence of extreme environmental factors.

3 Materials and research methods

The study of the possibility of creating effective methods for predicting the defining characteristics of CM in refined variational statements based on the combined use of necessary and sufficient conditions for an extremum and methods of exhaustive enumeration has been carried out. The methodology being developed is based on the techniques associated with the theoretical substantiation of the admissibility of the schemes of narrowing the set of admissible options, in which the desired globally optimal solution continues to be found on the resulting set of lower dimension [15, 16].

On the basis of a constructive study of the structure of efficiency indicators in the investigated variational formulations of inverse problems of predicting the residual resource of composites, it was found that optimal solutions that deliver an absolute minimum to the corresponding efficiency indicators can additionally satisfy a certain system of ratios. The development of a methodology for identifying such systems of additional relations made it possible to significantly reduce the dimension of the initial variational formulations of inverse problems of forecasting the residual resource of composites and, accordingly, significantly narrow the initial set of admissible variants of forecasting models M of the form

$$M = \left\{R(u;t)\colon u = (u_1, u_2, \ldots, u_n) \in U \subset E^n,\ 0 \le t \le T_{\max}\right\} \tag{4}$$

to a narrow compact set M^*, on which the problem of constructing a globally optimal solution can be effectively solved. In these designations: $R(u,t)$ is the predicted time dependence of the residual resource of the composite; $u = (u_1, u_2, \ldots, u_n)$ is the vector of undefined parameters from the admissible set of parameters U, t—time, $[0, T_{\max}]$ is the considered time interval.

Conducted constructive analysis of the investigated variational formulations of inverse problems of forecasting the residual resource made it possible to establish that the necessary conditions for an extremum for efficiency indicators for a number of parameters simultaneously coincide with sufficient conditions. Let us denote such parameters by $u_{j_1}, u_{j_2}, \ldots, u_{j_q}$. The necessary and sufficient conditions for an extremum for a given selected system of parameters $u_{j_1}, u_{j_2}, \ldots, u_{j_q}$ can be taken as a system of functional relationships g_j $(j = 1, 2, \ldots, q)$, which can serve as the basis for the subsequent significant reduction in the dimension of the original forecasting problem and simplification of its structure.

Based on the application of the modern mathematical apparatus of constructive analysis of the system of relations, to which the necessary and sufficient conditions for an extremum lead, the selected system of q parameters $u_{j_1}, u_{j_2}, \ldots, u_{j_q}$ can be expressed in terms of the remaining system of $r = n - q$ parameters u_l $(l = 1, 2, \ldots, n;\ l \ne j_i, i = 1, 2, \ldots, q)$ multiparameter forecasting model:

$$u_{j_i} = \varphi_i\big(u_1, u_2, \ldots, u_{j_1-1}, u_{j_1+1}, \ldots, u_{j_q-1}, u_{j_q+1}, \ldots, u_{n-1}, u_n\big), \quad i = 1, 2, \ldots, q. \tag{5}$$

As a result, the developed approach made it possible to obtain a variational problem of a much smaller dimension and, accordingly, a simpler structure compared to the original variational problem, for which, taking into account, the established features of its structure, effective methods of nonlocal optimization were developed.

A statement is formulated that substantiates the approach associated with an effective reduction in the dimension of the original variational problem of predicting the residual life of composites.

Statement. Let the functions $g_i(u_1, u_2, \ldots, u_n)$, $(i = 1, 2, \ldots, q)$ be continuous in their arguments, there exist and are continuous partial derivatives $\partial g_i/\partial u_j$, $(i = 1, 2, \ldots, q;$ $j = 1, 2, \ldots, n)$; $u^* = (u_1^*, u_2^*, \ldots, u_n^*)$ is the optimal solution to the original forecasting problem, and the condition

$$\begin{gathered} rang\left[\nabla g_1(u^*), \nabla g_2(u^*), \ldots, \nabla g_q(u^*)\right] = q, \\ \nabla g_i(u^*) = \big(\partial g_i(u^*)/\partial u_1, \partial g_i(u^*)/\partial u_2, \ldots, \partial g_i(u^*)/\partial u_n\big), \\ (i = 1, 2, \ldots, q). \end{gathered} \tag{6}$$

Then, $\tilde{u}_{j_l}$ $(l = q + 1, q + 2, \ldots, n)$ if there is an optimal solution of the generated auxiliary variational problem of dimension $r = n - q$, then the optimal solution of the original forecasting problem of dimension n is determined by the relations:

$$\begin{gathered} u_{j_i}^* = \varphi_i\left(\tilde{u}_1, \tilde{u}_2, \ldots, \tilde{u}_{j_1-1}, \tilde{u}_{j_1+1}, \ldots, \tilde{u}_{j_q-1}, \tilde{u}_{j_q+1}, \ldots, \tilde{u}_{n-1}, \tilde{u}_n\right), \\ (i = 1, 2, \ldots, q), \\ u_l^* = \tilde{u}_l, \quad l \ne j_i, \quad i = 1, 2, \ldots, q. \end{gathered} \tag{7}$$

The system of functional relations g_j $(j = 1, 2, \ldots, q)$, appearing in the formulation of Statement 1, reflects the necessary and sufficient conditions for an extremum in the forecasting problem under consideration.

According to the developed methods, based on the use of necessary and sufficient extremum conditions for effective constructive transformation of variational statements of inverse problems of predicting the residual resource of composites to variational problems of lower dimension, comparative computational experiments were carried out, which showed the effectiveness of the developed approach associated with the gradual complication of forecasting models and the corresponding sequential refinement optimal parameters of the model.

Computational experiments have shown the high efficiency of the developed approach for constructing optimal forecasting models of optimal complexity based on immersing the original forecasting problem into a parametric family of problems, in which the parameter is the number of uncertain model parameters.

4 Influence of the retrospection interval dimensions on the stability of the forecast accuracy assessment when applying the optimal generalized forecasting models

The study of the effect of the size of the retrospection interval $[0, T_{\min}]$, where the exposure is carried out, on the stability of the forecast accuracy assessment on the predicted time interval $[T_{\min}, T_{\max}]$ when using the optimal generalized models of forecasting the optimal complexity has been carried out.

The general case was considered when polymeric and composite materials can be simultaneously affected in various combinations by several different factors Fj associated with hardening processes, exposure to solar radiation, moisture saturation, exposure to ultraviolet radiation, exposure to extreme climatic factors, etc. A generalized model was investigated that describes the simultaneous impact of several factors. Under the assumptions that various physical factors exert an effect on a complex-built composite that is independent of the effect of other factors, and the changes caused in the composite are summarized, the studied generalized model of durability can be represented as:

$$R = R_0 + \sum_{j=1}^{p} F_j(u_{j,1}, u_{j,2}, \ldots, u_{j,l_j}; t). \tag{8}$$

Each of the functions $F_j(u_{j,1}, u_{j,2}, \ldots, u_{j,l_j}; t)$, $(j = 1, 2, \ldots, p)$ describing the effect of the j-th factor on a complexly constructed composite was presented in the form of a series expansion in a certain system of basis functions $\psi_{kj}(\beta_{kj}; t)$, $(k = 1, 2, 3, \ldots)$, which most fully characterize the features of the process of increasing material damage under the influence of extreme environmental factors.

$$F_j = \sum_{k=1}^{\infty} \alpha_{kj}(u_{j,1}, \ldots, u_{j,l_j})\, \psi_{kj}[\beta_{kj}(u_{j,1}, \ldots, u_{j,l_j}); t], \quad (j = 1, 2, \ldots, p). \tag{9}$$

In these designations: $\alpha_{kj}(u_{j,1}, \ldots, u_{j,l_j})$, $\beta_{kj}(u_{j,1}, \ldots, u_{j,l_j})$, $(j = 1, 2, \ldots, p;\ k = 0, 1, 2, \ldots)$ undefined parameters of the model describing the impact of the j-th factor. The system of Arrhenius functions $\psi_{kj}(\beta_{kj}; t)$, $(k = 1, 2, 3, \ldots)$ was taken as a system of basis functions.

The construction of optimal generalized models of durability of optimal complexity was carried out on the basis of the developed approach based on the principle of multiplicity of forecasting models within the framework of refined variational statements of inverse forecasting problems.

The regularities of dependence of optimal models of durability of optimal complexity $R^*_\tau(t) = R(u^*_\tau; t)$ $(0 \le \tau \le T_{\max}, 0 \le t \le T_{\max})$, built on the basis of the developed approach, based on the principle of multiplicity of forecasting models, have been investigated. Within the framework of the refined variational formulations of inverse forecasting problems, a parametric family of forecasting problems was considered, in which the parameter is the current size of the retrospection interval τ $(0 \le \tau \le T_{\max})$. u^*_τ—the optimal vector of uncertain parameters, which determines the optimal model of the optimal complexity corresponding to the parameter τ.

The important regularities of the structure of optimal models of durability of optimal complexity when changing the size of the retrospection interval have been established. Areas of stability and instability are highlighted, a qualitative characteristic of these areas is carried out, their structural features are investigated the behavior of optimal models of durability in these areas, the boundaries of these areas are estimated.

Figure 1 shows a graph of the dependence of the parametric family of optimal forecasting models of optimal complexity $R^*(\tau_r; t)$, depending on the parameter τ_r—the size of the time interval of retrospection $(0 \le tr \le T_{\max})$. The strength measured in Mpa is considered as the residual resource R of the investigated PVC type KM. The studied interval for studying the change in the residual resource of KM is $T_{\max} = 15$ years.

It was found that at the retrospection intervals the size of which τ_r is insufficient to identify the fundamental regularities of the micro- and macrostructure of the CM, which determine the nature of the change in the residual CM resource for a long period, the dependence of the parametric family of optimal models for predicting the optimal complexity $R^*(\tau_r; t)$ depending on the parameter τ_r has an unstable oscillatory character, while the amplitude fluctuations decreases as the parameter τ_r increases. At retrospection intervals, the size of which τ_r is already sufficient to highlight the fundamental regularities of the micro- and macrostructure of CM, which determine the nature of the change in the residual CM resource for a long period, the dependence of the parametric family of optimal models for predicting the optimal complexity $R^*(\tau_r; t)$ depending on the parameter τ_r acquires a monotonous stable character and smoothly approaches the real dependence residual resource as it increases τ_r (figure 1).

For the studied CM, already after seven years of observations $(\tau_r = 7)$, the dependence of the parametric family of optimal models for predicting the optimal complexity $R^*(\tau_r; t)$, depending on the parameter τ_r, acquires a monotonous stable character and does not significantly differ from the actual dependence of the residual resource.

Thus, it has been established that the success of the effective solution of the problems of predicting the residual life of composites largely depends on how adequately it is possible to take into account in the models of durability the fundamental regularities inherent in the processes occurring both at the micro- and at the macrolevel, which have a significant effect on the change in the residual resource of the composite. In accordance with this, an objective assessment of the minimum size of the retrospection interval τ_r, at which these fundamental regularities are already manifested in full, sufficient to solve the problems of effective forecasting, becomes important.

At the same time, it was found that the dependence of the parametric family of optimal forecasting models of optimal complexity $R^*(\tau_r; t)$ depending on the parameter τ_r characterizing the size of the retrospection interval is oscillatory. In this case, the nature of the oscillatory process largely depends on whether the fundamental regularities that characterize the processes occurring at the micro and macro levels, which have a decisive effect on the change in the residual resource of the composite, are sufficiently manifested in the studied time interval.

The given types of stable and unstable oscillatory processes, to which the parametric families of optimal models for predicting the optimal complexity $R^*(\tau_r; t)$ when the parameter τ_r

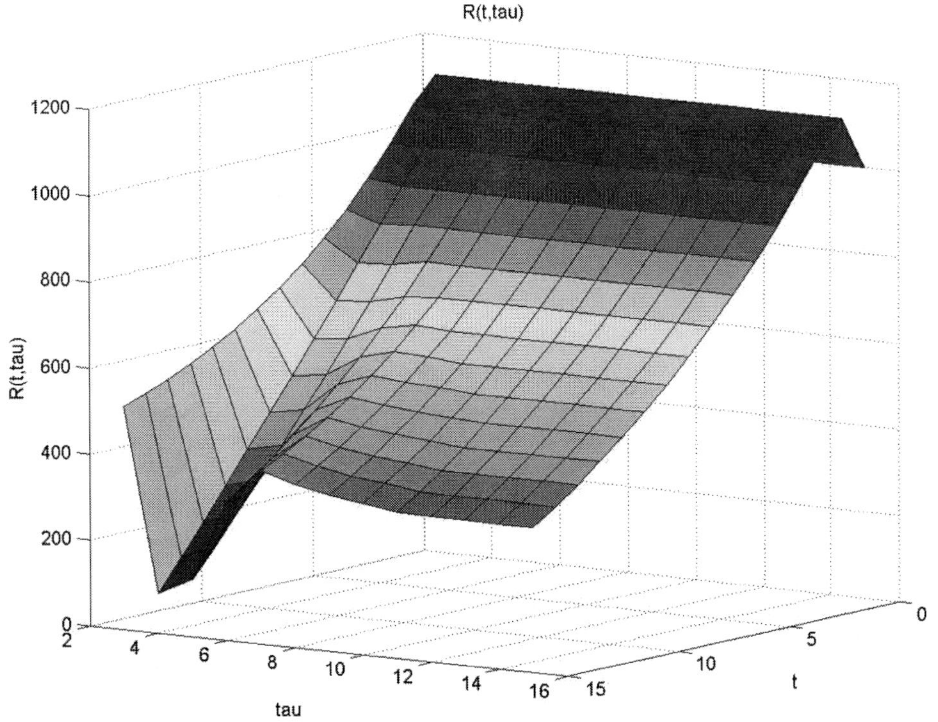

Figure 1. A graph of the dependence of the parametric family of optimal forecasting models of optimal complexity $R^*(\tau_r; t)$ depending on the parameter τ_r the size of the retrospection time interval $(0 \leq \tau_r \leq T_{\max})$

changes, are similar to oscillatory transient processes in stable and unstable automatic control systems (ACS).

5 Conclusion

Prospective ways of increasing the potentialities associated with solving the problems of predicting the residual resource, reliability, durability of polymer composites under the influence of extreme environmental factors are analyzed. A comparative analysis of existing approaches and their possible modifications made it possible to conclude that the most promising opportunities are associated with the development of a methodology for effectively narrowing search areas containing globally optimal solutions. The study of the influence of the size of the retrospection interval on the stability of the forecast accuracy estimate when using optimal generalized forecasting models of optimal complexity is carried out. Comparative computational experiments were carried out according to the developed methods based on the use of necessary and sufficient conditions for an extremum for an effective constructive transformation of variational statements of inverse problems of predicting the residual resource of composites to variational problems of lower dimension.

It is shown that the success of the effective solution of the problems of predicting the residual life of composites largely depends on how adequately it is possible to take into account in the models of durability the fundamental regularities inherent in the processes occurring both at the micro- and at the macro-level, and having a significant effect on the change in the residual resource of the composite. In accordance with this, an objective assessment of the minimum size of the retrospection interval at which these fundamental regularities are already manifested in full, sufficient for solving the problems of effective forecasting, becomes important.

The work was carried out within the framework of state tasks of the Institute of Oil and Gas Problems of the Federal Research Center, the Institute of Problems of Mechanics of the Russian Academy of Sciences.

References

[1] M.N. Stepanov, A.V. Zimin, *Prediction of fatigue resistance characteristics of materials and structural elements* (Innovatsionnoe Mashinostroenie, Moscow, 2007)

[2] S.A. Lurie, Yu.O. Solyaev, D.Q. Nguen et al., *Composites: Mechanics, Computations, Applications: An International Journal*, **3**, 1 (2015)

[3] E.L. Gusev, *Mathematical methods for the synthesis of layered structures* (Nauka, Novosibirsk, 1993)

[4] E.L. Gusev, *Qualitative regularities of the relationship of parameters in optimal structures in problems of optimal synthesis of inhomogeneous structures from a discrete set of materials under wave effects*, Reports of the Academy of Sciences, **346**, 324 (1996)

[5] E.L. Gusev, V.N. Bakulin, *Optimum design of structurally nonhomogeneous materials and constructions with required properties*, J. Eng. Phys. Thermophys., **89**, 260 (2016)

[6] E.L. Gusev, V.N. Bakulin, *Optimal control under a decrease in the thermal-field intensity based on selection of the heterogeneous-construction structure in the variational formulation*, Doklady physics, **63**, 213 (2018)

[7] E.L. Gusev, V.N. Bakulin, *Variation formulations of inverse problems in forecasting the residual life of composites*, Doklady physics, **63**, 388 (2018)

[8] A.I. Reutov, *Predicting the reliability of building products made of polymer materials* (Stroymaterialy, Moscow, 2007)

[9] L.D. Potapova, V.P. Yartsev, *Mechanics of materials under complex stress states: how are ultimate stresses predicted?* (Mashinostroenie, Moscow, 2005)

[10] A.I. Reutov, A.A. Sidorenko, I.S. Filatov, *Assessment of the climatic stability of polymer materials based on the probabilistic-statistical method*, Bulletin of the Tomsk Polytechnic University, **305**, 196 (2002)

[11] A.I. Reutov, *Forecasting to climatic stability polymeric materials, applicable in construction*, Vestnik Tomskogo gosudarstvennogo architecturno-stroitelnogo universiteta, **2**, 127 (2009)

[12] E.L. Gusev, *Using parallel procedures for the searching of the extremum for the decision of the inverse problems prediction of the defining characteristics of the composite materials*, Mater. Phys. Mech., **26**, 70 (2016)

[13] E.L. Gusev, V.N. Bakulin, V.D. Chernykh, *Development of combined search methods for efficiency indicator extreme in variation statement of forecasting tasks for determine characteristics of composite materials*, J. Phys.: Conf. Ser., **1392**, 012008 (2019)

[14] E.L. Gusev, F.I. Babenko, *Materials for technical devices and structures used in the Arctic* (All-Russian Institute of Aviation Materials, Moscow, 2015)

[15] E.L. Gusev, E.L. Research, *Methods of the Limit Possibilities of Structurally Inhomogeneous Structures with a Desired Set of Properties during Wave Actions in the Variation Statement*, J. Mach. Manuf. Reliab., 44, 148 (2015)

[16] E.L. Gusev, V.N. Bakulin, *Variation statement of the optimal design problem of composition constructions with the required complex of properties*, Mech. Compos. Mater., **51**, 637 (2015)

Mathematical methods for predicting the durability of polymer composites under extreme conditions

Evgeny Gusev[1,2,*], *Vladimir* Bakulin[3], and *Vera* Chernykh[1,2]

[1]Institute of the problems to Oils and Gas of Research Scientific Center "Yakut Scientific Center of the Siberian Branch RAS", 2, Petrovskogo st., Yakutsk, 677980, Russia
[2]Institute of Mathematics and Informatics, North-East Federal University, 48, Kulakovskogo st., Yakutsk, 677000, Russia
[3]Institute of Applied Mechanics of Russian Academy of Sciences, 7, Leningradsky prospekt, Moscow, 125040, Russia

Abstract. The paper investigates the issue of promising ways to predict the durability of polymer composites under extreme conditions. A study of the possibility of predicting the durability of the material, based on the information obtained in the first years of operation. Based on the modern provisions of molecular kinetic theory (MKT), the issue of matching the defining parameters of mathematical models calculated because of solving forecasting problems within the framework of the formulated variational formulation, considering the results of experimental measurements at the macro level, with the corresponding defining parameters of physical models describing molecular interactions at the microlevel, was investigated. The results of computational experiments are presented.

1 Introduction

In recent decades, increasing attention has been paid to the problem of creating reliable methods for quantifying the performance of products made of polymer and composite materials [1–3]. The importance of developing effective high-precision methods for long-term forecasting of the defining characteristics of composites (residual life, strength, reliability, durability) is since composite materials and structures made of them are an integral part of modern technology in such areas that determine scientific and technological progress as aviation and space technology, shipbuilding, oil, and gas industry, etc. [2–4].

Currently, the problems of mathematical modeling of problems of long-term forecasting of the durability of structures made of polymer composites under extreme conditions, under the influence of extreme climatic environmental factors, are solved in significantly simplified formulations that do not consider a significant number of factors that have a significant impact on the accuracy of the solution. As a rule, the applied forecasting models have a small number of uncertain parameters, which does not allow achieving the necessary degree of adequacy of the model of the real situation. At the same time, an integral part of the problem of developing effective methods for predicting the defining characteristics of composites is solving the problem of effectively constructing globally optimal solutions that evaluate the effectiveness of the constructed solutions to forecasting problems.

*e-mail: elgusev@mail.ru

The specificity of the studied problems of predicting the defining characteristics of polymer composite materials (PCM) under extreme conditions leads to the fact that for these problems the extrapolation approach in the traditional formulation is ineffective. As noted in several papers [5–8], a fundamental improvement of the extrapolation approach can be achieved based on the choice of a mathematical model with a focus on physical representations. The paper investigates the prospects of predicting the durability of polymer composites in extreme conditions. A study of the possibility of predicting the durability of materials, based on the information obtained in the first years of operation.

2 Application of the basic provisions of the molecular kinetic theory for the construction of generalized durability models

The development of generalized models of the durability of composites under extreme conditions is based on the main provisions of the modern molecular kinetic theory (MKT). This is because chemical transformations in composites can occur both at intermolecular and intramolecular levels.

It is known from the Boltzmann distribution for the kinetic energy of molecules that the number of molecules with energy E greater than the activation energy E_a is proportional to the exponent $\exp(-E_a/RT)$. As a result, the rate of a chemical reaction is represented by the Arrhenius equation obtained from thermodynamic considerations, which has the form in functional and differential forms:

$$k = k_0 \exp(-E_a/RT), \quad d(\ln k)/dT = E_a/RT^2. \tag{1}$$

Analysis of the Arrhenius equation (1) in differential form shows that the greater the activation energy E_a, the faster the reaction rate constant k increases with increasing temperature. In these notations, R is the Boltzmann constant. The frequency factor k_0 characterizes the collision frequency of reacting molecules (the number of active collisions of particles per unit volume in the reaction mixture). The frequency factor k_0 expresses the fraction of particles whose energy is equal to or greater than the activation energy E_a. The activation energy E_a is an excess of energy relative to the average energy of the particles at a given temperature, which is necessary for the reacting particles to enter a chemical reaction and is determined by the properties of the reacting particles, their energy state.

Following the main provisions of the MKT, the study was based on a physical model in which the nature of the change in time of the determining property of the composite R under the simultaneous influence of several extreme factors is determined by elementary chemical reactions at the molecular level described by the superposition of Arrhenius equations describing various types of molecular interactions initiated by the influence of various extreme factors $F_1, F_2, \ldots, F_p$ on the composite.

If various physical factors have an impact on the polymer composite, independent of the impact of other factors, and the changes caused in the composite are summed up, it can be assumed that a generalized durability model describing the simultaneous impact of several factors can be represented as:

$$R = R_0 + \sum_{j=1}^{p} F_j(u_{j,1}, u_{j,2}, \ldots, u_{j,l_j}; t). \tag{2}$$

By the current provisions of the MKT, it was assumed that the effect of each of the F_j factors on the composite activates at the micro-level a set of destructive elementary processes, the chemical reactions occurring in which can be described in the form of a superposition of Arrhenius equations. Each of the Arrhenius equations that make up the superposition describes a certain type of elementary chemical reaction, with its inherent parameters—the frequency factor characterizing the frequency of collisions of reacting molecules, and the activation energy.

3 Variational formulation of forecasting problems of defining characteristics of composites

Following the formulated provisions of the MKT, each factor F_j can be represented as a superposition of the Arrhenius equations in the following form:

$$F_j(u_{j,1},\dots,u_{j,l_j};t) = \sum_{k-1}^{N_j} \alpha_{kj}(u_{j,1},\dots,u_{j,l_j})\left[\exp\left(\beta_{kj}(u_{j,1},\dots,u_{j,l_j})t\right)-1\right], \quad (3)$$
$$j=1,2,\dots,p; \quad 0 \le t \le T_{\max}.$$

N_j is the number of elementary destructive processes, the course of chemical reactions in which is activated when the j-th factor is exposed to the composite, $u_{j_1}, u_{j_2}, \dots, u_{j,l_j}$—parameters describing the nature of the impact of the j-th factor on the composite.

The parameters $\alpha_{kj} = \alpha_{kj}(u_{j_1}, u_{j_2}, \dots, u_{j,l_j})$ express the fraction of particles whose energy is greater than or equal to the activation energy required to initiate the k-th elementary chemical reaction ($k = 1,2,\dots,N_j$) when exposed to the composite factor F_j; the parameters $\beta_{kj} = \beta_{kj}(u_{j_1}, u_{j_2}, \dots, u_{j,l_j})$ are related to the activation energy, i.e. with an excess of energy relative to the average energy of the particles at a given temperature, necessary for reacting particles to initiate the k-th elementary chemical reaction ($k = 1,2,\dots,N_j$) when exposed to the composite factor F_j. The excess energy associated with the activation energy is determined by the properties of the reacting particles, their energy state.

Variational formulation of forecasting problems of defining characteristics of composites: Based on short-term tests conducted on the time interval of retrospection $[0, T_{\min}]$, it is necessary to find optimal parameters $u_1^*, u_2^*, \dots, u_n^*$ of a generalized durability model $R(u^*;t)$ that deliver a global minimum to the efficiency criterion (4) on a set of multiparametric generalized forecasting models $\{R(u;t)\}$ that allow forecasting with an error not exceeding the predetermined maximum permissible error of the forecast $\gamma_R^{\max}$ (5):

$$\tilde{S} = S(u,Q) = \frac{1}{m}\sum_{i=1}^{m}\left[R(u_1,u_2,\dots,u_m;t)-\tilde{R}_i\right]^2 \Rightarrow \min_{u\in U}, \quad (4)$$

$$\max_{T_{\min}\le t\le T_{\max}} \left|\tilde{R}(t)-R^*(t)\right| \le \gamma_R^{\max}. \quad (5)$$

In these notations: $\tilde{R}_i = R_i(Q)$—measured values of the defining characteristic R at time points $t_1, t_2, \dots, t_m$, considering measurement errors $Q_1, Q_2, \dots, Q_m$. $R^*(t)$ real-time dependence. Vector of undefined parameters $u^*(Q)$ delivers the global minimum of the total standard error $\tilde{S}$ (4):

$$S(u^*(Q);Q) = \min_{u\in U} S(u;Q). \quad (6)$$

4 Problems of matching mathematical parameters and physical models within the framework of the formulated variational statement

Based on the modern provisions of the molecular kinetic theory, the question of matching the defining parameters of mathematical models calculated because of solving forecasting problems within the framework of the formulated variational formulation, considering the results of experimental measurements at the macro level, with the corresponding defining parameters of physical models describing molecular interactions at the micro-level, was investigated.

Several fundamental principles and concepts were introduced that allow for optimal coordination of the defining parameters calculated based on the constructed mathematical models at the macro level with the defining parameters of physical models describing molecular interactions at the micro-level.

Such concepts and principles as the principle of a multiplicity of forecasting models, forecasting models of optimal structure and complexity, maximum permissible accuracy of the forecast, etc. are introduced. Within the framework of the formulated variational statements of forecasting problems, a parametric family of generalized durability models $\{R^N(;t)\}$, $(1 \leq N < \infty)$ has been studied, in which the parameter is the number of terms N in the decomposition of the durability function into a series. Within the framework of the introduced parametric family of generalized durability models $\{R^N\}_{N=1}^{\infty}$, a forecasting model of optimal structure and complexity is constructed, which is understood as a model containing the optimal number of terms that allows solving the forecasting problem with the required accuracy.

The problem of constructing a model of optimal structure and complexity is reduced to solving the following extreme problem:

$$\max_{T_{\min}\leq t\leq T_{\max}} \left|R^{N^*}(u^{*N^*};t) - R^*(t)\right| = \min_{1\leq N<\infty} \max_{T_{\min}\leq t\leq T_{\max}} \left|R^N(u^{*N};t) - R^*(t)\right|,$$
$$\max_{T_{\min}\leq t\leq T_{\max}} \left|R^N(u^{*N};t) - R^*(t)\right| \leq \gamma_{\max}, \quad 1 \leq N < \infty. \tag{7}$$

In these notations, $N^*$is the optimal number of parameters of the prediction model of optimal structure and complexity R^{N^*}; u^{*N}—the optimal vector of parameters of the optimal forecasting model of the parametric family corresponding to the parameter N; $\gamma_{\max}$—the maximum allowable error in solving the prediction problem.

By the principle of the multiplicity of forecasting models, it is considered that if in each multiparametric family of generalized forecasting models there is a model that is most adequate to the real predicted time dependence $R^*(t)$, then this model in the considered parametric family is a model of optimal structure and complexity.

The physical problems of restoring the parameters of destructive chemical reactions initiated by the action of extreme environmental factors are formulated in mathematical form within the framework of refined variational formulations of inverse problems of predicting the defining characteristics of PCM. Following the formulated principle of the multiplicity of forecasting models and the introduced concept of optimal forecasting models of optimal structure and complexity in the variational model [9–11], the task of restoring uncertain parameters can be reduced to finding the global minimum of the functional J, assessing the degree of adequacy of the model describing the impact of extreme environmental factors on the composite

$$J\left(\left\{\left(u^*_{j,1}, u^*_{j,2}, \ldots, u^*_{j,l_j}\right)\right\}_{j=1}^{p}\right) = \min_{\{u_{jk}\}} J\left(\left\{\left(u_{j,1}, u_{j,2}, \ldots, u_{j,l_j}\right)\right\}_{j=1}^{p}\right). \tag{8}$$

To construct the global minimum in the extreme problem (8), combined methods of searching for the global extremum have been developed, based on the combined application of necessary and sufficient conditions of the extremum and methods of full search.

5 Results of computational experiments based on generalized durability models

Figure 1 shows the results of computational experiments on predicting the residual life of polymer fiber composites (PFC) within the framework of the developed methodology. The simultaneous effect of two factors was considered: the hardening factor and the aging factor.

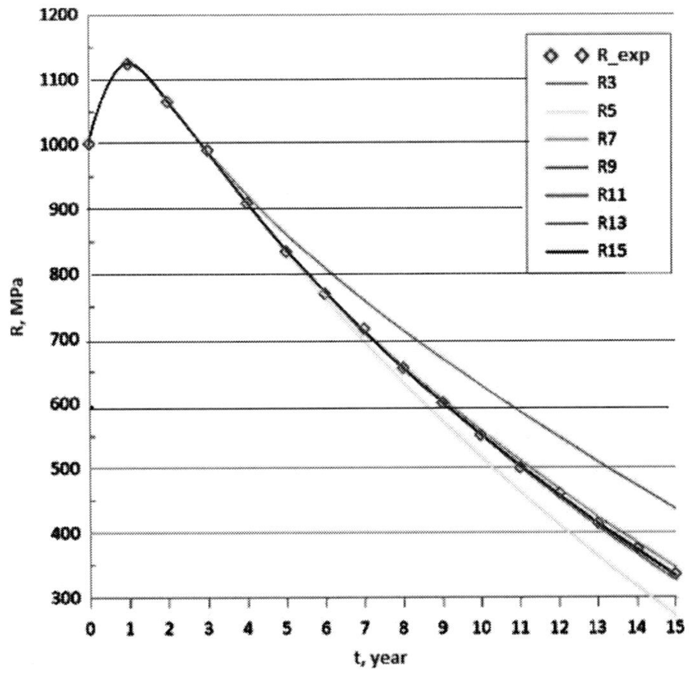

Figure 1. Optimal durability functions constructed based on optimal generalized forecasting models for retrospection intervals, the sizes of which are sequentially 3, 5, 7, 9, 11, and 13 years

As initial data for the construction of generalized models of the durability of the optimal structure and complexity based on the Arrhenius equations, the results of physical experiments on measuring the residual life of PFC were used.

Based on the computational experiments carried out according to the developed methodology, optimal multiparametric generalized models of the durability of the optimal structure and complexity were constructed with a sequential increase in the size of the time interval of retrospection.

The graph shows simultaneously all the optimal durability functions calculated based on the developed methodology, respectively, for the retrospection intervals, the sizes of which are sequentially 3, 5, 7, 9, 11, and 13 years. Experimental values are marked with diamonds on the graph. In the given designations: R3—forecast of the residual resource according to the data for the first 3 years of exposure, R5—according to the data for the first 5 years, etc.

Conducting computational experiments based on the developed methodology can help achieve a balance between exposure time and the accuracy of the forecast results.

6 Conclusion

The developed methodology for matching the parameters of mathematical and physical models at the micro and macro levels allowed solving the problem of restoring the parameters of physics-chemical processes occurring at the molecular level and leading to destructive changes in composites and deterioration of their characteristics over time.

An objective assessment of the parameters of destructive elementary chemical reactions in composites based on the conducted physical experiments made it possible to build effective generalized forecasting models for the long term, to conduct a constructive analysis of the influence of individual extreme factors on the durability of the composite, to give a comparative assessment of the degree of influence of one or another extreme factor on the durability of the composite. Knowledge of the parameters of destructive elementary chemical reactions in the composite initiated by the action of extreme environmental factors, their comparative constructive analysis, can allow us to develop a methodology for the synthesis of new materials with increased durability.

The work was carried out within the framework of tasks of the Institute of Oil and Gas Problems of the Federal Research Center and the Institute of Problems of Mechanics of the Russian Academy of Sciences.

References

[1] A.Y. Bashkarev, V.I. Vettegren, M.A. Suslov, *Longevity polymeric composites* (Publishing House of Saint Petersburg Polytechnic University, Saint Petersburg, 2016)

[3] A.I. Reutov, *Forecasting to climatic stability polymeric materials, applicable in construction*, Vestnik Tomskogo gosudarstvennogo architecturno-stroitelnogo universiteta, **2**, 127 (2009)

[3] S.A. Lurie, Yu.O. Solyaev, D.Q. Nguen et al., *Composites: Mechanics, Computations, Applications: An International Journal*, **3**, 1 (2015)

[4] A.M. Dumansky, L.P. Tairova, *Time-dependent behavior of carbon fiber reinforced laminates*, in Proceedings of the Second Int. Conference on Advanced Composite Materials and Technologies for Aerospace Applications, 11–13 June 2012, Wrexham, North Wales, United Kingdom, 75 (2012)

[5] U.S. Urzhumtsev, I.N. Chersky, *Scientific foundations of engineering climatology of polymer and composite materials*, Mechanics of Composite Materials, **4**, 708 (1985)

[6] V.N. Bulmanis, V.A. Yartsev, V.V. Krivonos, *Efficiency design from polymeric composites at influence of the steady-state loads and climatic factors*, Mechanics of Composite Materials, **5**, 915 (1987)

[7] O.N. Karpuhin, *Determining the service life of a polymer material as a physics-chemical problem*, Uspehi himii, **8**, 1523 (1980)

[8] I.S. Filatov, R.N. Bochkarev, *Some problems of assessment and forecasting of climatic stability of polymer materials*, Methods for assessing climatic stability of polymer materials, 11 (YF AN SSSR, Yakutsk, 1986)

[9] E.L. Gusev, V.N. Bakulin, *Variation formulations of inverse problems in forecasting the residual life of composites*, Doklady physics, **63**, 388 (2018)

[10] E.L. Gusev, *Using parallel procedures for the searching of the extremum for the decision of the inverse problems prediction of the defining characteristics of the composite materials*, Material Physics and Mechanics, **26**, 70 (2016)

[11] E.L. Gusev, V.N. Bakulin, *Optimal control under a decrease in the thermal-field intensity based on selection of the heterogeneous-construction structure in the variational formulation*, Doklady physics, **63**, 213 (2018)

Application of spike neural network for stabilizing pendulum in the nonlinear formulation

Anastasia Ivanova[1,*], *Regina* Khamatnurova[2], *Nikita* Kharin[1], *Tatyana* Baltina[1], and *Oskar* Sachenkov[1]

[1]Kazan Federal University, 18, Kremlyovskaya st., Kazan, 420008, Russia
[2]Goethe-Universität Frankfurt am Main, 6 (PEG Building), Theodor-W.-Adorno-Platz, Frankfurt am Main, 60323, Germany

Abstract. The article describes the solution to the problem of stabilizing a non-linear system using machine learning methods. Neural networks are one of the promising directions in this area. The article describes a model of spiking neural network, which differs from previous generations of networks by its similarity to biological neurons. A pendulum on an elastic foundation was chosen as a dynamic system for the study. The input layer of the neural network is the so-called sensory neuron, and information about the deviation of the pendulum from the equilibrium position was received on it. The Leaky Integrate-and-Fire model of the spiking neural network was used. The article shows the process of stabilization of a pendulum on an elastic foundation. The closed system was built and a method for a numerical solution was implemented. Two configurations of control functions have been considered. It is shown that the time required to bring the system into a steady equilibrium state depends on the choice of the control function.

1 Introduction

Despite the existence of classical approaches for solving the optimal control problem, the need for other approaches still remains. Nowadays, solving this class of problems using artificial intelligence has become popular. One of the promising areas in this field is neural networks. A spiking neural network is an artificial neural network of the third generation, its mathematical model is as close as possible to biological processes. The process of generating a spike (neuron output value) depends on the sequence of signal pulses received at the input [1, 2]. Despite the complexity of the model and the need to solve a system of differential equations at each calculation step, the advantage of spiking neural networks over multilayer perceptrons has been shown [3]. Due to their higher computational power, such networks require fewer neurons in their architecture.

The application of spiking neural networks in control problems is gaining popularity [4–9]. Such an approach makes it possible to depart from classical methods for solving optimal control problems: Pontryagin's maximum principle and Bellman's dynamic programming. Since these methods require complex computational processes, they are not always conveniently implemented in practice, especially in cases of discontinuous control function

*e-mail: ivanovaanastasya0449@gmail.com

or complex dynamic systems. It was shown in [10] the example of modeling of a virtual insect, how the potential of neural networks can be used in the future to solve practical problems. Other works [11–16] show how a spiking neural network can be used to build a self-learning control system of the virtual agent. However, the aforementioned articles used simplified mathematical models (e.g., do not take into account inertia force), which led to our study. The purpose of this paper is to apply spiking neural networks to complicated dynamic systems.

2 Materials and methods

The design of a spiking neural network is fundamentally different from a multilayer network by the mathematical model of the neuron itself. From a biological point of view, neurons are complex structures, due to which a number of scientific studies have been devoted to their mathematical description. Because of this, at the moment there are several corresponding mathematical models. In the research Leaky Integrate-and-Fire (LIF) model of a neuron was used. The LIF model can be described by an equation:

$$\tau_{mem}\frac{dV}{dt} = E_{leak} - V + \sum_{m=1}^{N_e} g_{e,m}(E_e - V) + \sum_{n=1}^{N_i} g_{i,n}(E_i - V), \tag{1}$$

where τ_{mem} is the membrane time constant, V is the membrane voltage, E_{leak} is the reversal potential for the leak, E_e is the reversal potential for excitatory (depolarizing) inputs, E_i is the reversal potential for inhibitory inputs, g_e is the excitatory conductance, g_i is the inhibitory conductance, N_e and N_i is the quantity of excitatory and inhibitory neurons, respectively.

Excitatory conduction can be described by the equation:

$$\frac{dg_{e,m}}{dt} = -\frac{g_{e,m}}{\tau_e} + w_{e,m}\sum_{s=1}^{N_m}\delta(t - t_s), \tag{2}$$

where τ_e is the time constant of postsynaptic potential, w_e is the strength of the excitatory synapse, t_s is the time of an excitatory input spike, δ is the Dirac delta function.

Inhibitory conduction can be described by the equation:

$$\frac{dg_{i,n}}{dt} = -\frac{g_{i,n}}{\tau_i} + w_{i,n}\sum_{p=1}^{N_p}\delta(t - t_p), \tag{3}$$

where τ_i is the time constant of postsynaptic potential, w_i is the strength of the inhibitory synapse, t_p is the time of an inhibitory input spike.

In this approach, each neuron has characterizing value—potential. The potential is compared to the neuron threshold value. And if the potential exceeds the threshold value, the neuron sends an impulse to the next layer and its potential drops to a relaxing level. Such a process calls spike. Otherwise, potential accumulation occurs.

Let us consider a pendulum on an elastic foundation. To derive the equations of motion we use Lagrange's equation, the angle of deviation from the equilibrium position φ and the vertical displacement x were taken as generalized coordinates (figure 1).

The kinetic energy of the system can be written as follows:

$$T = m\dot{x}^2 + \frac{m}{2}(\dot{\varphi}^2 l^2 + 2\dot{x}\dot{\varphi} l \sin\varphi), \tag{4}$$

where m is rod mass; l is rod length; x, φ are generalized coordinates.

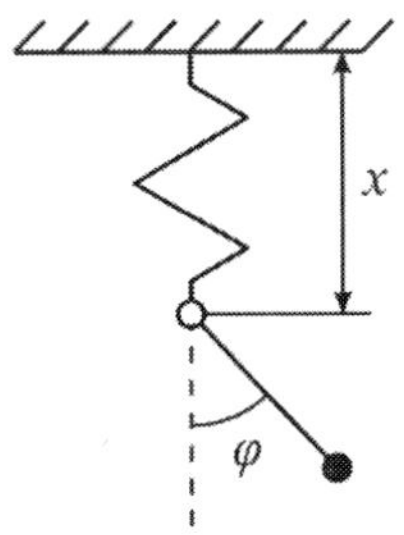

Figure 1. Pendulum with two degrees of freedom and generalized coordinates φ and x

The generalized forces can be written as follows:

$$
\begin{aligned}
Q_x &= -mg + kx, \\
Q_\varphi &= -\frac{mgl}{2}\sin\varphi.
\end{aligned}
\tag{5}
$$

Then differential equations of motion can be rewritten as follows:

$$
\begin{cases}
\ddot{\varphi} + \dfrac{kl\dot{x}\sin\varphi + 0.5l^2m\dot{\varphi}^2\sin 2\varphi}{\sin^2\varphi - 2l^2m} = U_\varphi, \\
\ddot{x} + \dfrac{0.5glm\sin^2\varphi + klx + m\dot{\varphi}^2l^2\cos\varphi}{lm(2-\sin^2\varphi)} = U_x,
\end{cases}
\tag{6}
$$

where U_φ, U_x—are generalized control functions.

Consider the problem of stabilization of a pendulum. The behavior of a dynamical system is described by two generalized coordinates corresponding to angular and translational oscillations. Let us introduce control functions for the corresponding oscillations and assume that they depend on the corresponding pairs of generalized coordinates. These control functions can be force factors for the corresponding character of motion, namely, moment for angular oscillations and force for translational oscillations (figure 2(a)), or two forces (figure 2(b)).

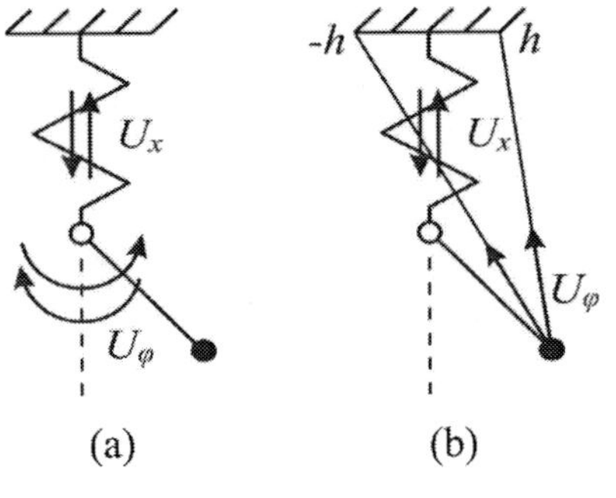

Figure 2. (a) The system with stabilizing forces and moments. (b) The system with stabilizing forces, here h is the distance from the suspension point of the pendulumto the point of thread fastening

At the initial moment of time, the pendulum is in a certain state: initial velocity and position. It is required to stabilize the system by the control functions. It was decided to use four networks at once, each consisting of one neuron responsible for each of the four control functions. The input layer for the first two neurons is the sensory neuron, which receives a discrete signal of constant amplitude with a frequency proportional to the deviation of the generalized coordinates from the equilibrium position. When the deviation is zero, no signal is applied. So, if the output spike from the corresponding neuron denotes as L, the generalized control functions can be introduced as follows:

$$\begin{cases} \dot{U}_\lambda = -R_\lambda \cdot U_\lambda, \\ U_\lambda(0) = C_\lambda \cdot L_\lambda, \end{cases} \tag{7}$$

where λ defines generalized coordinate—x or φ, R—corresponding damping coefficient, C—corresponding scale factor.

Equations (1)–(3), (6), and (7) with the initial states for the dynamic systems are the closed system for solving a stabilization problem.

Two options for generalized control functions are considered for angular coordinate:

$$\begin{aligned} U_\varphi^1(0) &= C_\varphi^1 \cdot L_\varphi, \\ U_\varphi^2(0) &= C_\varphi^2 \cdot L_\varphi \cdot \|\vec{e}_\varphi \times \vec{e}_h\|, \end{aligned} \tag{8}$$

where e_φ—direction cosines of the pendulum, e_h—direction cosines of corresponding force direction. Let's call them case 1 and case 2.

3 Results and discussion

The numerical simulation of stabilization of pendulum on an elastic foundation using spike neural network performed. In calculations constants for equations (1)–(3) were as follows: $\tau_{mem} = 0.2$ ms, $E_{leak} = -60$ mV, $E_e = 0$, $E_i = -80$ mV, $\tau_e = 0.05$ ms, $N_e = 1$, $N_i = 0$. In calculations constants for equations (6) were as follows: $m = 0.8$ kg, $l = 0.5$ m, $k = 100$ N/m. In calculations constants for equations (7) were as follows: $R_x = R_\varphi = 1$, $C_x = C_\varphi = 1$. Equations (1)–(3), (6), and (7) were solved numerically. Let's consider the results of stabilization in more detail.

Let's focus on results for case 1. Moreover, in figure 3(a) a time point about 250 sec should be mentioned. At this time the first jump of the angle law envelope appears. The second jump appears after 1500 sec. From figure 3(b) it can be noted that the stabilizing moment reduces up to zero and there is a sharp decrease in the amplitude of the oscillation in figure 3(a) at the same time. It means that the system stabilized instantaneously. As soon as a constant amplitude of the applied force is established in figure 3(d), the cessation of oscillation can be observed in figure 3(c). It is also worth noting that the damping of oscillations in different directions occurs at the same point in time. Sharp decrease of control function's amplitude explains by the neuron's spike "capturing" the near-equilibrium area. It's like the control system "catch" the dynamic system near equilibrium position. The weak point of such control system is amplitude–jumping pattern. In practice such behavior provide revolute joint wobble.

Going over to results for case 2 some difference occurs. In figures 3(e), (f) it can be noted a smoother envelope of the vibration wave profile during attenuation, in contrast to the previous case. It can be explained by the fact that in this case the control function is not a constant, but depends on the coordinates of the phase space.

The time required for stabilizing the system decreases. The control function became continuous; as a result, the behavior of the generalized coordinates is smoother. In contrast to the first case, there are no envelope jumps. Despite specific embodiment of control function (in mechanical terms), received results illustrate more suitable mode of dynamic system.

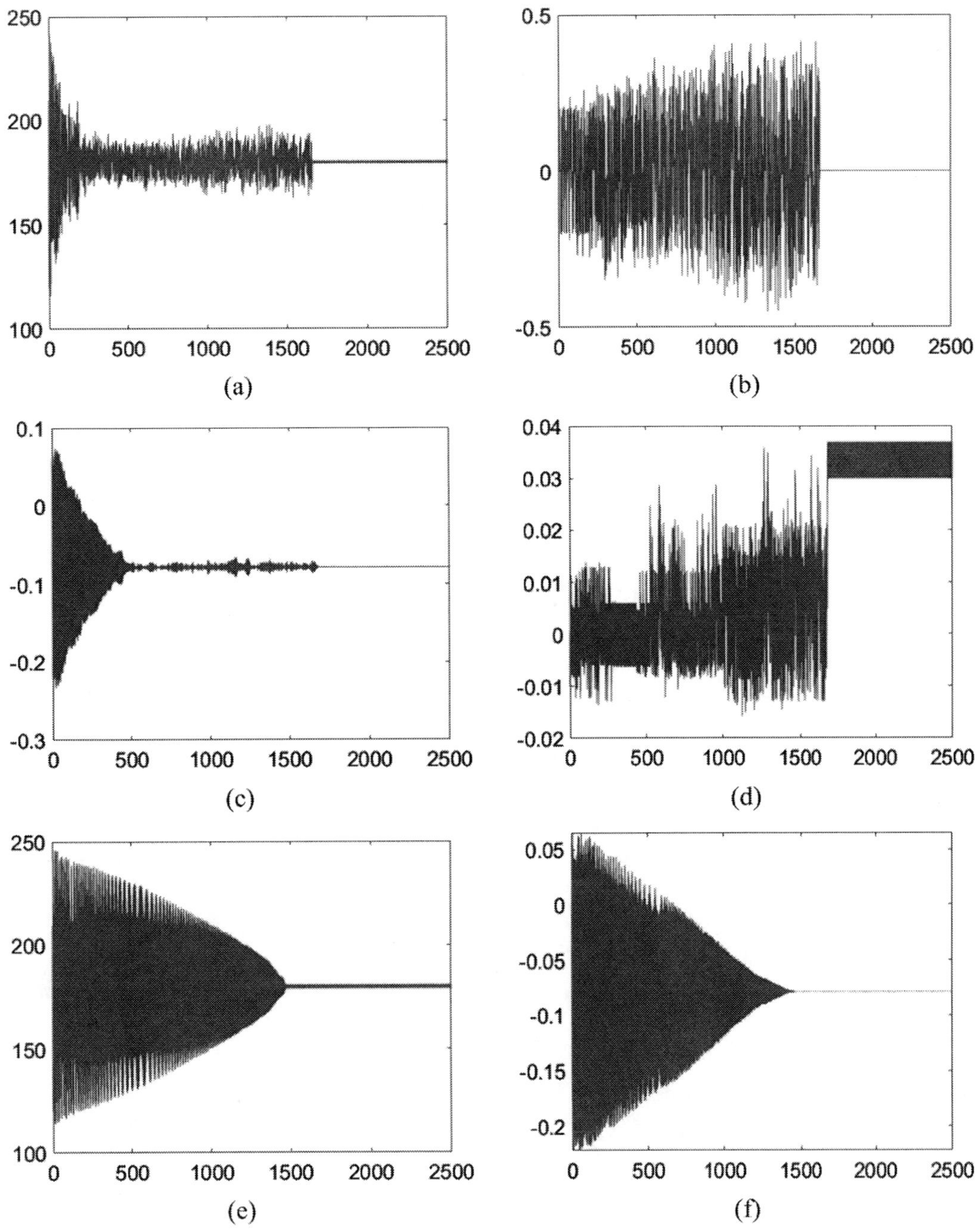

Figure 3. (a) The graph of the angle (deg) φ versus time (s). (c) The graph of the vertical displacement (m) x versus time (s). (b), (d) Graphs of the generalized control function for moments and forces, respectively versus time (s). (e) The graph of the angle (deg) φ versus time (s). (f) The graph of the vertical displacement (m) x versus time (s)

4 Conclusion

The article presents a method for stabilizing a nonlinear dynamic system using machine-learning tools. The LIF model of the spiking neural network was used. The article shows the process of stabilization of a pendulum on an elastic foundation. The closed system was built and a method for a numerical solution was implemented. Two configurations of control functions have been considered. It is shown that the time required to bring the system into a steady equilibrium state depends on the choice of the control function. Further development of this approach and its application to complex systems with a large number of degrees of freedom is envisaged.

This research was funded by RFBR, grant number 20-01-00535.
The results are obtained under support from the Kazan Federal University Strategic Academic Leadership Program ("PRIORITY-2030").

References

[1] C. Jimenez-Romero, J. Johnson, Neural. Comput. Appl., **28**, 755 (2017)
[2] S. Haykin, *Neural Networks and Learning Machines: Third Edition* (Prentice Hall, New York, 2009), pp. 674–711
[3] W. Maass, Neural Netw., **10**, 1659 (1997)
[4] N. Kasabov, K. Dhoble, N. Nuntalid, G. Indiveri, Neural Netw., **41**, 188 (2013)
[5] C. Jimenez-Romero, J. Johnson, Neural. Comput. Appl., **28**, 755 (2017)
[6] V. Timchenko, O. Ukhin, D. Lebedev, J. Autom. Inf. Sci., **10**, 33 (2017)
[7] S. Andropov, A. Guirik, M.Yu. Budko, M.B. Budko, Sci. Tech. J. Inf. Technol. Mech. Opt., **16**, 796 (2016)
[8] V. Deryabin, Vestnik Gosudarstvennogo universiteta morskogo i rechnogo flota imeni admiral S.O. Makarova, **10**, 665 (2018)
[9] A. Bakefayat, M. Tabrizi, *Lyapunov stabilization of the nonlinear control systems via the neural networks*, Appl. Soft Comput., **42**, 459 (2016)
[10] X. Zhang, Z. Xu, C. Henriquez, S. Ferrari, *Spike-Based Indirect Training of a Spiking Neural Network-Controlled Virtual Insect*, in Proceedings of the IEEE Conference on Decision and Control, 52nd IEEE Conference on Decision and Control, Institute of Electrical and Electronics Engineers, Florence, Italy, 6798 (2013)
[11] N. Fremaux, H. Sprekeler, W. Gerstner, J. Neurosci., **30**, 13326 (2010)
[12] O. Sinyavskiy, A. Kobrin, Russian Journal of Nonlinear Dynamics, **7**, 859 (2011)
[13] X. Li, Y. Li, *Neural networks optimized learning control of state constraints systems*, Neurocomputing, **453**, 512 (2021)
[14] F. Ponulak, A. Kasiński, Neural. Comput., **22**, 467 (2010)
[15] A. Perrusquia, W. Yu, *Identification and optimal control of nonlinear systems using recurrentneural networks and reinforcement learning: An overview*, Neurocomputing, **438**, 145 (2021)
[16] P. Manoonpong, F. Woegoetter, F. Pasemann, The International Journal of Applied Biomedical Engineering (IJABME), **3**, 1 (2010)

Self-oscillations in a certain Belousov–Zhabotinsky model

Liudmila Kondratieva[1,*] and *Aleksandr* Romanov[2]

[1]Moscow Aviation Institute (National Research University), 4, Volokolamskoe shosse, Moscow, 125993, Russia
[2]School of Applied Mathematics, HSE University, 34, Tallinskaya st., Moscow, 123458, Russia

Abstract. We consider the dynamic properties of a system of three differential equations known as the oreganator model. This model depends on four external parameters and describes one of the periodic Belousov–Zhabotinsky reactions. We obtain broad conditions for the parameters that ensure the existence of non-stationary steady-state regimes in oregonator model. With classical values of the parameters, the localization of the limit (at a long time) dynamics in the phase space has been improved. In fact, using numerical analysis, we significantly narrow the bounded region of the phase space containing the trajectories of the system. An iterative procedure is proposed for the approximate localization of closed trajectories (cycles) of the system on algebraic surfaces in R^3. A promising problem of theoretical substantiation of the numerical convergence of this procedure is posed.

1 Introduction

The problem of the existence of periodic regimes in problems of mechanics, chemistry, biology, economics, etc. is highly relevant. Let us consider in this connection one of the models of the Belousov–Zhabotinsky reaction [1–4], known in the literature as the "oregonator". This model corresponds to the system of ODEs

$$\begin{aligned}\frac{dx}{dt} &= s(y - xy + x - qx^2),\\ \frac{dy}{dt} &= s^{-1}(-y - xy + sfz),\\ \frac{dz}{dt} &= wx - wz\end{aligned} \tag{1}$$

with positive constants s, w, f, q. According to [1–3], system (1) describe oscillations in a real chemical reaction at $s = 77.27$, $q = 8.375 \cdot 10^{-6}$ and appropriate constraints on w, f. Various aspects of this model were considered in [5–7]. An interesting work [8] demonstrates the possibility of applying the oregonator model to problems not related to chemistry.

For $q < 1$, system (1) admits [2] a bounded positively invariant domain $D \subset R^3_+$ containing a single stationary point $\bar{p}$. We find broad (different from the well-known [2]) conditions on the parameters s, w, q and $f \leq 1$, under which a given point is unstable. Under these conditions, almost all solutions of system (1) go to nonstationary regimes when $t \to +\infty$. An

*e-mail: liudmila.kondratieva@inbox.ru

unstable invariant manifold $M^u(\bar{p})$ of a unstable point $\bar{p}$ is two-dimensional, and a stable manifold $M^s(\bar{p})$ is one-dimensional. System (1) generates in D a smooth phase semiflow $\{\Phi_t\}_{t\geq 0}$ and for each point $p \in D\backslash M^s(\bar{p})$ the limit set $\omega(p)$ does not contain $\bar{p}$, and therefore has a nontrivial geometric structure. Acting as in [2], the existence of a closed trajectory (cycle) $\Gamma \subset D$ of system (1) can be established using the so-called torus principle [9, § 18].

We analyze the instability of the system at $q < 1$, $f \leq 1$. It is convenient to consider these two parameters as control ones. In principle, reducing (1) to a two-dimensional manifold $M^u(\bar{p})$ and applying the Poincaré–Bendixson theory can provide asymptotic orbital stability of the cycle Γ. A similar technique was applied in [10–12] to the problem of satellite motion, as well as to one biochemical model of dynamics in cellular processes.

In the case $s = 77.27$, $q = 8.375 \cdot 10^{-6}$, $w = 41.86$, $f = 1$, we significantly narrow the positively invariant domain of system (1). The new domain still contains a stationary point $\bar{p}$and an unstable manifold $M^u(\bar{p})$. Thus, we improve the localization of the limit (as $t \to +\infty$) phase dynamics of system (1). Numerical analysis using the Maple package demonstrates that the system enters the self-oscillation mode ($M^u(\bar{p})$ contains a cycle Γ) and this cycle "well placed" on some algebraic surface of the fifth order in R^3.

2 Stationary point instability

As seen from (1), the positive octant R^3_+ is positively invariant. It is easy to show (see [2]) that the parallelepiped

$$D = D(f,q)\colon \left\{1 < x < \frac{1}{q},\ \frac{fq}{1+q} < y < \frac{f}{2q},\ 1 < z < \frac{1}{q}\right\}$$

has the same property.

We find the only stationary point $\bar{p} = (\bar{x}, \bar{y}, \bar{z}) \in R^3_+$, $\bar{p} = \bar{p}(f,q)$, of system (1) from the relations

$$y - xy + x - qx^2 = 0, \quad -y - xy + fz = 0, \quad x - z = 0.$$

We have $q\bar{x}^2 + (q + f - 1)\bar{x} - 1 - f = 0$ and

$$\bar{x} = \bar{z} = \frac{1}{2q}\left(\sqrt{(q+f-1)^2 + 4q(1+f)} - (q+f-1)\right),$$

$$\bar{y} = \frac{f\bar{x}}{1+\bar{x}} = \frac{1}{2}(1 + f - q\bar{x}), \quad \bar{y} < f.$$

The convex bounded positively invariant domain $D \subset R^3_+$ contains at least one stationary point; therefore, $\bar{p} \in D$. The identity $q\bar{x}(1+\bar{x}) = 2f + (1-f)(1+\bar{x})$ implies the estimates:

$$1 + \bar{x} > \sqrt{\frac{2}{q}}, \quad f = 1; \quad 1 + \bar{x} > \frac{1-f}{q}, \quad f < 1. \tag{2}$$

Consider the Jacobi matrix

$$J'(\bar{p}) = \begin{pmatrix} s(1 - \bar{y} - 2q\bar{x}) & s(1-\bar{x}) & 0 \\ -s^{-1}\bar{y} & -s^{-1}(1+\bar{x}) & s^{-1}f \\ w & 0 & -w \end{pmatrix}$$

and its characteristic polynomial $\det(\lambda E - J'(\bar{p})) = \lambda^3 + \alpha\lambda^2 + \beta\lambda + \gamma$, where E is the unit matrix. Calculate:

$$\gamma = -\det J'(\bar{p}) = w(1 + q\bar{x} + 2\bar{y}) > 0$$

at all $q, f, s, w > 0$. Further,

$$\alpha = \frac{3}{2}sq\bar{x} + s^{-1}(1+\bar{x}) + w + \frac{s}{2}(f-1) > 0$$

at $f = 1$. The estimate $\alpha > 0$ is also valid [2, p. 681] for all $0 < f < 1$.

Since $\gamma > 0$, $\alpha > 0$, then according to the Routh–Hurwitz criterion the instability of a stationary point provides a condition $\gamma > \alpha\beta$, which therefore the condition $\beta \leq 0$. The only stationary point $\bar{p} \in D$ is non-degenerate, its Poincare index is 1, and the Jacobi matrix $J'(\bar{p})$ with all permissible parameter values has in $\operatorname{Re}\lambda > 0$ either two (with multiplicity) of eigenvalues or none. So, with $\beta \leq 0$ an unstable manifold of the point $\bar{p}$ two-dimensional.

The value β is the sum of the main 2×2 minors of the matrix $J'(\bar{p})$ and

$$\beta = \bar{y}(2 + sw) + (2q\bar{x} - 1)(1 + \bar{x} + sw) + s^{-1}w(1 + \bar{x}).$$

Put $1 + \bar{x} = A$, then $\bar{y} = \dfrac{f\bar{x}}{1+\bar{x}} = f - \dfrac{f}{A}$ and $2q\bar{x} - 1 = 1 + 2f - 4\bar{y} = 1 - 2f + \dfrac{4f}{A}$. As $\bar{y} < f$,

$$\beta < 6f + sw(1-f) + (1-2f)A + \frac{3swf}{A} + \frac{w}{s}A.$$

In particular, the condition $\beta \leq 0$ is fair if

$$\left(2f - 1 - \frac{w}{s}\right)A^2 - 2\theta A - 3swf > 0$$

at $\theta = 3f + \dfrac{sw}{2}(1-f)$. Taking into account estimates (2), we obtain sufficient conditions for the instability of the stationary point $\bar{p} \in D$ as

$$f = 1, \quad w < s, \quad \sqrt{\frac{2}{q}} > \frac{3 + \sqrt{9 + 3w(s-w)}}{1 - s^{-1}w} = \mu_0, \tag{3.1}$$

$$\frac{1}{2}\left(1 + \frac{w}{s}\right) < f < 1, \quad \frac{1-f}{q} > \frac{\theta + \sqrt{\theta^2 + 3sw(2f - 1 - s^{-1}w)}}{2f - 1 - s^{-1}w} = \mu_1. \tag{3.2}$$

It is convenient to record estimates (3.1), (3.2) in the form of $q < \dfrac{2}{\mu_0^2}$ at $f = 1$ and $q < \dfrac{1-f}{\mu_1}$ at $f < 1$. Under conditions (3), the stationary point $\bar{p}$ is a saddle with a one-dimensional stable and two-dimensional unstable manifold.

3 Localization and visualization of limit dynamics

After normalization of variables $x \to s^{-1}x$, $y \to sy$, $z \to s^{-1}z$, the system (1) acquires the form

$$\begin{aligned} \frac{dx}{dt} &= s^{-1}y - xy + sx - s^2qx^2, \\ \frac{dy}{dt} &= -s^{-1}y - xy + sfz, \\ \frac{dz}{dt} &= wx - wz \end{aligned} \tag{4}$$

with a positively invariant domain

$$\Omega = \Omega(f, q, s)\colon \left\{s^{-1} < x < \frac{s^{-1}}{q},\ \frac{sfq}{1+q} < y < \frac{sf}{2q},\ s^{-1} < z < \frac{s^{-1}}{q}\right\}.$$

We preserve the designations of the coordinates x, y, z and the stationary point $\bar{p}$ when moving from the system (1) to the system (4). We assume $f = 1$ and try to reduce the domain Ω. Let us show that on the surface

$$S_1: \quad z = g_1(x, y) \doteq \frac{s}{4} + \frac{1}{s}(x + y)^2$$

the vector field $\bar{F}$ of system (4) is directed to the domain. Indeed, if we take the field of normals on S_1 in the form $\bar{n} = (2s^{-1}(x + y), 2s^{-1}(x + y), -1)$, then the scalar product

$$\bar{F} \cdot \bar{n}\Big|_{S_1} = 2s^{-1}(x + y)(s^{-1}y - xy + sx + s^2qx) + 2s^{-1}(x + y)(-s^{-1}y - xy + sz) + w(z - x) =$$
$$= 2(x + y)(z - 2s^{-1}xy) + 2x(x + y)(1 - sqx) + w(z - x) > 0$$

since all expressions in parentheses are positive. Since $s^{-1} < s/4 + s^{-2}$ for all $s > 0$, the lower edge of the parallelepiped Ω is contained in the domain Ω_1.

We further assume that $w = 41.86$ and (as in [1–3]) $s = 77.27$, $q = 8.375 \cdot 10^{-6}$. Consider the surface

$$S_2: \quad z = g_2(x, y) \doteq -s^{-1}(x + 48)(y - 263) + 333$$

and take the field of normals on on the surface in the form $\bar{n} = (s^{-1}(y - 263), s^{-1}(x + 48), 1)$. Numerical analysis shows that on the surface S_2 the vector field of system (4) is directed to the domain $\Omega_2: z < g_2(x, y)$, i.e. $\bar{F} \cdot \bar{n}|_{S_2} < 0$.

The plane $z = z_0$, $z_0 > s/4$, intersects the surfaces S_1 and S_2 in the straight line $x + y = \sqrt{s(z_0 - s/4)}$ and in the hyperbola $(x + 48)(y - 263) = s(333 - z_0)$, respectively. We accept $z_0 = 251$, then (also, numerically) it can be shown that on a part of the plane $z = z_0$, $(x, y, z_0) \in \Omega_1 \cap \Omega_2$, the vector field $\bar{F}$ is directed to the domain $\Omega_3: z < z_0$, i.e. $\bar{F} \cdot \bar{n} = w(x - z) < 0$ for the normal $\bar{n} = (0, 0, 1)$.

As we can see, the domain

$$\Omega_0 = \Omega \cap \Omega_1 \cap \Omega_2 \cap \Omega_3$$

is positively invariant for system (4) and homeomorphic to a ball. The domain Ω_0 contains a single stationary point $\bar{p}$, its two-dimensional unstable manifold $M^u(\bar{p})$, and a stable (as numerical analysis shows) limit cycle $\Gamma \subset M^u(\bar{p})$. At the same time, domain Ω_0 is significantly less than Ω. Let the y^+, y_0 and z^+, z_0 be the maximum coordinate values of y and z in the domains Ω and Ω_0, respectively. Then we have $y^+ \approx 4.6 \cdot 10^6$, $y_0 \approx 789.9$ and $z^+ \approx 1543.3$, $z_0 = 251$.

Let us pose the problem of approximate localization of the cycle Γ on some algebraic set (algebraic surface). For an arbitrary polynomial $Q \neq \text{const}$ in R^3, we put

$$S(Q) = \{(x, y, z) \in R^3 : Q(x, y, z) = 0\}.$$

If $S(Q) \supset \Gamma$ and gradient $\nabla Q \neq 0$ on Γ then the vector $\bar{F}$ lies in the tangent plane of the surface $S(Q)$ for all points of the cycle, i.e. $\nabla Q \cdot \bar{F} = 0$ on Γ. Let us introduce a linear differential operator $L: Q \to \nabla Q \cdot \bar{F}$. We have

$$(LQ)(x(t), y(t), z(t)) = \frac{d}{dt} Q(x(t), y(t), z(t))$$

for an arbitrary solution $(x(t), y(t), z(t))$ of system (4) in Ω_0, i.e. LQ is the derivative of the function $Q(x, y, z)$ by virtue of (4). For the polynomial Q_0, we set $Q_1 = LQ_0$ and, in general, $Q_k = LQ_{k-1}$ for $k \geq 1$. At each iteration, the degrees of the polynomials are increased by one. The sets $S(Q_k)$, $k \geq 1$, contain stationary points of system (4) in R^3.

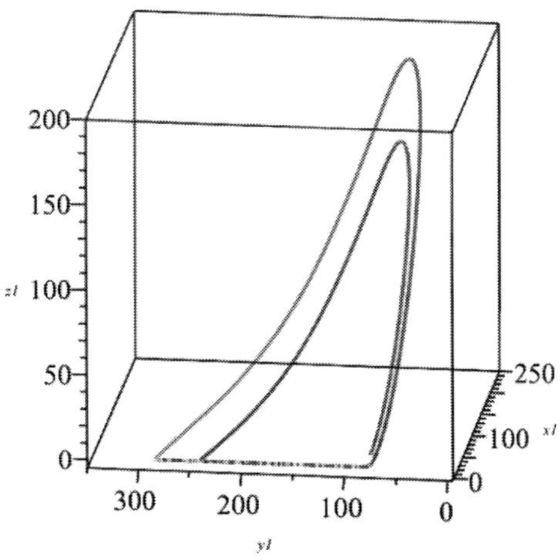

Figure 1. Limit cycle image

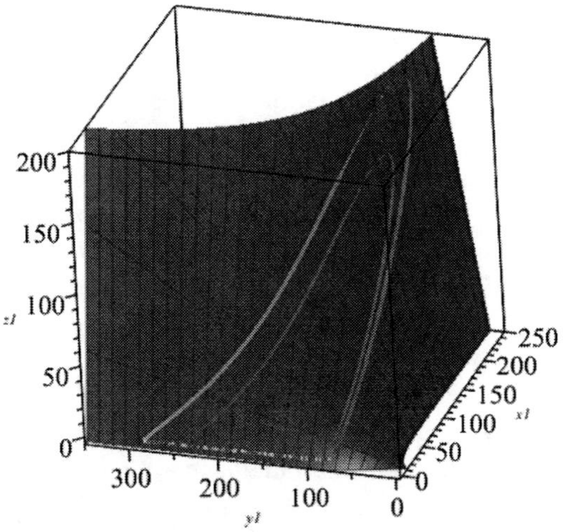

Figure 2. The hyperbolic paraboloid $S(Q_0)$ and limit cycle Γ

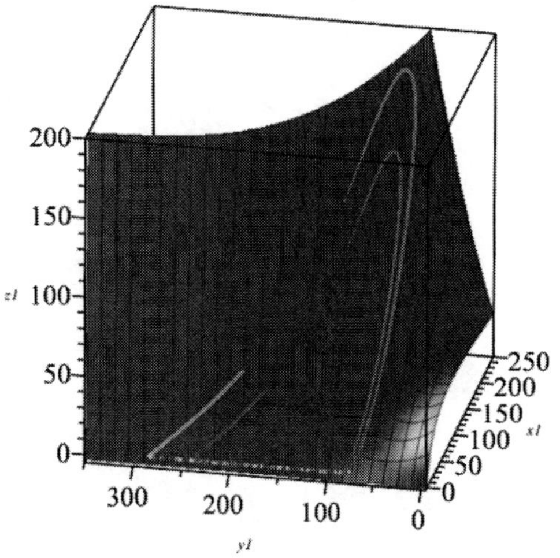

Figure 3. The 3rd order surface $S(Q_1)$ and limit cycle Γ

Let us try to choose an initial approximation Q_0 such that, with increasing k, the distances (deviations) from the cycle Γ to $S(Q_k)$ decrease. Figure 1 shows the limit cycle Γ and the trajectory approaching it.

Using computer visualization, we select Q_0 such that the surface $S(Q_0)$ repeats the outline Γ. For example, put $Q_0(x, y, z) = xy - sz$, i.e. $S_0 = S(Q_0)$ is the hyperbolic paraboloid $z = s^{-1}xy$. At the first iteration, we find the polynomial $Q_1 = LQ_0 = \nabla Q_0 \cdot \bar{F}$ of degree 3 and the corresponding surface

$$S(Q_1):\ z(w + x)s^2 = syx^2(qs^2 + 1) + x(ws^2 - ys^2 + sy^2 + y) - y^2.$$

Figures 2 and 3 show the localization of the limit cycle Γ near surfaces $S(Q_0)$ and $S(Q_1)$, respectively.

At the second iteration, we find the polynomial $Q_2 = LQ_1 = \nabla Q_1 \cdot \bar{F}$ of degree 4 and the surface $S(Q_2)$. At the third iteration, we obtain the polynomial $Q_3 = LQ_2 = \nabla Q_2 \cdot \bar{F}$ of

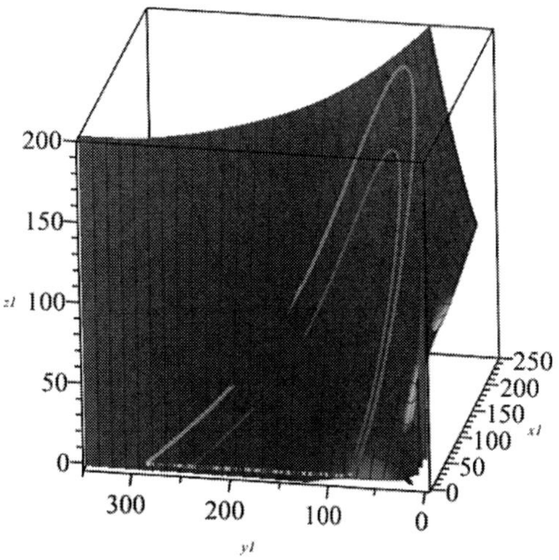

Figure 4. The 4th order surface $S(Q_2)$ and limit cycle Γ

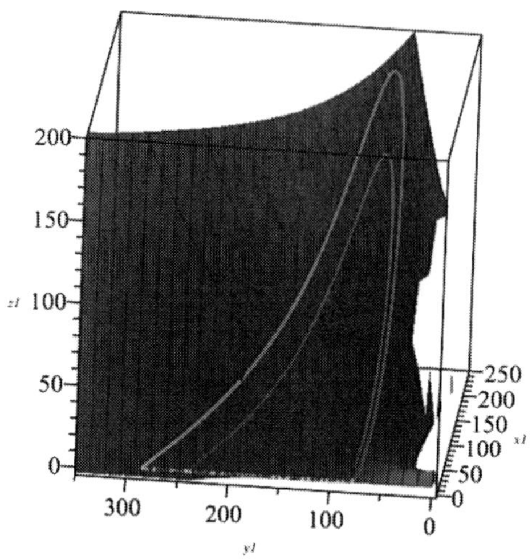

Figure 5. The 5th order surface $S(Q_3)$ and limit cycle Γ

degree 5 and the surface $S(Q_3)$. Figures 4 and 5 show the localization of the limit cycle Γ near surfaces $S(Q_2)$ and $S(Q_3)$, respectively.

One can see (visually) that the deviations $\rho_0, \rho_1, \rho_2, \rho_3$ are sequentially decrease and the cycle Γ "well placed" on the surface of the fifth order $S(Q_3) \subset R^3$. It is interesting to find out under what conditions for the considered iterative procedure the deviations $\rho_k \to 0$ at $k \to \infty$. This task looks very promising and may become the subject of further research.

4 Conclusion

The area of variation of parameters $s, w, f, q > 0$ is described, at which non-stationary limiting modes arise in system (1) and in its equivalent system (4). In the case $s = 77.27$, $q = 8.375 \cdot 10^{-6}$, $w = 41.86$, $f = 1$, the known positively invariant domain is substantially reduced to one containing a single stationary point and a stable limit cycle Γ. An iterative procedure for the approximate localization of the cycle Γ on a suitable two-dimensional algebraic surface in three-dimensional phase space is described.

References

[1] R.J. Field, R.M. Noyes, J. Chem. Phys., **60**, 1877 (1974)

[2] S.P. Hastings, J.D. Murray, SIAM J. Appl. Math., **28**, 678 (1975)

[3] G. Nicolis, I. Prigogine, *Self-Organization in Nonequilibrium Systems* (Wiley, New York, 1977)

[4] I.R. Epstein, J.A. Pojman, *An introduction to nonlinear chemical dynamics* (Oxford University Press, New York, 1998)

[5] A.K. Dutt, AIP Advances, **1**, 042147 (2011)

[6] R.D. Ikramov, S.A. Mustafina, International Research Journal (Phys. and Math.), **9-2**, 124 (2016)

[7] A. Adamatzky, Int. J. Bifurc. Chaos, **27(3)**, 1750041 (2017)

[8] A. Adamatzky, N. Phillips, R. Weerasekera, M.A. Tsompanas, G.Ch. Sirakoulis, arXiv:1803.01632 (2018)

[9] V.A. Pliss, *Nonlocal Problems of the Theory of Oscillations* (Academic Press, New York, 1966)
[10] L.A. Kondratieva, Aerospace MAI Journal, **19**, 75 (2012)
[11] L. Kondratieva, *Computational model for satellite periodic motion*, AIP Conf. Proc., **2181**, 020002 (2019)
[12] L.A. Kondratieva, A.V. Romanov, Elect. J. Qual. Th. Differ. Eq., **96**, 1 (2019)

Stability Analysis of helicopter dynamics with incomplete information using MPC

Irina Kudryavtseva[1,*] and *Kirill* Petrov[1,**]

[1]Moscow Aviation Institute (National Research University), 4, Volokolamskoe shosse, Moscow, 125993, Russia

Abstract. The paper is devoted to the problem of seeking a helicopter motion control using MPC provided that the state vector is not fully observable. It is supposed that the helicopter motion is described by a dynamic system linearized near regimes with different fixed values of the forward speed. To solve the problem for an arbitrary value of the forward speed the interpolation strategy when the system matrices are decomposed in the basis of the finite functions is used. The proposed technique is applied to analyze dynamic behavior of three models of helicopters: ZD559-Lynx, SA330-Puma, S123-Bo105. The MPC approach permits to gain the stability of helicopters with complete and incomplete information in varying a time interval on which the system is under control and an interval over which a desired output is needed to achieve, initial conditions of a motion and values of the forward speed.

1 Introduction

A helicopter as a complex dynamic system requires to find such a control strategy that permits to attain desired trajectories and to make a rapid response to appearing disturbances. The flight control system design faces to a variety of difficulties related to different flight regimes for better performance of tasks, presence of external impacts or appearing effects caused by practical implementation of a helicopter control [1–3]. There are a few flight control design algorithms developed to enhance helicopter dynamics and handling qualities: linear quadratic optimal control (LQR/LQG), $H_2\backslash H_\infty$ optimal control, backstepping control, gain-scheduling control, adaptive control, model predictive control, observer feedback control, fuzzy control, intelligent control [4, 5]. The MPC approach has a set of benefits such as an intuitive implementation, applicability to both SISO and MIMO systems, taking into account constraints in their "natural form" that make this scheme attractive for practical using [6–8]. In the paper the helicopter motion control problem is proposed to solve using MPC.

*e-mail: kudryavtseva.irina.a@gmail.com
**e-mail: kpetroff12@yandex.ru

2 The problem formulation

Consider the helicopter flight dynamic model including the force equations and the moment equations in the form [1, 2]:

$$\begin{aligned} \dot{u} &= -(wq - vr) + \frac{X}{M} - g\sin\theta, \\ \dot{v} &= -(ur - wp) + \frac{Y}{M} + g\cos\theta\sin\varphi, \\ \dot{w} &= -(vp - uq) + \frac{Z}{M} + g\cos\theta\cos\varphi, \end{aligned} \tag{1}$$

$$\begin{aligned} I_{xx}\dot{p} &= (I_{yy} - I_{zz})qr + I_{zz}(\dot{r} + pq) + L, \\ I_{yy}\dot{q} &= (I_{zz} - I_{xx})rp + I_{xz}(r^2 - p^2) + R, \\ I_{zz}\dot{r} &= (I_{xx} - I_{yy})pq + I_{xz}(\dot{p} - qr) + N, \end{aligned} \tag{2}$$

and also completed with the equations for the Euler angles [1, 2]:

$$\begin{aligned} \dot{\varphi} &= p + q\sin\varphi\,\mathrm{tg}\,\theta + r\cos\varphi\,\mathrm{tg}\,\theta, \\ \dot{\theta} &= q\cos\varphi - r\sin\varphi, \\ \dot{\psi} &= q\frac{\sin\varphi}{\cos\theta} + r\frac{\cos\varphi}{\cos\theta}, \end{aligned} \tag{3}$$

where u, v, w, p, q, r are components of the translational and angular velocities respectively; φ, θ, ψ are the Euler angles defining an orientation of the helicopter with respect to the fixed coordinate system (relative to the earth); X, Y, Z and L, R, N are components of applied external forces and moments respectively; g is the gravitational acceleration; M is mass of the aircraft; I_{xx}, I_{yy}, I_{zz} are moments of inertia about x-, y- and z-axes; I_{xz} is a moment of inertia about the x- and z-axes. The last equation of the presented model can be dropped due to the fact of no contribution in the helicopter dynamics in the considered flight mode.

The system (1)–(3) is gone through linearization provided that the helicopter moves under conditions corresponding to sea level with zero sideslip and turn rate at a fixed value of the forward speed from 0 to 140 kts: $V_i = (i-1)\Delta V_i$, $\Delta V_i = 20$ kts, $i = 1, \ldots, 8$

$$\dot{x} = A(V_i)x + B(V_i)u, \quad i = 1, \ldots, 8, \tag{4}$$

where $x = (u, w, q, \theta, v, p, \varphi, r)^T$ is the system state, $u = (\theta_0, \theta_{1s}, \theta_{1c}, \theta_{0T})^T$ is the control input, θ_0 and θ_{0T} are the main and tail rotor collective pitch angles; θ_{1s} and θ_{1c} are longitudinal and lateral cyclic pitch; i is a number of a flight mode associated with a certain value of the forward speed in the mentioned range: 0 kts, 20 kts, …, 140 kts; $A(V_i)$, $B(V_i)$ are known matrices given in [1].

The system state x is assumed to be observable in accordance with the rule:

$$y = C(V_i)x, \quad i = 1, \ldots, 8, \tag{5}$$

where y is the system output, $C(V_i)$ is a known matrix of an appropriate dimension. The rank of the matrix C might be less than dimension of the state vector x that means a deficiency of information about x.

The paper aims to find the control input that allows to achieve the trajectory stability of the aircraft in case of incomplete information about the state vector. It is supposed to analyze the aircraft dynamics at an arbitrary forward speed not exceeding 140 kts. For this purpose the interpolation strategy based on the decomposition in finite function is used. To solve the problem the Model Predictive Control approach is utilized.

3 Solution method

Running the system (4)–(5) through discretization leads to

$$\begin{aligned} x_{j+1} &= (hA_i - E)x_j + hB_i u_j, \\ y_j &= C_i x_j, \\ x_j &= x(t_j), \quad u_j = u(t_j), \quad t_j = jh, \quad j = 0, 1, \ldots; \quad i = 1, \ldots, 8, \end{aligned} \tag{6}$$

where h is a time step, j is utilized for moments of time, i specifies a helicopter's flight mode.

For further exposition rewriting system (6) in the difference form will be needed [6]

$$\Delta x_{j+1} = \tilde{A}_i \Delta x_j + \tilde{B}_i \Delta u_j, \quad j = 0, 1, \ldots,$$

where $\Delta x_{j+1} = x_{j+1} - x_j$, $\Delta u_j = u_j - u_{j-1}$, $\tilde{A}_i = hA_i - E$, $\tilde{B}_i = hB_i$, $i = 1, \ldots, 8$.

The difference analogue of the second equation of system (6) has the form [6]:

$$y_j - y_{j-1} = C_i \Delta x_j, \quad j = 0, 1, \ldots, \quad i = 1, \ldots, 8.$$

That implies

$$\begin{aligned} X_{j+1} &= A'_i X_j + B'_i \Delta u_j, \\ y_j &= C'_i X_j, \end{aligned} \tag{7}$$

where $X_j = (\Delta x_j \quad y_j)^T$, $A'_i = \begin{pmatrix} \tilde{A}_i & O \\ C_i\tilde{A}_i & I \end{pmatrix}$, $B'_i = \begin{pmatrix} \tilde{B}_i \\ C_i\tilde{B}_i \end{pmatrix}$, $C'_i = (O \quad I)$.

Choose two time intervals (the prediction horizon N_p and the control horizon N_c: $N_c \leq N_p$) over which system trajectories and control trajectories are predicted. Assume that the system information X_j is available with the control Δu_j at t_j. Then for moments of time within the time interval of length determined by the prediction horizon N_p the predicted vectors $X_{j+1}, \ldots, X_{j+N_p}$ and the predicted outputs $y_{j+1}, \ldots, y_{j+N_p}$ can be expressed as

$$\begin{aligned} X_{j+1} &= A'_i X_j + B'_i \Delta u_j, \\ y_{j+1} &= C'_i A'_i X_j + C'_i B'_i \Delta u_j, \\ X_{j+2} &= A'^2_i X_j + A'_i B'_i \Delta u_j + B'_i \Delta u_{j+1}, \\ y_{j+2} &= C'_i A'^2_i X_j + C'_i A'_i B'_i \Delta u_j + C'_i B'_i \Delta u_{j+1}, \\ &\ldots \\ X_{j+N_p} &= A'^{N_p}_i X_j + A'^{N_p-1}_i B'_i \Delta u_j + \cdots + A'^{N_p-N_c} B'_i \Delta u_{j+N_c-1}, \\ y_{j+N_p} &= C'_i A'^{N_p}_i X_j + C'_i A'^{N_p-1}_i B'_i \Delta u_j + \cdots + C'_i A'^{N_p-N_c} B'_i \Delta u_{j+N_c-1}. \end{aligned}$$

Thus, system (7) can be transformed into the system:

$$Z = FX_j + \Phi\,\Delta U, \quad j = 1, \ldots, N_p,$$

where $Z = (y_{j+1}, \dots, y_{j+N_p})^T$, $\Delta U = \left(\Delta u_j, \dots, \Delta u_{j+N_c-1}\right)^T$,

$$F = \begin{pmatrix} C_i' A_i' \\ C_i' A_i'^2 \\ \dots \\ C_i' A_i'^{N_p} \end{pmatrix}, \quad \Phi = \begin{pmatrix} C_i' B_i' & 0 & \dots & 0 \\ C_i' A_i' B_i' & C_i' B_i' & \dots & 0 \\ \dots & \dots & \dots & \dots \\ C_i' A_i'^{N_p-1} B_i' & C_i' A_i'^{N_p-2} B_i' & \dots & C_i' A_i'^{N_p-N_c} B_i' \end{pmatrix}.$$

The optimal control ΔU can be found by minimizing the cost function:

$$I = (R - Z)^T (R - Z) + \Delta U^T Q \Delta U \to \min_{\Delta U},$$

where $R_{\dim(x_j)\cdot N_p \times 1} = (r_j \quad \dots \quad r_j)^T$ defines a final desired position of the system; $Q = qE$ is a positive definite symmetric matrix; E is an identity matrix of an appropriate dimension; q is a parameter that serves to account the control effect. The cost function I takes its minimum value at

$$\Delta U^* = (\Phi^T \Phi + Q)^{-1} \Phi^T (R - F X_j).$$

Inspite of the fact that the control ΔU^* is being obtained over whole time interval defined by the control horizon N_c only the first entry of the vector ΔU^* takes part in calculation. Then it should be modified using a new observation.

Hence, the predicted state x_{j+1} can be determined from obtained Δu_j^* and the known u_{j-1} as follows

$$x_{j+1} = \tilde{A}_i x_j + \tilde{B}_i (u_{j-1} + \Delta u_j^*).$$

In the case when the helicopter moves with a constant forward speed of the value from 0 to 140 kts but differing from the claimed earlier eight modes the matrices included in the linearized dynamic model (4)–(5) can be found via the decomposition in a set of finite functions [9, 10]:

$$S_p^*(\tau) = \begin{cases} 2^{p-1}(1+\tau)^p, & \tau \in \left[-1, -\frac{1}{2}\right], \\ 1 - 2^{p-1}|\tau|^p, & \tau \in \left[-\frac{1}{2}, \frac{1}{2}\right], \\ 2^{p-1}(1-\tau)^p, & \tau \in \left[\frac{1}{2}, 1\right], \\ 0, & \tau \notin [-1, 1], \end{cases}$$

where $p = 2, 3, 4$ is an interpolation parameter. The decomposition is performed in accordance with the relations:

$$A = \sum_i A_i S_p^* \left(\frac{V}{\Delta V} - (i-1)\right), \quad B = \sum_i B_i S_p^* \left(\frac{V}{\Delta V} - (i-1)\right),$$

$$C = \sum_i C_i S_p^* \left(\frac{V}{\Delta V} - (i-1)\right).$$

It should be noticed, the finite functions take zero values outside the interval $[-1, 1]$.

4 Numerical results

The presented approach is proposed to apply to investigate dynamical behavior of three models of helicopters: ZD559-Lynx, the twin-engine multi-purpose helicopter for military tasks; SA330-Puma, the twin-engine medium transport/utility helicopter; S123-Bo105, the twin-engine multi-purpose helicopter.

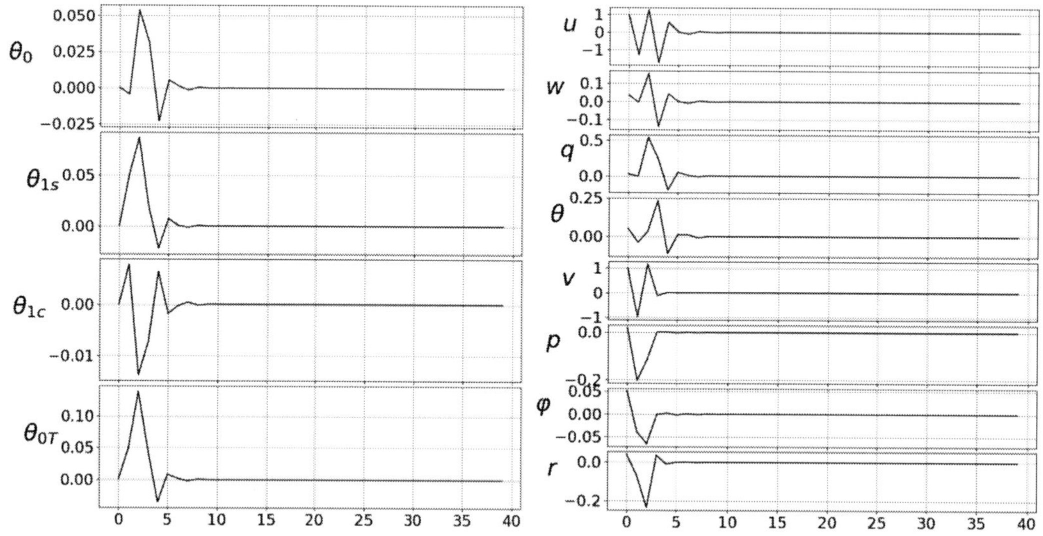

Figure 1. The control and system states trajectories for the ZD559-Lynx: $N_p = N_c = 8$, $q = 0.1$

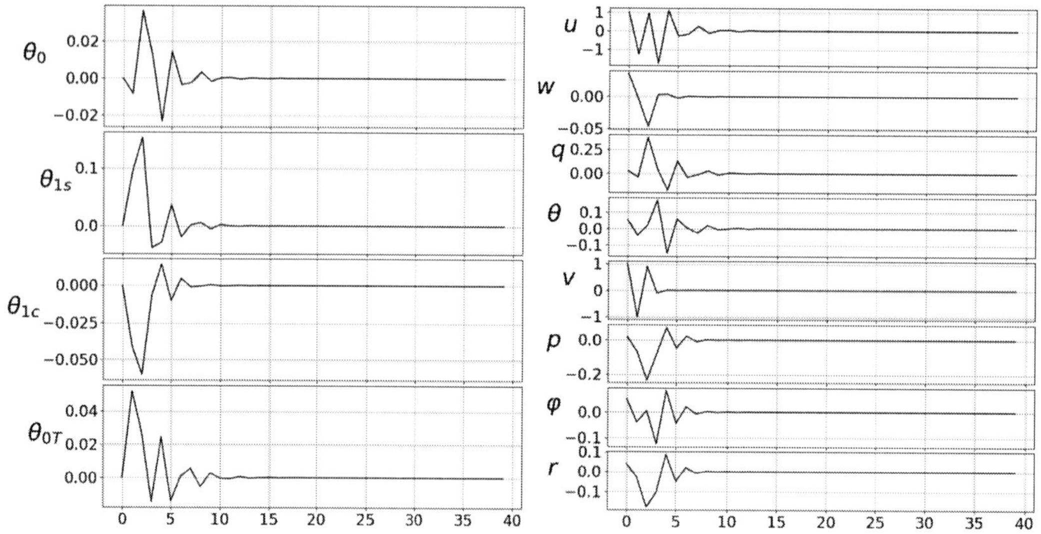

Figure 2. The control and system states trajectories for the SA330-Puma: $N_p = N_c = 8$, $q = 0.1$

During numerical experiments effects of the model and method parameters on the motion stability are analyzed in dependence on completeness of system information. In figures 1–3 the evolution of control trajectories and system states trajectories are shown provided that the control and the prediction horizons are equal: $N_p = N_c = 8$. It is also supposed that the system

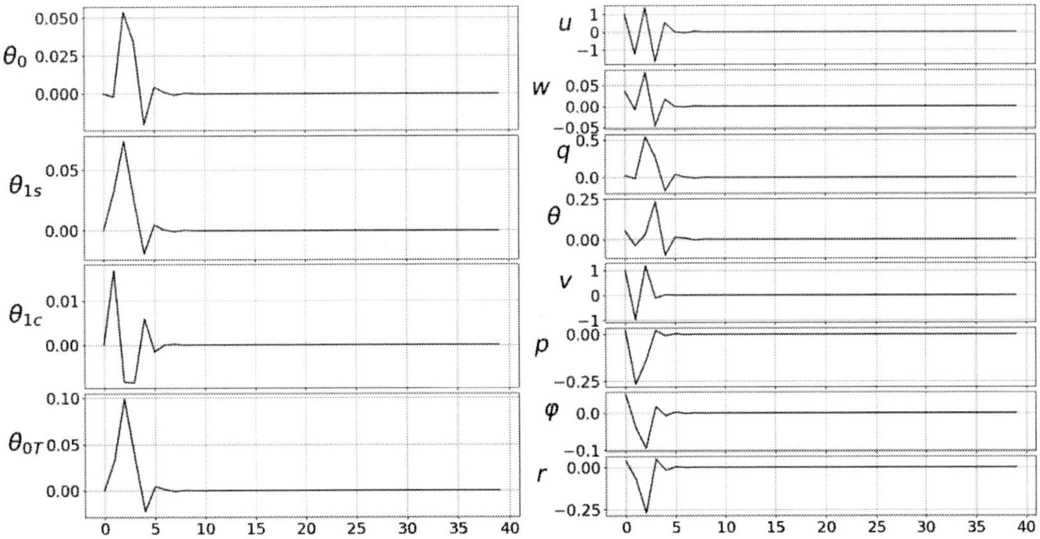

Figure 3. The control and system states trajectories for the S123-Bo105: $N_p = N_c = 8$, $q = 0.1$

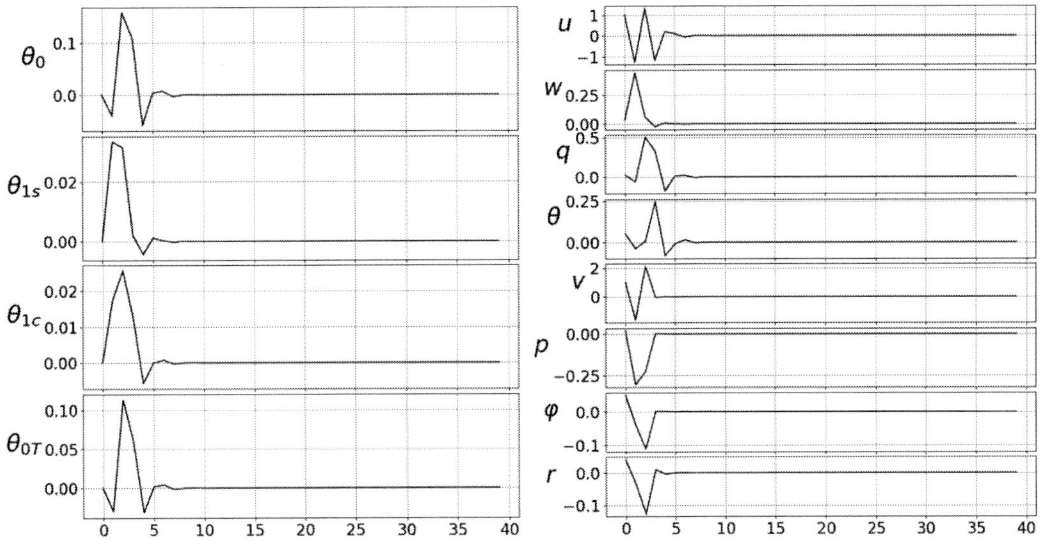

Figure 4. The control and system states trajectories for the S123-Bo105: $N_p = N_c = 8$, $q = 0.1$, $V = 55$ kts

state is completely observable. The helicopters perform the hover flight. It is clearly seen that the motion of all of the considered helicopters' models is stabilized within the interval given by the prediction horizon N_p. It should be noticed that different kind of transients for the SA330-Puma takes more time in comparison with the other models. It might be linked with poor controllability of the mentioned helicopter.

Utilizing the decomposition on a basis of the finite functions permits to obtain trajectories for an arbitrary values of the forward speed from 0 to 140 kts. Figure 4 demonstrates trajectories of the S123-Bo105 for the flight mode with $V = 55$ kts. As in the previous case

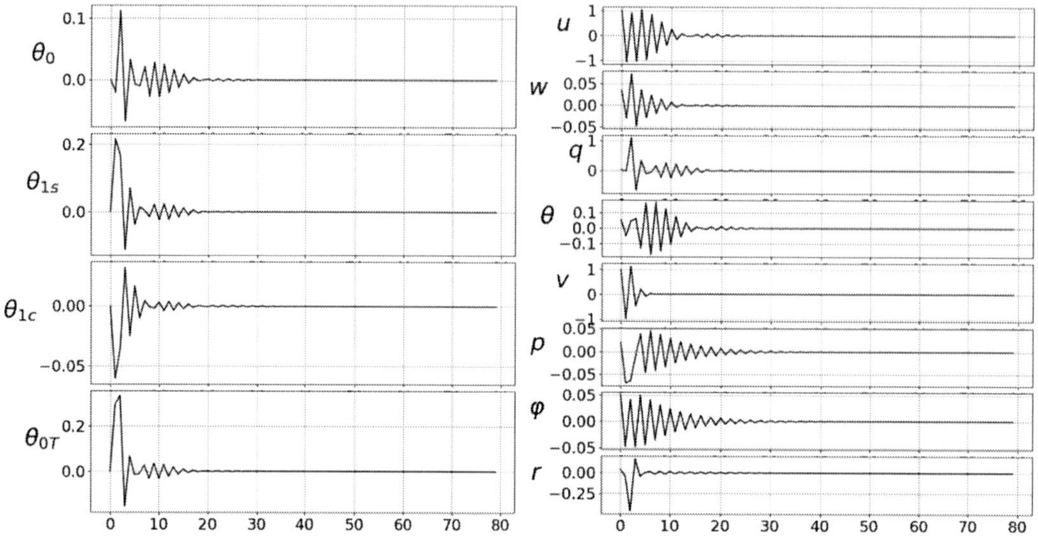

Figure 5. The control and system states trajectories for the S123-Bo105: $N_p = N_c = 16$, $q = 0.1$

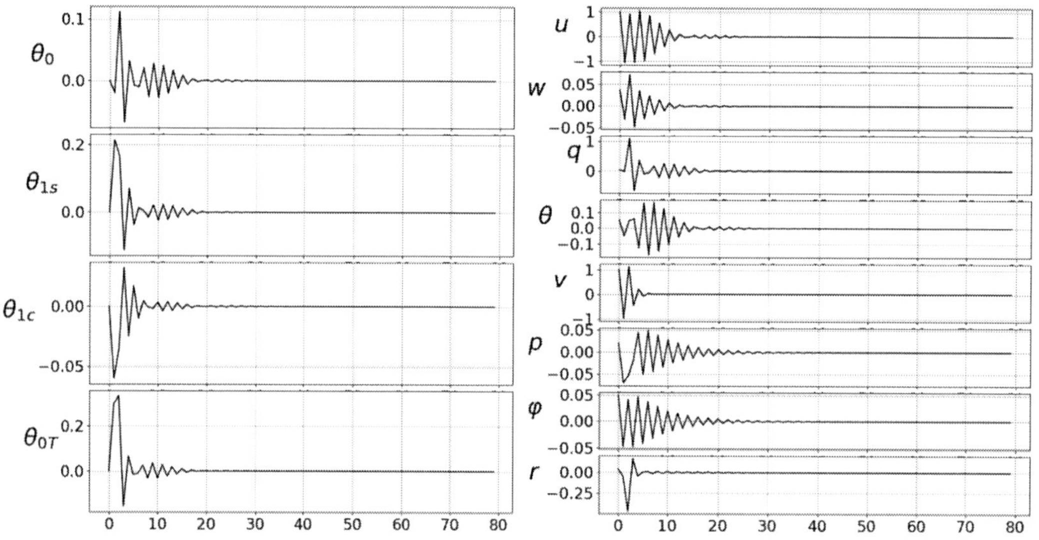

Figure 6. The control and system states trajectories for the S123-Bo105: $N_p = 164$, $N_c = 8$, $q = 0.1$

all fluctuations dampen within the prediction horizon. Significant distinctions from the cases observed before are not identified.

It's worthwhile to say that increasing the prediction and control horizons forces to decrease the time step of discretization. Figure 5–6 prove a benefit of the MPC strategy. Reducing the control horizon doesn't much influence achievement of the desired final position.

However it should be added that increasing the prediction and control horizons should be convoyed with decreasing the time step. Otherwise it may lead to the instability. Figures 7–9 demonstrate evolution of the state vector and the control in time when the state vector is

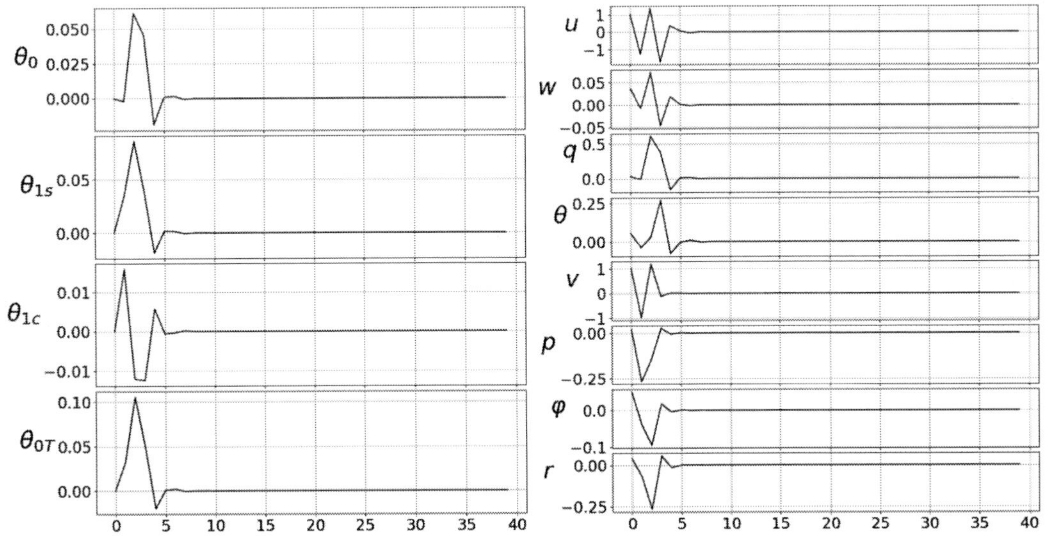

Figure 7. The control and system states trajectories for the S123-Bo105: $N_p = 8$, $N_c = 4$, $q = 0.01$, rank $C_i = 8, i = 1, \ldots, 8$

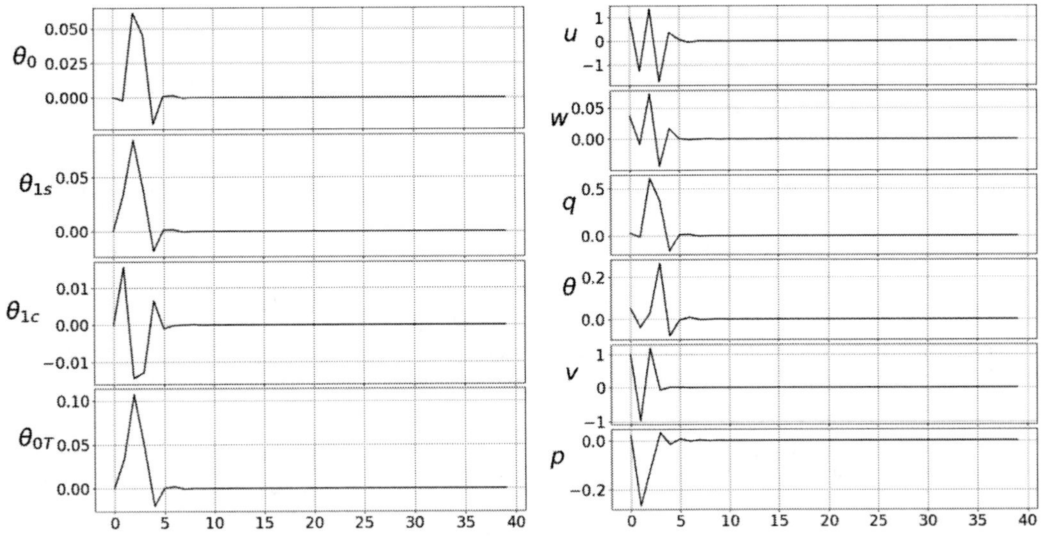

Figure 8. The control and system states trajectories for the S123-Bo105: $N_p = 8$, $N_c = 4$, $q = 0.01$, rank $C_i < 8, i = 1, \ldots, 8$

completely and partly observable. First it is supposed to select such a relation between the prediction horizon and the control horizon that makes the process stable. Then the effect of loss of information is investigated.

It's clearly seen that significant cutting of the output vector implies substantial changes in processes behavior. It can be traced when comparing figure 8 with figure 9. Nevertheless all trajectories approach to occupy a wished position. The suitable choice of the parameters N_c and q permits to gain a wished result when all transition processes fade.

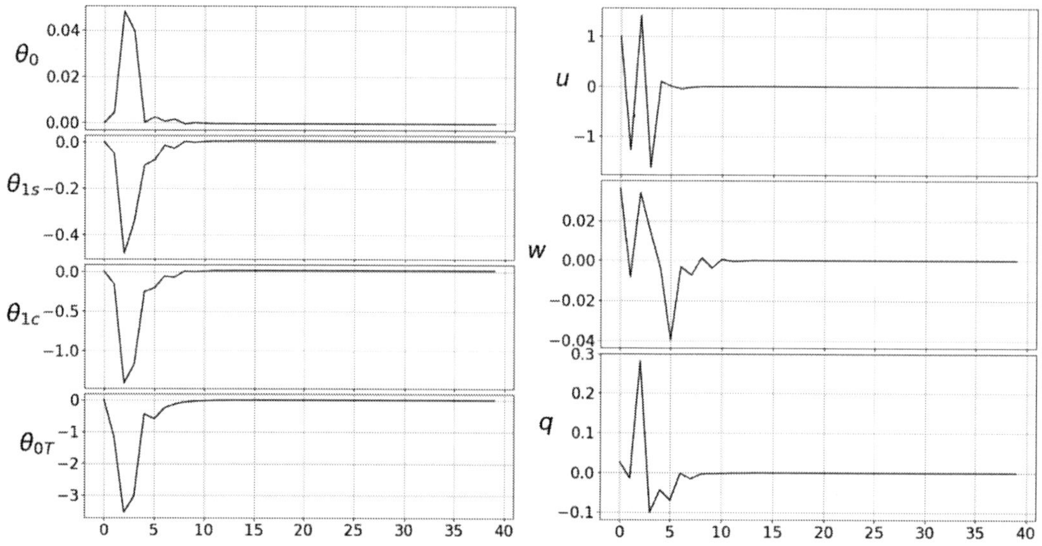

Figure 9. The control and system states trajectories for the S123-Bo105: $N_p = 8$, $N_c = 4$, $q = 0.01$, rank $C_i < 8, i = 1, \ldots, 8$

5 Conclusion

In the article the problem of a helicopter control with use of MPC is considered. The proposed strategy is utilized to find control and state trajectories of three helicopters models: ZD559-Lynx, SA330-Puma and S123-Bo105. It is supposed that the helicopters perform flight on sea level with zero sideslip and turn rate with a fixed value of the forward speed.

The described approach permits to damp all transferring processes caused by initial disturbances. The obtained results demonstrate the robustness of the considered method in the case of changing intensity of initial disturbances, the control and prediction horizons, the forward speed of the helicopters and completeness of system state information. The suitable relation between the control and prediction horizons leading to the motion stabilization over a time interval defined by the prediction horizon is determined. It is revealed that to reach a desired trajectory it is needed to reduce the discretization step for the proposed numerical scheme when switching from the given continuous model to the discrete one. Moreover significant changes in control trajectories behavior are observed in the case of substantial cutting of the dimension of the system output.

References

[1] G.D. Padfield, *Helicopter flight dynamic. The theory and application of flying qualities and simulation modelling* (Blackwell Publishing, Oxford, 2007)

[2] G.D. Padfield, *Helicopter flight dynamics: Including a treatment of tiltrotor aircraft* (John Wiley & Sons, New York, 2018)

[3] T. Oktay, C. Sultan, Aircraft Engineering and Aerospace Technology, **85(1)**, 32 (2013)

[4] J. Hu, H. Gu, International Journal of Aerospace Engineering, **2017** (2017)

[5] C.C. Luo, R.F. Liu, C.D. Yang, Y.H. Chang, Aerospace Science and Technology, **7(2)**, 159 (2003)

[6] L. Wang, *Model predictive control system design and implementation using MATLAB* (Springer-Verlag, Melbourne, 2009)

[7] J.B. Rawlings, *Model predictive control: Theory and design* (Nob Hill Publishing, Madison, 2009)

[8] I.A. Kudryavtseva, A.V. Panteleev, Smart Innovation, Systems and Technologies, **274**, 363 (2022)

[9] K.A. Rybakov, Differential Equations and Control Processes, **2**, 39 (2016)

[10] I. Kudryavtseva, A. Efremov, A. Panteleev, *Optimization of helicopter motion control based on the aggregated interpolation model* AIP Conf. Proc., **2181**, 020008 (2019)

Numerical methods in the training of aerospace engineers

Elena Kuznetsova[1,*], *Victoria* Kindinova[1], and *Oleg* Krinetsky[1]

[1]Moscow Aviation Institute (National Research University), 4, Volokolamskoe shosse, Moscow, 125993, Russia

Abstract. The paper analyzes the experience of teaching numerical methods in the preparation of engineering students at the Moscow Aviation Institute (National Research University). The necessity of linking the numerical methods studied in the basic course with the topic of a particular specialty is substantiated. The disadvantages of using Ms Excel spreadsheets in the implementation of numerical methods are explained. The necessity of teaching software implementation of numerical methods on a computer is shown. The new American Strategic Computer Initiative is analyzed, as well as the current state of computing power in Russia. The need for the development and implementation of specialized courses in engineering areas for teaching complex methods of numerical analysis is shown. Such learning is recommended to be carried out within the framework of profile physical tasks. It is proposed to acquaint future engineers with the problem of parallelizing numerical algorithms for solving them on a superecomputer. The essence of the engineering approach to the parallelization of numerical algorithms is presented.

1 Introduction

Numerical methods are a powerful tool for solving complex scientific and technical problems using computers. For decades, the Department of Computational Mathematics and Programming has been providing teaching of the discipline Numerical Methods at engineering faculties (institutes) Moscow Aviation Institute (National Research University). The basic course of the discipline includes basic numerical methods for solving linear and nonlinear algebra problems, numerical methods for approximating functions, numerical integration and numerical differentiation, numerical methods for solving ordinary differential equations, as well as numerical methods for solving the Cauchy problem and the boundary value problem for partial differential equations. The discipline is studied in the third year of bachelor's degree and specialist degree of engineering faculties. Teaching includes a lecture course (a two-hour lecture per week) and practical classes in the same volume for one semester. Lectures are based on knowledge from the sections of higher mathematics obtained in junior courses. At the lectures, students study the mathematical formulation of problems, the numerical methods used, the issues of numerical error, convergence, as well as the application of numerical methods for calculating physical systems in a critical state (at points of bifurcation, poor conditionality, transition to another phase state). In practical classes, students get acquainted with examples of problem solving, perform training exercises. As students study the discipline, they perform calculation and graphic work on an individual task, the assessment for

*e-mail: oekrin@mail.ru

which is decisive when they put down the final examination grade for the discipline. In the course of training, methodological and teaching aids on the discipline are used, developed by the staff of the Department of Computational Mathematics and Programming [1, 2], as well as textbooks of Russian and foreign specialists recommended by the scientific and technical community [3, 4].

The purpose of research is to analyze the experience of teaching numerical methods in the preparation of engineering students at the Moscow Aviation Institute. Consider the shortcomings and possible prospects for the development of the discipline. As a criterion for the effectiveness of teaching, we will take the competitiveness of graduates of engineering faculties at the international level.

2 Analysis of teaching a basic course of numerical methods

Currently, there is an understanding that it is necessary to link the numerical methods studied in the basic course with the topics of a particular institute. In other words, before describing numerical methods in a mathematical formulation, it is first necessary to present a physical formulation of an engineering problem in which these methods are used. Such an approach, on the one hand, contributes to the development of the student's creative potential, develops in him the ability to find objective cause-and-effect relationships in the surrounding world, the ability to prognostic activity. On the other hand, it increases his motivation, facilitates the process of transition to practical activity in the acquired specialty. This approach was used at the department before [5, 6], but was undeservedly forgotten. It seems appropriate to begin the study of numerical methods with a description of the physical (engineering) tasks inherent in a specific specialty. Let us consider as an example the presentation of numerical methods for students of the Institute No. 2 "Aviation, rocket engines and power plants". When studying the topic of solving nonlinear equations, it will be useful to describe in advance the physical problem of calculating the steady-state surface temperature of an aircraft. And only after that to present iterative methods for solving algebraic and transcendental equations and systems: the method of simple iteration, the secant method, the Newton method. When studying the topic of solving systems of nonlinear equations, it will be useful to first consider a simplified scheme of the main hydraulic subsystem of a rocket engine. And only after that to present iterative methods of numerical solution of systems of nonlinear equations: the method of simple iteration, the Seidel method, the Newton method.

In addition, it will be useful to supplement the training course of a particular institute with a mathematical apparatus that is most adequate for solving the practical problems of this institute. So, for example, for students of Institute No. 2, it is advisable to devote more time to studying the theory of difference schemes that allow solving problems of strength and thermal conductivity, to numerical methods of hydrodynamics. For students of Institute No. 3 "Control Systems, Informatics and Power Engineering", it is desirable to pay more attention to the study of numerical optimization methods, statistical numerical methods, statistical methods for processing test results. For students of the Institute No. 1 "Aviation Engineering", the most interesting are statistical methods such as the Monte Carlo method, which allow you to effectively solve complex integro-differential systems that cannot be solved by classical analytical methods. At present, in this regard, it is practiced by students to write essays on this topic and to defend them in the framework of seminars. However, abstracts are clearly not enough.

Initially, solving seminar tasks and performing settlement and graphic work involved performing calculations manually. In recent years, it has become fashionable to perform these calculations using MS Excel spreadsheets. Many students know the technique of working with the universal MS Excel application from the school course, others master it on their own. The use of MS Excel is welcome, because, on the one hand, it reduces routine computational work and increases the accuracy of calculations, on the other hand, it expands the tools that students master. In addition, students can use the graphical capabilities of MS Office to build flat graphs and spatial surfaces. In MS Excel, there are simple options for varying parameters, facilitating the possibility of conducting numerical experiments. But it is necessary to focus students' attention on the fact that the use of MS Excel has serious drawbacks, the most important of which are the following.

- The logic of the program loses visibility. Formulas expressed in terms of cell addresses, with a large number of them, are difficult to perceive, since cell addresses themselves do not carry any semantic load. This fact makes it much more difficult to check the correctness of the program and understand the formulas out of context.
- Programs in tabular processors are highly dependent on their dimensionality and are rigidly tied to the grid. When adding and deleting rows or columns, it is often necessary to make many changes to various parts of the table. Adding and removing dimensions is so difficult and inconvenient that when developing a tabular program, the author is actually forced to choose the required dimensions in advance and plan far ahead, which contradicts the idea of using a tabular processor as a means of rapid development and modeling.
- It is difficult to differentiate access to data. A user who has access to the table may accidentally or intentionally make changes to it that may disrupt the operation of the program. It is difficult to find out what changes, by whom and when were made to the program.
- The pace of development of tabular programs is significantly reduced due to the fact that the developer has to work at the cell level, which significantly increases the risk of errors. It is difficult to control the input: it is easy to enter data in the wrong format and crash the calculations.
- It is difficult to maintain data integrity. The inability to control corrections increases the risk of errors arising from the inability to track, test and isolate changes.
- Joint development of applications by a team of developers is difficult, it is impossible to cooperate with external systems, there are no notifications, etc.

The reason for these shortcomings is the low level of abstraction of the table processor, when the program developer is forced to operate with such low-level objects as a cell and a formula fixed on a grid. It is necessary to explain to students that it is a mistake to consider spreadsheets as the only means of implementing numerical methods on a computer. In practice, they will inevitably face the need for software implementation of numerical methods using algorithmic languages on modern computers.

It seems appropriate to introduce students to the implementation of these numerical methods programmatically using modern algorithmic languages such as Fortran, Pascal, Delphi, C++, C# on personal computers and workstations. The software implementation of the numerical methods under study is in most cases a non-trivial task. In previous years, the department had a practice of teaching methods of programming engineering problems [7]. Currently, within the framework of the one-semester educational process, it is not possible to introduce students to the most important issues in the field of numerical methods programming. This is regrettable. Nevertheless, in order to prepare students for the difficulties that they will inevitably face when implementing numerical methods programmatically, the following is being undertaken. Students are invited to independently prepare and debug programs for computational tasks using algorithmic languages that they studied as part of the educational process of their institute. Errors that occur during the independent implementation of numerical algorithms are considered and corrected at seminars. Students who have not received sufficient training in programming are given the opportunity to study and debug ready-made programs for typical tasks from the relevant textbooks [8, 9].

3 Current state of computer power

Currently, the computing capabilities of modern computers are rapidly developing. In May 2005, the U.S. government reviewed an analytical report of the Advisory Committee on Information Technology under the President of the United States entitled: “Computational Science: Ensuring America’s Competitiveness” [10]. The report analyzed the potential opportunities for the development of science, industry and the economy, their relationship with achievements in the field of Computational Science. Analysts have identified Computational Science as one of the most important areas providing innovative development of society in the XXI century. The report highlighted the basic components of Computational Science, including:

- models, computational algorithms and modeling programs for solving scientific, technical and humanitarian problems;
- optimization and development of hardware, software and network data processing tools that provide solutions to these computational tasks;
- computer infrastructure to support scientific research and solve technical problems.

The report identified measures aimed at maintaining US national security, including:

- creating promising areas and maintaining a high level of research in the field of computational sciences;
- creation of infrastructure for exploratory research;
- creation of a national repository of data and software products;
- creation and support of national high-performance computing centers.

The new American strategic computer initiative entailed a new stage of competitive struggle in the field of science, education and information technology, determined the direction of development of all countries without exception.

Under the influence of these global processes, the Program “Development of supercomputers and Grid technologies” [11] was adopted in Russia in 2009 in accordance with the decision of the Commission under the President of the Russian Federation on Modernization and Technological Development of the Economy. Its goal is to create a supercomputer computing industry in Russia, including a basic range of supercomputers of various capacities, high-speed communication channels, and training of specialized specialists. The implementation of this program is entrusted to the Rosatom State Corporation. In accordance with this program, supercomputers were installed in the largest Russian state universities, including Moscow (MSU), Tomsk, Saint Petersburg Polytechnic, Nizhny Novgorod, South Ural, Far Eastern and Novosibirsk to support scientific research. In November 2021, Russia entered the top twenty in the world Top 500 ranking, presenting seven powerful supercomputers, one of which is installed at Moscow State University [12]. Currently, there is a tendency to concentrate high-performance computing resources in collective use centers and the development of remote access infrastructure using telecommunications. The Moscow Supercomputer Center of the Russian Academy of Sciences (MSC RAS), the supercomputer center and the grid system in Sarov are examples of such high-performance computing centers with remote access.

The aerospace industry, in the light of the scientific and defense challenges facing it at the present stage, is one of the main consumers of high-performance computing power. Several compact supercomputers created within the framework of the Program "Development of Supercomputers and Grid technologies" are operated in the design bureau of Sukhoi Airlines. On these computers, Sukhoi engineers calculate a significant part of the medium-scale tasks that arise when creating new models of military and civil aircraft. With their help, processes related to aerodynamics, gas dynamics, heat and mass transfer, strength, and safety of aircraft operation are investigated. Large-scale computational problems (aeroelasticity of the structure, flutter, aeroacoustics, connected and conjugate heat and mass transfer, optimization problems) requiring more detailed computer models and large computing resources are solved in remote access mode on a powerful supercomputer installed at the Russian Federal Nuclear Center in Sarov [11].

The rapid growth of computing power, the need to calculate more complex models pose new challenges for universities that train engineering personnel in the aerospace industry, and force them to take a fresh look at teaching Numerical Methods.

4 Learning in complex numerical methods

New high-performance computing technology, large-scale realistic computer models require the use of much more complex methods of numerical analysis. As it became clear at the end of the last century: "the development of mathematics in a certain period of time reflects the nature of computing tools available at that moment to a much greater extent than one might assume" [13, p. 152]. Complex methods of numerical analysis include numerical methods for solving elliptic, hyperbolic and parabolic problems [14]. It makes sense to consider the methods of this class within the framework of the physical formulation of tasks corresponding to each specific field of study.

For example, when training engineers in the field of strength, it makes sense to first determine the task of calculating the strength of the structure. This can be a strength calculation under static and dynamic loading, calculation of natural frequencies and forms of vibrations, calculation of critical loads and forms of loss of stability, two-dimensional and three-dimensional calculation of volumetric, thin-walled and rod structures. Then you need to show which mathematical model corresponds to a specific physical problem. And then you need to describe the numerical methods used: the finite element method, the boundary element method, the spectral element method and their possible variants.

When training engineers in the field of hydroaerodynamics, you first need to describe a physical problem. This may be a transient mode in the boundary layer, separation and attachment of the flow, reduction of aerodynamic drag during takeoff, the interaction of a shock wave with a turbulent boundary layer in a transonic flow. Then you need to show the corresponding mathematical model. Finally, it is necessary to describe the numerical methods used: the finite difference method, the finite volume method, the finite element method, the smoothed particle method, the method using the probability distribution function, as well as the application features of each method.

Particular attention should be paid to the implementation of complex numerical methods on a computer. The implementation of each of them is carried out in three stages:

- preparatory stage of forming the geometry of the model, determining the physical properties, boundary and initial conditions;
- stage of direct calculation;
- stage of analysis and visualization of calculation results.

Both the preparatory phase and the analysis phase are very resource-intensive and time-consuming.

There are foreign and domestic commercial software packages that implement these complex numerical methods for problems in the aerospace industry. However, using these packages without understanding the methods they use can lead to erroneous decisions.

Complex numerical methods can be taught within the discipline, which is a natural continuation and development of the discipline Numerical Methods. Building such a discipline is a difficult task. However, similar specialized courses on numerical methods are already taught at some universities in Russia: at Moscow State University, at the Moscow Institute of Physics and Technology (National Research University), at the South Ural State University. It seems promising to develop and introduce such profile courses for the training of engineers at the Moscow Aviation Institute, although some elements of the study of special numerical methods can be found in the curriculum of a number of institutes of the Moscow Aviation Institute within the framework of the specialist degree and master's degree.

5 Preparation of parallel applications

At the present time, a further increase in the performance of computers only by improving the characteristics of electronic elements has reached the limit determined by physical laws. A further increase in the productivity of computers is possible only through the parallelization of information processing processes. The construction of parallel computing processes that ensure maximum performance is an important independent problem.

Porting a conventional program to a multiprocessor computing system may not provide the expected performance gain. Moreover, as a result of such a transfer, the program may run even slower. One of the authors of the article faced the problem of transferring a sequential program performing strength calculations by the finite element method to a supercomputer back in the 90s of the last century [15, 16].

It is extremely difficult to write an effective parallel program or adapt an existing serial program for parallel computing on multiprocessor systems. This requires, on the one hand, to perform a structural analysis of the algorithm, to identify its internal parallelism. Such an analysis requires a deep understanding of the essence of the task. On the other hand, when developing parallel algorithms for composing effective code, it is necessary to know the features of the computing system on which the algorithm will be implemented. In particular, knowledge of the topology of connections, the type of processor, the relative memory volumes of caches of various levels, etc. is required.

The problem is that there are not enough specialists who are equally proficient in both the subject of research in their "own" field of knowledge and knowledge in the field of high-performance computing systems. Therefore, an approach is practiced that allows the task to be divided into two independent tasks. The first task is the task of parallel computing. The task is to perform a structural analysis of the algorithm, identify the parallelism in it and make a time diagram of its execution. This task should be performed by a highly qualified engineer, a specialist in the subject area. The second task is the task of parallel programming. The task is to compile effective code and debug the program according to the time diagram developed by the engineer. This task is performed by a programmer who knows the architecture of the computing system well.

Moreover, the inverse problem can be posed. If there is a choice of calculators, then to build the most efficient parallel algorithm, you can set the task of determining the calculators that most fully implement the features of the parallel algorithm.

The approach under consideration allows us to study the construction of parallel algorithms and programming issues independently. When studying parallel computing, the issues of analyzing the effectiveness of the parallel structure of the algorithm, building models and schemes of parallel computing, as well as the analysis of achievable acceleration and efficiency for various applied computing tasks are considered. The solution of the parallel computing problem is a description of a parallel algorithm in the form of flowcharts, graphs and diagrams that are understandable to the programmer and can be used by him to write a parallel program.

In recent years, efforts have been made to expand the training of specialists in the field of problem solving on multiprocessor supercomputers. Since the solution of the problem of parallel computing is assigned to an engineer, let's consider the essence of the engineering approach [17].

To solve the problem of parallel computing, it is necessary to first study the architecture of a parallel computing system.

Currently, the vast majority of parallel computers are implemented in the MIMD architecture class (multiple instruction streams and multiple data streams) [17, 18]. Among them, the following main subclasses are distinguished:

- Vector-conveyor computers. This subclass uses a set of vector commands that perform operations with arrays of independent data in one clock cycle. Example—CRAY.
- Massively parallel with distributed memory. In this subclass, microprocessors, each having its own local memory, are connected through some communication medium.
- Symmetric multiprocessors (SMP), consist of a set of processors having divide shared memory with a single address space and operating under the control of a single operating system.
- Clusters are formed from computing modules of any of the types discussed above, united by a system of connections or by means of shared external memory.

To describe the information dependencies of the algorithm for solving the problem, an algorithm model is used in the form of an acyclic oriented graph (algorithm graph). In this model, the set of algorithm operations and the information dependencies existing between the operations describe the set of vertices and arcs of the graph. The vertices represent the operations of the algorithm, and the arcs of the graph establish a partial order of operations. The arcs of the graph are assigned weights reflecting the intensity of information exchange between the corresponding vertices—operations.

The computing system is also represented as a graph. In the graph of a computing system, vertices are a set of processors, arcs represent communication lines between processors. The bandwidth of communication lines is characterized by the weights of the arcs of the graph.

The effectiveness of implementing specific calculations on a specific system depends on how much the structure of the algorithm corresponds to the architecture of the computing system. The central problem in the planning of computing resources is the problem of mapping a parallel algorithm and its corresponding program to the architecture of a multiprocessor computing system.

The second task is to build a schedule for the parallel implementation of the algorithm, in which each operation is assigned the number of the processor used for its execution and the start time of the operation.

In order for the schedule to be realizable, it is necessary that the same processor was not assigned to different operations at the same time; by the designated time of the operation, all the required data has already been calculated. A set of parallelization methods and techniques that do not require a strict solution to the problem of mapping the graph of an algorithm to the graph of a computing system was called an engineering approach.

The algorithm's ability to parallelize is potentially related to one of two internal properties of the algorithm, which are characterized as task parallelism (message passing) and data parallelism (data parallel). When task parallelism, a computational task is divided into several relatively independent subtasks, each of which is loaded into its own processor and implemented independently, but uses common data and/or exchanges the results of its work with other subtasks. If there is a data parallelism property in the algorithm, one operation can be performed on all elements of the data array at once. In this case, different fragments of the array can be processed independently on different processors. A number of algorithms have both task parallelism and data parallelism to one degree or another. The following types of algorithms are distinguished:

- Algorithms using Data Parallelism. They are characteristic of numerical algorithms for processing large arrays represented as vectors and matrices.
- Algorithms with data distribution (Data Partitioning). Data parallelism is used, in which the data space can be divided into disjoint areas. Independent processes are associated with each area. A rare exchange between processes is required.
- Relaxation algorithms (Relaxed Algorithm). The algorithm can be represented as independent processes without synchronization of communication between them, but processors must have access to shared data.
- Algorithms with synchronization of iterations (Synchronous Iteration). Many of the standard numerical iterative parallel algorithms require synchronization at the end of each iteration. Synchronization means that permission to start the next iteration is given after all processors have completed the previous iteration.
- Self-replicating tasks (Replicated Workers). For tasks of this class, a central pool (storage) of similar computing tasks is created and maintained. This technology is typical for graph or tree research.
- Pipelined Computing. This type of computation is typical for processes that can be represented in the form of some regular structure.

With the engineering approach, the following stages of parallel applications preparation technology are recommended:

- decomposition of the task into subtasks that are implemented independently;
- definition of information interactions for the generated set of subtasks;
- scaling of subtasks, determining the number of processors;
- determining the architecture of the system, assigning subtasks to processors, scheduling.

The execution time of the parallel algorithm is determined by the maximum time value in the schedule. For a fixed graph of calculations, it is advisable to build a schedule that ensures the minimum execution time of the algorithm. To reduce the implementation time of the algorithm, various computational schemes are considered. Each scheme has its own schedule.

The considered approach is implemented in the educational process of Samara State Aerospace University named after S.P. Korolev within the framework of an independent discipline on parallel computing. Such a discipline is a natural continuation and development of the discipline of Numerical Methods. It also seems useful at the Moscow Aviation Institute at the engineering faculties to supplement the discipline of Numerical methods with parallel computing sections.

6 Conclusion

The Moscow Aviation Institute (National Research University) is one of the leading universities in Russia that train engineers for the aerospace industry. The discipline Numerical methods at engineering faculties allows students to systematize the knowledge acquired in the study of computer modeling and algorithmization, programming languages and methods, modern information technologies, as well as apply the knowledge gained to solving problems of computational mechanics. A critical analysis of the experience of teaching the discipline at engineering faculties allows us to assess the compliance of teaching results with the requirements of state policy in the field of training competitive engineering personnel for the aerospace industry. The paper suggests possible prospects for the development of the discipline. Movement in this direction will contribute to the development of students' creative abilities, love for their chosen profession, and the opportunity to fully realize their creative potential. Familiarization of students with supercomputer technologies in the educational process is an effective way to involve students in the real practice of scientific research in the aerospace industry.

References

[1] U.G. Pirumov, *Numerical methods* (MAI Publishing House, Moscow, 1998)

[2] M.A. Cherkasov, *Numerical methods. Problem solving* (MAI Publishing House, Moscow, 2007)

[3] B.P. Demidovich, I.L. Maron, *Fundamentals of computational mathematics*, 3rd ed. (Nauka, Moscow, 1966)

[4] N.V. Kopchenova, I.A. Maron, *Computational mathematics in examples and problems* (Pal, Saint Petersburg, 2008)

[5] Yu.N. Badlevsky, P.V. Biryukov, T.Ya. Kosheleva et al., *Numerical methods and algorithms for solving problems of rocket technology*, ed. by T.Ya. Koshelevoy (MAI Publishing House, Moscow, 1990)

[6] V.I. Kireev, *Numerical methods for solving problems of algebra and approximation theory* (MAI Publishing House, Moscow, 1991)

[7] V.E. Zaitsev, O.I. Korchagina, E.A. Kuzmenkova, M.V. Rotanina, Yu.A. Shebeko, *Methods of programming engineering problems in the FORTRAN language* (MAI Publishing House, Moscow, 1988)

[8] Yu.V. Vasilkov, N.N. Vasilkova, *Computer computing technologies in mathematical modeling* (Finance and Statistics, Moscow, 2004)

[9] A.M. Mudrov, *Numerical methods for PCs in Basic, Fortran and Pascal languages* (RASKO MP, Tomsk, 1991)

[10] *Computational Science: Ensuring America's Competitiveness* (President's Information Tehnology Advisory Committee, May 27, 2005)

[11] *Supercomputers for aircraft*, All-Russian weekly newspaper VPK, 19.12.2012, URL `https://vpk.name/news/81078_superkompyutery_dlya_samoletov.html`

[12] Rating_supercomputers_Russia_and_SNG_Top50,04.01.2020, URL `https://www.tadviser.ru/index.php`

[13] T. Toffoli, N. Margolus, *Machines of cellular automata* (Mir, Moscow, 1991)

[14] V.F. Formalev, D.L. Reviznikov, *Numerical methods* (FIZMATLIT, Moscow, 2006)

[15] E.V. Kuznetsova, *A vectorized algorithm for solving systems of linear algebraic equations by the frontal method in the FEM package "Composite"* (Mobile software, Tver, 1991)

[16] E.V. Kuznetsova, *Research and development of a transportable version of the fortran package for strength calculations by the finite element method*, in Candidate of Physical and Mathematical Sciences: 05.13.11, 05.07.03

[17] V.P. Gergel, V.A. Fursov, *Lectures on parallel computing* (SSAU Publishing house, Samara, 2009)

[18] J.L. Hennessy, D.A. Patterson, *Computer architecture. Quantitative approach*, 5th ed. (Elsevier, 2012)

Optimization of the mean-square approximation procedures for iterated Stratonovich stochastic integrals of multiplicities 1 to 3 with respect to components of the multi-dimensional Wiener process based on Multiple Fourier–Legendre series

Dmitriy Kuznetsov[1,*] and *Mikhail* Kuznetsov[2]

[1]Institute of Physics and Mechanics, Peter the Great Saint-Petersburg Polytechnic University, 29, Polytechnicheskaya st., 195251, Saint-Petersburg, Russia
[2]Faculty of Computer Science and Technologies, Saint-Petersburg Electrotechnical University, 5, Professora Popova st., 197376, Saint-Petersburg, Russia

Abstract. The article is devoted to approximation of iterated Ito and Stratonovich stochastic integrals of multiplicities 1 to 3 by the method of multiple Fourier–Legendre series. The mentioned stochastic integrals are part of strong numerical methods with convergence order 1.5 for Ito stochastic differential equations with multidimensional noncommutative noise. These numerical methods are based on the so-called Taylor–Ito and Taylor–Stratonovich expansions. We calculate the exact lengths of sequences of independent standard Gaussian random variables required for the mean-square approximation of iterated Stratonovich stochastic integrals. Thus, the computational cost for the implementation of numerical methods can be significantly reduced.

1 Strong Taylor–Ito and Taylor–Stratonovich numerical schemes with convergence order 1.5

The relevance of the problem of numerical integration of Ito stochastic differential equations (SDEs) is explained by a wide range of their applications related to the construction of adequate mathematical models of dynamical systems of various physical nature under random perturbations. Among these models, we note models in financial mathematics, biology, epidemiology, medicine, aerospace industry, hydrology, seismology, geophysics, genetics, electrodynamics, chemical kinetics [1–10]. Moreover, Ito SDEs are used to solve various mathematical problems such as signals filtering with random noises, stochastic optimal control, stochastic stability and bifurcation analysis, parameter estimation of stochastic systems [1, 2, 9]. In addition, exact solutions of Ito SDEs are known in rare cases, and besides, knowing the exact solution of Ito SDE does not always allow us to simulate it numerically without using special numerical methods.

One of the effective approaches to the numerical integration of Ito SDEs is an approach based on the Taylor–Ito and Taylor–Stratonovich expansions [1, 2, 9, 10]. The important feature of these expansions is iterated Ito and Stratonovich stochastic integrals with respect to components of the multi-dimensional Wiener process.

*e-mail: sde_kuznetsov@inbox.ru

The article continues the research [11] and is devoted to the development of an effective method of iterated Ito and Stratonovich stochastic integrals approximation based on generalized multiple Fourier series and proposed by the first author of the article in [10]. More precisely, we consider mean-square approximations of iterated Stratonovich stochastic integrals of multiplicities 1 to 3 with respect to components of the multi-dimensional Wiener process by the method of multiple Fourier–Legendre series. It should be noted that the iterated Stratonovich stochastic integrals are more convenient to use than the iterated Ito stochastic integrals due to simplicity of their approximations [1, 2, 9, 10, 12–22] (compare formulas (18) and (19)).

Among advantages of the method based on generalized multiple Fourier series [10, 12] over the methods [1, 2, 8, 9, 13–21] of mean-square approximation of iterated Ito and Stratonovich stochastic integrals we note the following. The well-known approach based on the Karhunen–Loeve expansion of the Brownian bridge process [1, 8] (also see [17]) leads to iterated application of the operation of limit transition. At the same time the operation of limit transition is implemented only once in theorem 1 (see below). This feature is more appropriate for the approximation. Moreover, the method [10, 12] allows one to calculate the exact lengths of sequences of independent standard Gaussian random variables individually for different iterated Ito or Stratonovich stochastic integrals. Thus, the computational cost for the implementation of numerical methods for Ito SDEs can be significantly reduced. For more details see chapter 6 of monograph [12].

Let (Ω, F, P) be a complete probability space, let $\{\mathrm{F}_t, t \in [0, T]\}$ be a nondecreasing right-continuous family of σ-algebras of F and let $\mathbf{w}_t$ be a standard m-dimensional Wiener process, which is F_t-measurable for any $t \in [0, T]$. We assume that the components $\mathbf{w}_t^{(i)}$ $(i = 1, \ldots, m)$ of this process are independent. Consider an Ito SDE in the integral form

$$\mathbf{x}_t = \mathbf{x}_0 + \int_0^t \mathbf{a}(\mathbf{x}_\tau, \tau)\, d\tau + \sum_{i=1}^m \int_0^t B_i(\mathbf{x}_\tau, \tau)\, d\mathbf{w}_\tau^{(i)}, \quad \mathbf{x}_0 = \mathbf{x}(0, \omega), \tag{1}$$

where $\omega \in \Omega$, $\mathbf{a}(\mathbf{x}, t)\colon \mathbf{R}^n \times [0, T] \to \mathbf{R}^n$, $B_i(\mathbf{x}, t)$ is the i-th column of $B(\mathbf{x}, t)\colon \mathbf{R}^n \times [0, T] \to \mathbf{R}^{n\times m}$, $\mathbf{x}_0$ is F_0-measurable, $\mathbf{M}|\mathbf{x}_0|^2 < \infty$ ($\mathbf{M}$ is an expectation), $\mathbf{x}_0$ and $\mathbf{w}_t - \mathbf{w}_0$ are independent $(t > 0)$.

Let us consider the following differential operators

$$L = \frac{\partial}{\partial t} + \sum_{i=1}^n \mathbf{a}^{(i)}(\mathbf{x}, t) \frac{\partial}{\partial \mathbf{x}^{(i)}} + \frac{1}{2} \sum_{j=1}^m \sum_{l,i=1}^n B^{(lj)}(\mathbf{x}, t) B^{(ij)}(\mathbf{x}, t) \frac{\partial^2}{\partial \mathbf{x}^{(l)} \partial \mathbf{x}^{(i)}}, \quad \bar{L} = \frac{\partial}{\partial t} + \sum_{i=1}^n \bar{\mathbf{a}}^{(i)}(\mathbf{x}, t) \frac{\partial}{\partial \mathbf{x}^{(i)}}, \tag{2}$$

$$\bar{\mathbf{a}}(\mathbf{x}, t) = \mathbf{a}(\mathbf{x}, t) - \frac{1}{2} \sum_{j=1}^m G_0^{(j)} B_j(\mathbf{x}, t), \quad G_0^{(i)} = \sum_{j=1}^n B^{(ji)}(\mathbf{x}, t) \frac{\partial}{\partial \mathbf{x}^{(j)}}, \quad i = 1, \ldots, m, \tag{3}$$

where $\mathbf{a}^{(i)}(\mathbf{x}, t)$, $\bar{\mathbf{a}}^{(i)}(\mathbf{x}, t)$ are i-th components of $\mathbf{a}(\mathbf{x}, t)$, $\bar{\mathbf{a}}(\mathbf{x}, t)$ respectively and $B^{(ij)}(\mathbf{x}, t)$ is the ij-th element of $B(\mathbf{x}, t)$.

Consider the following iterated Ito and Stratonovich stochastic integrals [10–12]

$$I_{(l_1 \ldots l_k)s,t}^{(i_1 \ldots i_k)} = \int_t^s (t - t_k)^{l_k} \ldots \int_t^{t_2} (t - t_1)^{l_1} d\mathbf{w}_{t_1}^{(i_1)} \ldots d\mathbf{w}_{t_k}^{(i_k)}, \tag{4}$$

$$I_{(l_1 \ldots l_k)s,t}^{*(i_1 \ldots i_k)} = \int_t^s (t - t_k)^{l_k} \ldots \int_t^{t_2} (t - t_1)^{l_1} \circ d\mathbf{w}_{t_1}^{(i_1)} \ldots \circ d\mathbf{w}_{t_k}^{(i_k)}, \tag{5}$$

where $\circ d\mathbf{w}_s^{(i_1)}$ is the Stratonovich differential, $l_1, \ldots, l_k = 0, 1, 2, \ldots$, and $i_1, \ldots, i_k = 1, \ldots, m$.

Assume that $\mathbf{a}(\mathbf{x}, t)$ and $B(\mathbf{x}, t)$ are enough smooth functions with respect to the variables $\mathbf{x}$ and t. Consider the partition $\{\tau_q\}_{q=0}^N$ such that $0 = \tau_0 < \tau_1 < \ldots < \tau_N = T$, $\Delta_N = \max_{0 \le q \le N-1} |\tau_{q+1} - \tau_q|$.

We will say [1] that a numerical scheme $\mathbf{y}_{\tau_q}$, $q = 0, 1, \ldots, N$ converges strongly with order $\gamma > 0$ at time moment T to the process $\mathbf{x}_t$, $t \in [0, T]$ if there exist a constant $C > 0$, which does not depend on Δ_N, and a $\delta > 0$ such that $\mathbf{M}|\mathbf{x}_T - \mathbf{y}_T| \le C(\Delta_N)^\gamma$ for each $\Delta_N \in (0, \delta)$.

Consider the strong Taylor–Ito and Taylor–Stratonovich schemes for Ito SDE (1) with convergence order 1.5 [1, 8–10, 12]:

$$\mathbf{y}_{q+1} = \mathbf{y}_q + \sum_{i_1=1}^{m} B_{i_1} \hat{I}^{(i_1)}_{(0)\tau_{q+1},\tau_q} + \Delta\mathbf{a} + \sum_{i_1,i_2=1}^{m} G_0^{(i_1)} B_{i_2} \hat{I}^{(i_1 i_2)}_{(00)\tau_{q+1},\tau_q} +$$
$$+ \sum_{i_1=1}^{m} \left[G_0^{(i_1)} \mathbf{a} \left(\Delta \hat{I}^{(i_1)}_{(0)\tau_{q+1},\tau_q} + \hat{I}^{(i_1)}_{(1)\tau_{q+1},\tau_q} \right) - L B_{i_1} \hat{I}^{(i_1)}_{(1)\tau_{q+1},\tau_q} \right] + \sum_{i_1,i_2,i_3=1}^{m} G_0^{(i_1)} G_0^{(i_2)} B_{i_3} \hat{I}^{(i_1 i_2 i_3)}_{(000)\tau_{q+1},\tau_q} + \frac{\Delta^2}{2} L\mathbf{a}, \tag{6}$$

$$\mathbf{y}_{q+1} = \mathbf{y}_q + \sum_{i_1=1}^{m} B_{i_1} \hat{I}^{*(i_1)}_{(0)\tau_{q+1},\tau_q} + \Delta\bar{\mathbf{a}} + \sum_{i_1,i_2=1}^{m} G_0^{(i_1)} B_{i_2} \hat{I}^{*(i_1 i_2)}_{(00)\tau_{q+1},\tau_q} +$$
$$+ \sum_{i_1=1}^{m} \left[G_0^{(i_1)} \bar{\mathbf{a}} \left(\Delta \hat{I}^{*(i_1)}_{(0)\tau_{q+1},\tau_q} + \hat{I}^{*(i_1)}_{(1)\tau_{q+1},\tau_q} \right) - \bar{L} B_{i_1} \hat{I}^{*(i_1)}_{(1)\tau_{q+1},\tau_q} \right] + \sum_{i_1,i_2,i_3=1}^{m} G_0^{(i_1)} G_0^{(i_2)} B_{i_3} \hat{I}^{*(i_1 i_2 i_3)}_{(000)\tau_{q+1},\tau_q} + \frac{\Delta^2}{2} L\mathbf{a}, \tag{7}$$

where the functions B_i, $\mathbf{a}$, $G_0^{(i)} B_j$, $\ldots$ are calculated at the point $(\mathbf{y}_q, \tau_q)$, $\mathbf{y}_q$ is an approximation of solution to Ito SDE (1) at moment τ_q, $\hat{I}^{(i_1 \ldots i_k)}_{(l_1 \ldots l_k)\tau_{q+1},\tau_q}$, $\hat{I}^{*(i_1 \ldots i_k)}_{(l_1 \ldots l_k)\tau_{q+1},\tau_q}$ are approximations of the iterated Ito and Stratonovich stochastic integrals $I^{(i_1 \ldots i_k)}_{(l_1 \ldots l_k)\tau_{q+1},\tau_q}$, $I^{*(i_1 \ldots i_k)}_{(l_1 \ldots l_k)\tau_{q+1},\tau_q}$ respectively, and L, $\bar{L}$, $\bar{\mathbf{a}}$, $G_0^{(i)}$, $i = 1, \ldots, m$ are defined by equalities (2) and (3). Note that the first four terms on the right-hand side of scheme (6) correspond to the strong Milstein scheme [8] with convergence order 1.0. Among the standard conditions ensuring the convergence of schemes (6) and (7), we note the conditions for approximation of iterated Ito and Stratonovich stochastic integrals [1]:

$$\mathbf{M}\left(I^{(i_1 \ldots i_k)}_{(l_1 \ldots l_k)\tau_{q+1},\tau_q} - \hat{I}^{(i_1 \ldots i_k)}_{(l_1 \ldots l_k)\tau_{q+1},\tau_q} \right)^2 \le C\Delta^4, \tag{8}$$

$$\mathbf{M}\left(I^{*(i_1 \ldots i_k)}_{(l_1 \ldots l_k)\tau_{q+1},\tau_q} - \hat{I}^{*(i_1 \ldots i_k)}_{(l_1 \ldots l_k)\tau_{q+1},\tau_q} \right)^2 \le \bar{C}\Delta^4, \tag{9}$$

where $q = 0, 1, \ldots, N-1$, constants C and $\bar{C}$ are independent of Δ.

2 Method of approximation of iterated Ito stochastic integrals based on generalized multiple Fourier series

Let us consider the effective method of expansion and mean-square approximation of iterated Ito stochastic integrals [10, 12] (other approaches can be found in [1, 2, 8, 9, 13–21]):

$$J[\psi^{(k)}]_{T,t} = \int_t^T \psi_k(t_k) \ldots \int_t^{t_2} \psi_1(t_1)\, d\mathbf{w}^{(i_1)}_{t_1} \ldots d\mathbf{w}^{(i_k)}_{t_k}, \tag{10}$$

where $i_1, \ldots, i_k = 0, 1, \ldots, m$, every $\psi_l(\tau)$ $(l = 1, \ldots, k)$ is a continuous non-random function on $[t, T]$, $\mathbf{w}^{(i)}_\tau$ $(i = 1, \ldots, m)$ are independent standard Wiener processes, and $\mathbf{w}^{(0)}_\tau = \tau$.

Let $K(t_1,\ldots,t_k)=\psi_1(t_1)\ldots\psi_k(t_k)\mathbf{1}_{\{t_1<\ldots<t_k\}}$ for $t_1,\ldots,t_k\in[t,T]$ $(k\ge 2)$ and $K(t_1)\equiv\psi_1(t_1)$ for $t_1\in[t,T]$, where $\mathbf{1}_A$ is the indicator of the set A. Suppose that $\{\phi_j(x)\}_{j=0}^{\infty}$ is a complete orthonormal system of functions in the space $L_2([t,T])$. It is well known that the generalized multiple Fourier series of $K(t_1,\ldots,t_k)\in L_2([t,T]^k)$ is converging to $K(t_1,\ldots,t_k)$ in the mean-square sense, i.e.

$$\lim_{p_1,\ldots,p_k\to\infty}\left\|K(t_1,\ldots,t_k)-\sum_{j_1=0}^{p_1}\ldots\sum_{j_k=0}^{p_k}C_{j_k\ldots j_1}\prod_{l=1}^{k}\phi_{j_l}(t_l)\right\|=0,$$

where

$$C_{j_k\ldots j_1}=\int_{[t,T]^k}K(t_1,\ldots,t_k)\prod_{l=1}^{k}\phi_{j_l}(t_l)\,dt_1\ldots dt_k \tag{11}$$

is the Fourier coefficient and $\|\cdot\|$ is the $L_2([t,T]^k)$-norm.

Theorem 1 [10], [12]. Suppose that every $\psi_l(\tau)$ $(l=1,\ldots,k)$ is a continuous non-random function on $[t,T]$ and $\{\phi_j(x)\}_{j=0}^{\infty}$ is a complete orthonormal system of continuous functions in the space $L_2([t,T])$. Then

$$\mathbf{M}\left(J[\psi^{(k)}]_{T,t}-J[\psi^{(k)}]_{T,t}^{p_1,\ldots,p_k}\right)^2\le k!\left(I_k-\sum_{j_1=0}^{p_1}\ldots\sum_{j_k=0}^{p_k}C_{j_k\ldots j_1}^2\right)\to 0 \tag{12}$$

if $p_1,\ldots,p_k\to\infty$, where $I_k^{1/2}$ is the $L_2([t,T]^k)$-norm of $K(t_1,\ldots,t_k)$,

$$J[\psi^{(k)}]_{T,t}^{p_1,\ldots,p_k}=\sum_{j_1=0}^{p_1}\ldots\sum_{j_k=0}^{p_k}C_{j_k\ldots j_1}\left(\prod_{l=1}^{k}\zeta_{j_l}^{(i_l)}-\underset{N\to\infty}{\text{l.i.m.}}\sum_{(l_1,\ldots,l_k)\in G_k}\phi_{j_1}(\tau_{l_1})\Delta\mathbf{w}_{\tau_{l_1}}^{(i_1)}\ldots\phi_{j_k}(\tau_{l_k})\Delta\mathbf{w}_{\tau_{l_k}}^{(i_k)}\right),$$

$i_1,\ldots,i_k=1,\ldots,m$ for $T-t\in(0,\infty)$ and $i_1,\ldots,i_k=0,1,\ldots,m$ for $T-t\in(0,1)$, $J[\psi^{(k)}]_{T,t}$ is defined by formula (10), l.i.m. is a limit in the mean-square sense, $\zeta_j^{(i)}=\int_t^T\phi_j(s)\,d\mathbf{w}_s^{(i)}$ $(i=0,1,\ldots,m)$ are i.i.d. $N(0,1)$-r.v.'s for various i or j if $i\ne 0$, $C_{j_k\ldots j_1}$ is defined by formula (11), $G_k=H_k\backslash L_k$, $H_k=\{(l_1,\ldots,l_k):\ l_1,\ldots,l_k=0,1,\ldots,N-1\}$, $L_k=\{(l_1,\ldots,l_k):\ l_1,\ldots,l_k=0,1,\ldots,N-1;\ l_g\ne l_r\ (g\ne r);\ g,r=1,\ldots,k\}$, $\Delta\mathbf{w}_{\tau_q}^{(i)}=\mathbf{w}_{\tau_{q+1}}^{(i)}-\mathbf{w}_{\tau_q}^{(i)}$, $\{\tau_q\}_{q=0}^{N}$ is a partition of $[t,T]$ such that $t=\tau_0<\ldots<\tau_N=T$, $\max_{0\le q\le N-1}(\tau_{q+1}-\tau_q)\to 0$ if $N\to\infty$.

Note that a number of generalizations and modifications of theorem 1 can be found in [12].

Let $J[\psi^{(k)}]_{T,t}^{p}$ be $J[\psi^{(k)}]_{T,t}^{p_1,\ldots,p_k}$ for $p_1=\ldots=p_k=p$ and let $E_k^p=\mathbf{M}\left(J[\psi^{(k)}]_{T,t}-J[\psi^{(k)}]_{T,t}^{p}\right)^2$. Combining estimates (8) and (12) for $p_1=\ldots=p_k=p$ and $k=3$, we obtain

$$3!\left(I_3-\sum_{j_1,j_2,j_3=0}^{p}C_{j_3j_2j_1}^2\right)\le C(T-t)^4. \tag{13}$$

It is not difficult to see that the multiplier factor 3! on the left-hand side of inequality (13) leads to a significant increase of computational costs for approximation of iterated Ito stochastic integrals. The mentioned problem can be overcome if we calculate the mean-square approximation error E_k^p exactly.

Theorem 2 [12]. Suppose that the conditions of theorem 1 are fulfilled. Then

$$E_k^p=I_k-\sum_{j_1,\ldots,j_k=0}^{p}C_{j_k\ldots j_1}\mathbf{M}\left(J[\psi^{(k)}]_{T,t}\sum_{(j_1,\ldots,j_k)}\int_t^T\phi_{j_k}(t_k)\ldots\int_t^{t_2}\phi_{j_1}(t_1)\,d\mathbf{w}_{t_1}^{(i_1)}\ldots d\mathbf{w}_{t_k}^{(i_k)}\right), \tag{14}$$

where $i_1,\ldots,i_k = 1,\ldots,m$; expression $\sum_{(j_1,\ldots,j_k)}$ means the sum with respect to all possible permutations $(j_1,\ldots,j_k)$. At the same time if j_r swapped with j_q in the permutation $(j_1,\ldots,j_k)$, then i_r swapped with i_q in the permutation $(i_1,\ldots,i_k)$; another notations are the same as in theorem 1.

For the further consideration, we note that

$$\mathbf{M}\left(J[\psi^{(k)}]_{T,t}\int_t^T \phi_{j_k}(t_k)\ldots\int_t^{t_2}\phi_{j_1}(t_1)\,d\mathbf{w}_{t_1}^{(i_1)}\ldots d\mathbf{w}_{t_k}^{(i_k)}\right) = C_{j_k\ldots j_1}. \tag{15}$$

Theorem 3 [12]**.** Suppose that $\{\phi_j(x)\}_{j=0}^{\infty}$ is a complete orthonormal system of Legendre polynomials or trigonometric functions in the space $L_2([t,T])$. Then, for the iterated Stratonovich stochastic integral of third multiplicity $I^{*(i_1i_2i_3)}_{(000)T,t}$ the following expansion

$$I^{*(i_1i_2i_3)}_{(000)T,t} = \underset{p_1,p_2,p_3\to\infty}{\text{l.i.m.}}\sum_{j_1=0}^{p_1}\sum_{j_2=0}^{p_2}\sum_{j_3=0}^{p_3} C^{000}_{j_3j_2j_1}\zeta_{j_1}^{(i_1)}\zeta_{j_2}^{(i_2)}\zeta_{j_3}^{(i_3)} \quad (i_1,i_2,i_3 = 1,\ldots,m) \tag{16}$$

that converges in the mean-square sense is valid, where

$$C^{000}_{j_3j_2j_1} = \int_t^T \phi_{j_3}(t_3)\int_t^{t_3}\phi_{j_2}(t_2)\int_t^{t_2}\phi_{j_1}(t_1)\,dt_1\,dt_2\,dt_3; \tag{17}$$

another notations are the same as in theorem 1.

3 Approximation of iterated Ito and Stratonovich stochastic integrals using Legendre polynomials

Using theorems 1 and 3 and the complete orthonormal system of Legendre polynomials in the space $L_2([t,T])$, we obtain the following formulas for numerical modeling of iterated Ito and Stratonovich stochastic integrals from numerical schemes (6) and (7) [10, 12]

$$I^{(i_1)}_{(0)T,t} = \sqrt{T-t}\,\zeta_0^{(i_1)}, \quad I^{(i_1)}_{(1)T,t} = -\frac{(T-t)^{3/2}}{2}\left(\zeta_0^{(i_1)} + \frac{1}{\sqrt{3}}\zeta_1^{(i_1)}\right),$$

$$I^{(i_1i_2)q}_{(00)T,t} = \frac{T-t}{2}\left(\zeta_0^{(i_1)}\zeta_0^{(i_2)} + \sum_{i=1}^{q}\frac{1}{\sqrt{4i^2-1}}\left(\zeta_{i-1}^{(i_1)}\zeta_i^{(i_2)} - \zeta_i^{(i_1)}\zeta_{i-1}^{(i_2)}\right) - \mathbf{1}_{\{i_1=i_2\}}\right),$$

$$I^{(i_1i_2i_3)q_1}_{(000)T,t} = \sum_{j_1,j_2,j_3=0}^{q_1} C^{000}_{j_3j_2j_1}\Big(\zeta_{j_1}^{(i_1)}\zeta_{j_2}^{(i_2)}\zeta_{j_3}^{(i_3)} - \mathbf{1}_{\{i_1=i_2\}}\mathbf{1}_{\{j_1=j_2\}}\zeta_{j_3}^{(i_3)} - \mathbf{1}_{\{i_2=i_3\}}\mathbf{1}_{\{j_2=j_3\}}\zeta_{j_1}^{(i_1)} - \mathbf{1}_{\{i_1=i_3\}}\mathbf{1}_{\{j_1=j_3\}}\zeta_{j_2}^{(i_2)}\Big), \tag{18}$$

$$I^{*(i_1)}_{(0)T,t} = \sqrt{T-t}\,\zeta_0^{(i_1)}, I^{*(i_1)}_{(1)T,t} = -\frac{(T-t)^{3/2}}{2}\left(\zeta_0^{(i_1)} + \frac{1}{\sqrt{3}}\zeta_1^{(i_1)}\right),$$

$$I^{*(i_1i_2)q}_{(00)T,t} = \frac{T-t}{2}\left(\zeta_0^{(i_1)}\zeta_0^{(i_2)} + \sum_{i=1}^{q}\frac{1}{\sqrt{4i^2-1}}\left(\zeta_{i-1}^{(i_1)}\zeta_i^{(i_2)} - \zeta_i^{(i_1)}\zeta_{i-1}^{(i_2)}\right)\right),$$

$$I^{*(i_1i_2i_3)q_1}_{(000)T,t} = \sum_{j_1,j_2,j_3=0}^{q_1} C^{000}_{j_3j_2j_1}\zeta_{j_1}^{(i_1)}\zeta_{j_2}^{(i_2)}\zeta_{j_3}^{(i_3)}, \tag{19}$$

where $\mathbf{1}_A$ is the indicator of the set A; $i_1,\ldots,i_3 = 1,\ldots,m$,

$$C^{000}_{j_3j_2j_1} = \frac{1}{8}L_{j_1j_2j_3}(T-t)^{3/2}\bar{C}^{000}_{j_3j_2j_1}, \quad L_{j_1\ldots j_k} = \left(\prod_{l=1}^{k}(2j_l+1)\right)^{1/2},$$

$$\bar{C}^{000}_{j_3j_2j_1} = \int_{-1}^{1}P_{j_3}(z)\int_{-1}^{z}P_{j_2}(y)\int_{-1}^{y}P_{j_1}(x)\,dx\,dy\,dz,$$

$P_j(x)$ is the Legendre polynomial; other notations are the same as in theorem 1.

Let $E_k^{(l_1\ldots l_k)p}$ be the left-hand side of equality (14) for inetgral (4) and $E_k^{*(l_1\ldots l_k)p}$ be the analogous value for integral (5).

4 Main results

This section is devoted to the optimization of approximation procedures for iterated Stratonovich stochastic integrals, i.e. we discuss how to essentially minimize the values q and q_1 from the previous section for iterated Stratonovich stochastic integrals.

From theorem 2 we obtain (equality (15)) [11, 12]

$$E_2^{(00)p} = E_2^{*(00)p} = \frac{(T-t)^2}{2}\left(\frac{1}{2} - \sum_{i=1}^{p}\frac{1}{4i^2-1}\right), \quad i_1 \neq i_2, \tag{20}$$

$$E_3^{(000)p} = E_3^{*(000)p} = (T-t)^3\left(\frac{1}{6} - \frac{1}{64}\sum_{j_1,j_2,j_3=0}^{p} L_{j_1j_2j_3}^2 (\bar{C}_{j_3j_2j_1}^{000})^2\right), \quad i_1 \neq i_2, \quad i_1 \neq i_3, \quad i_2 \neq i_3, \tag{21}$$

$$E_3^{(000)p} = (T-t)^3\left(\frac{1}{6} - \frac{1}{64}\sum_{j_1,j_2,j_3=0}^{p} L_{j_1j_2j_3}^2 \left(\left(\bar{C}_{j_3j_2j_1}^{000}\right)^2 + \bar{C}_{j_3j_1j_2}^{000}\bar{C}_{j_3j_2j_1}^{000}\right)\right), \quad i_1 = i_2 \neq i_3, \tag{22}$$

$$E_3^{(000)p} = (T-t)^3\left(\frac{1}{6} - \frac{1}{64}\sum_{j_1,j_2,j_3=0}^{p} L_{j_1j_2j_3}^2 \left(\left(\bar{C}_{j_3j_2j_1}^{000}\right)^2 + \bar{C}_{j_2j_3j_1}^{000}\bar{C}_{j_3j_2j_1}^{000}\right)\right), \quad i_1 \neq i_2 = i_3, \tag{23}$$

$$E_3^{(000)p} = (T-t)^3\left(\frac{1}{6} - \frac{1}{64}\sum_{j_1,j_2,j_3=0}^{p} L_{j_1j_2j_3}^2 \left(\left(\bar{C}_{j_3j_2j_1}^{000}\right)^2 + \bar{C}_{j_3j_2j_1}^{000}\bar{C}_{j_1j_2j_3}^{000}\right)\right), \quad i_1 = i_3 \neq i_2. \tag{24}$$

Obviously, conditions (20)–(24) do not contain the multiplier factors 2!, 3! in contrast to estimate (12). However, the number of the mentioned conditions is quite large, which is inconvenient for practice. In [11] we proposed and confirmed the hypothesis that all formulas (20)–(24) can be replaced by formulas (20) and (21) in which we can suppose that $i_1, \ldots, i_3 = 1, \ldots, m$. At that we have no noticeable loss of the mean-square approximation accuracy of iterated Ito stochastic integrals. In this article, we propose and confirm similar hypothesis for iterated Stratonovich stochastic integrals.

It should be noted that unlike the method based on theorem 1, existing approaches to the mean-square approximation of iterated stochastic integrals [1, 2, 8, 9, 13–21] do not allow to choose different numbers p for approximations of different iterated stochastic integrals with multiplicities $k = 2, 3, \ldots$ Moreover, the noted approaches exclude the possibility for obtaining of approximate and exact expressions similar to formulas (12), (14). The detailed comparison of theorem 1 with methods from [1, 2, 8, 9, 13–21] is given in chapter 6 of monograph [12].

Consider the analogs of formulas (22)–(24) for iterated Stratonovich stochastic integrals.

Theorem 4. Suppose that $\{\phi_j(x)\}_{j=0}^{\infty}$ is a complete orthonormal system of Legendre polynomials in the space $L_2([t, T])$. Then, the following relations are hold

$$E_3^{*(000)p} = E_3^{(000)p} + (T-t)^3\left(\frac{1}{12} - \frac{1}{16}\sum_{j_1=0}^{p} L_{j_1}^2\left(\bar{C}_{0j_1j_1}^{000} + \bar{C}_{1j_1j_1}^{000}\right) + \frac{1}{64}\sum_{j_3=0}^{p} L_{j_3}^2\left(\sum_{j_1=0}^{p} L_{j_1}^2\bar{C}_{j_3j_1j_1}^{000}\right)^2\right),$$
$$i_1 = i_2 \neq i_3, \tag{25}$$

$$E_3^{*(000)p}=E_3^{(000)p}+(T-t)^3\left(\frac{1}{12}-\frac{1}{16}\sum_{j_3=0}^{p}L_{j_3}^2\left(\bar{C}_{j_3j_30}^{000}-\bar{C}_{j_3j_31}^{000}\right)+\frac{1}{64}\sum_{j_1=0}^{p}L_{j_1}^2\left(\sum_{j_3=0}^{p}L_{j_3}^2\bar{C}_{j_3j_3j_1}^{000}\right)^2\right),$$

$$i_1\neq i_2=i_3, \tag{26}$$

$$E_3^{*(000)p}=E_3^{(000)p}+\frac{(T-t)^3}{64}\sum_{j_2=0}^{p}L_{j_2}^2\left(\sum_{j_1=0}^{p}L_{j_1}^2\bar{C}_{j_1j_2j_1}^{000}\right)^2, \quad i_1=i_3\neq i_2. \tag{27}$$

Proof. Consider the following three cases $i_1=i_2\neq i_3$, $i_1\neq i_2=i_3$, $i_1=i_3\neq i_2$. First, consider the case $i_1=i_2\neq i_3$. From the standard relations between Stratonovich and Ito stochastic integrals [1] and formulas (18), (19) we obtain

$$E_3^{*(000)p}=\mathbf{M}\left(I_{(000)T,t}^{(i_1i_2i_3)}-I_{(000)T,t}^{(i_1i_2i_3)p}+\frac{1}{2}\int_t^T\int_t^\tau ds\,d\mathbf{f}_\tau^{(i_3)}-\sum_{j_1,j_3=0}^{p}C_{j_3j_1j_1}^{000}\zeta_{j_3}^{(i_3)}\right)^2. \tag{28}$$

According to formulas (1.39), (1.83) [12], the quantity $I_{(000)T,t}^{(i_1i_2i_3)}-I_{(000)T,t}^{(i_1i_2i_3)p}$ includes only iterated Ito stochastic integrals of multiplicity 3. At the same time, the quantity

$$\frac{1}{2}\int_t^T\int_t^\tau ds\,d\mathbf{f}_\tau^{(i_3)}-\sum_{j_1,j_3=0}^{p}C_{j_3j_1j_1}^{000}\zeta_{j_3}^{(i_3)}$$

contains only iterated Ito stochastic integrals of multiplicity 1. This means that from (28) we get

$$E_3^{*(000)p}=E_3^{(000)p}+\mathbf{M}\left(\frac{1}{2}\int_t^T(\tau-t)\,d\mathbf{f}_\tau^{(i_3)}-\sum_{j_1,j_3=0}^{p}C_{j_3j_1j_1}^{000}\zeta_{j_3}^{(i_3)}\right)^2. \tag{29}$$

Theorem 2 implies that

$$E_3^{(000)p}=\frac{(T-t)^3}{6}-\sum_{j_1,j_2,j_3=0}^{p}\left(C_{j_3j_2j_1}^{000}\right)^2-\sum_{j_1,j_2,j_3=0}^{p}C_{j_3j_1j_2}^{000}C_{j_3j_2j_1}^{000}, \quad i_1=i_2\neq i_3. \tag{30}$$

We have

$$\mathbf{M}\left(\frac{1}{2}\int_t^T(\tau-t)\,d\mathbf{f}_\tau^{(i_3)}-\sum_{j_1,j_3=0}^{p}C_{j_3j_1j_1}^{000}\zeta_{j_3}^{(i_3)}\right)^2=$$

$$=\frac{1}{4}\int_t^T(\tau-t)^2d\tau-\sum_{j_1,j_3=0}^{p}C_{j_3j_1j_1}^{000}\int_t^T(\tau-t)\,\phi_{j_3}(\tau)\,d\tau+\sum_{j_3=0}^{p}\left(\sum_{j_1=0}^{p}C_{j_3j_1j_1}^{000}\right)^2, \tag{31}$$

where $\phi_{j_3}(\tau)$ is the Legendre polynomial. Moreover, we obtain

$$\int_t^T(\tau-t)\,\phi_{j_3}(\tau)\,d\tau=\frac{(T-t)^{3/2}}{2}\begin{cases}1, & j_3=0,\\ 1/\sqrt{3}, & j_3=1,\\ 0, & j_3\geq 2.\end{cases} \tag{32}$$

Combining (29)–(32), we get

$$E_3^{*(000)p}=\frac{(T-t)^3}{4}-\sum_{j_1,j_2,j_3=0}^{p}\left(C_{j_3j_2j_1}^{000}\right)^2-\sum_{j_1,j_2,j_3=0}^{p}C_{j_3j_1j_2}^{000}C_{j_3j_2j_1}^{000}-$$

$$-\frac{(T-t)^{3/2}}{2}\sum_{j_1=0}^{p}\left(C_{0j_1j_1}^{000}+\frac{1}{\sqrt{3}}C_{1j_1j_1}^{000}\right)+\sum_{j_3=0}^{p}\left(\sum_{j_1=0}^{p}C_{j_3j_1j_1}^{000}\right)^2, \tag{33}$$

where $i_1 = i_2 \neq i_3$. The case $i_1 = i_2 \neq i_3$ is proved. The cases $i_1 \neq i_2 = i_3$ and $i_1 = i_3 \neq i_2$ can be considered by analogy to the case $i_1 = i_2 \neq i_3$. Theorem 4 is proved.

Let $q(\alpha)$ be numbers p from formulas (20), (21), (25)–(27), where α are numbers of these formulas. Let

$$E_2^{*p} \le (T-t)^4, E_3^{*p} \le (T-t)^4, \tag{34}$$

where E_2^{*p}, E_3^{*p} mean the left-hand sides of equalities (20), (21), (25)–(27) respectively.

Let us show by numerical experiments that in most situations the following inequalities are fulfilled (under condition (34)): $q(21) \ge q(25)$, $q(21) \ge q(26)$, $q(21) \ge q(27)$, where all numbers in these inequalities are minimal natural numbers satisfying to condition (34). In tables 1–3 we can see the results of numerical experiments. In table 3 numbers $p(21)$, $p(25)$–$p(27)$ means numbers p corresponding to formulas (21), (25)–(27) respectively. These results confirm the hypothesis proposed in this paper.

Table 1. Condition (34)

$T-t$	0.011	0.008	0.0045	0.0035	0.0027	0.0025
$q(21)$	12	16	28	36	47	50
$q(25)$	6	8	14	18	23	25
$q(26)$	6	8	14	18	23	25
$q(27)$	12	16	28	37	47	51

Table 2. Strong scheme with order 1.5. Condition (34)

$T-t$	2^{-1}	2^{-3}	2^{-5}	2^{-8}
$q(20)$	1	8	128	8192
$q(21)$	0	1	4	32
$q(25)$	0	0	2	16
$q(26)$	0	0	2	16
$q(27)$	0	1	4	33

Table 3. Values $E_3^{*(000)p} \cdot (T-t)^{-3} \overset{\text{def}}{=} E_p^*$

$T-t$	0.011	0.008	0.0045	0.0035	0.0027	0.0025
$p = p(21)$	12	16	28	36	47	50
E_p^*	0.010154	0.007681	0.004433	0.003456	0.002652	0.002494
$p = p(25)$	12	16	28	36	47	50
E_p^*	0.005102	0.003855	0.002221	0.001731	0.001328	0.001248
$p = p(26)$	12	16	28	36	47	50
E_p^*	0.005102	0.003855	0.002221	0.001731	0.001328	0.001248
$p = p(27)$	12	16	28	36	47	50
E_p^*	0.010407	0.007845	0.004500	0.003501	0.002680	0.002519

Let q_1 be minimal natural number satisfying to $E_3^{*(000)q_1} \le (T-t)^4$ (the left-hand side of these inequality is defined by formula (21)). Let p_1 be minimal natural number satisfying to $3!\,E_3^{*(000)p_1} \le (T-t)^4$, where the value $E_3^{*(000)p_1}$ is defined by formula (21).

Table 4. Comparison of numbers q_1 and p_1

$T-t$	2^{-1}	2^{-2}	2^{-3}	2^{-4}	2^{-5}	2^{-6}
q_1	0	0	1	2	4	8
p_1	1	3	6	12	24	48

In table 4 we can see the numerical comparison of numbers q_1 and p_1. Obviously, excluding of the multiplier factor 3! essentially (in many times) reduces the calculation costs for the mean-square approximations of iterated Stratonovich stochastic integrals. Note that in this paper we use the exactly calculated Fourier–Legendre coefficients using the Python programming language [23].

5 Conclusion

As we mentioned above, existing approaches [1, 2, 8, 9, 13–21] to the mean-square approximation of iterated stochastic integrals do not allow to choose different numbers p (see theorem 2) for approximations of different iterated stochastic integrals with multiplicity $k = 2, 3, \ldots$ and exclude the possibility for obtaining of approximate and exact expressions similar to formulas (12) and (14). This leads to unnecessary terms usage in the expansions of iterated Ito and Stratonovich stochastic integrals and, as a consequence, to essential increase of computational costs for the implementation of numerical methods for Ito SDEs. In this article, we have optimized method based on theorems 1–3, which makes it possible to correctly choose the lengths of sequences of standard Gaussian random variables required for the approximation of iterated Stratonovich stochastic integrals. Thus, the computational costs for the implementation of numerical methods for Ito SDEs based on the Taylor–Stratonovich expansion are significantly reduced. The analogous optimization for iterated Ito stochastic integrals is carried out in [11].

On the base of the obtained results we can recommend the following conditions for correct choosing the minimal natural numbers q and q_1: $E_2^{*(00)q} \le \overline{C}(T-t)^4$, $E_3^{*(000)q_1} \le \overline{C}(T-t)^4$ (for the strong scheme (7) with order 1.5). Here the left-hand sides of the above inequalities are defined by relations (20) and (21), correspondingly; $\overline{C}$ is a constant from condition (9).

References

[1] P.E. Kloeden, E. Platen, *Numerical Solution of Stochastic Differential Equations* (Springer, Berlin, 1995)

[2] E. Platen, N. Bruti-Liberati, *Numerical Solution of Stochastic Differential Equations with Jumps in Finance* (Springer, Berlin–Heidelberg, 2010)

[3] X. Han, P.E. Kloeden, *Random Ordinary Differential Equations and their Numerical Solution* (Springer, Singapore, 2017)

[4] Z. Zhang, A. Zeb, S. Hussain, E. Alzahrani, *Dynamics of COVID-19 mathematical model with stochastic perturbation*, Adv. Differ. Equ., **451** (2020), DOI: 10.1186/s13662-020-02909-1

[5] A. Alzahrani, A. Zeb, *Detectable sensation of a stochastic smoking model*, Open Math., **18**, 1045 (2020), DOI: 10.1515/math-2020-0068

[6] G. Li, Y. Liu, *The dynamics of a stochastic SIR epidemic model with nonlinear incidence and vertical transmission*, Discr. Dyn. Nat. Soc., **2021**, 4645203 (2021), DOI: 10.1155/2021/4645203

[7] Y. Jiao, C. Ma, S. Scotti, C. Zhou, *The Alpha-Heston stochastic volatility model*, Math. Finance, **31(3)**, 943 (2021), DOI: 10.1111/mafi.12306

[8] G.N. Milstein, *Numerical Integration of Stochastic Differential Equations* (Ural University Press, Sverdlovsk, 1988)

[9] G.N. Milstein, M.V. Tretyakov, *Stochastic Numerics for Mathematical Physics* (Springer, Berlin, 2004)

[10] D.F. Kuznetsov, *Numerical Integration of Stochastic Differential Equations 2* (Polytechnical University Publishing House, Saint-Petersburg, 2006), DOI: 10.18720/SPBPU/2/s17-227

[11] D.F. Kuznetsov, M.D. Kuznetsov, *Optimization of the mean-square approximation procedures for iterated Ito stochastic integrals based on multiple Fourier–Legendre series*, J. Phys.: Conf. Ser., **1925**, 012010 (2021), DOI: 10.1088/1742-6596/1925/1/012010

[12] D.F. Kuznetsov, *Strong approximation of iterated Ito and Stratonovich stochastic integrals based on generalized multiple Fourier series. Application to numerical solution of Ito SDEs and semilinear SPDEs*, arXiv:2003.14184v27 [math.PR], 1-869 (2022), DOI: 10.48550/arXiv.2003.14184

[13] P.E. Kloeden, E. Platen, I.W. Wright, *The approximation of multiple stochastic integrals*, Stoch. Anal. Appl., **10(4)**, 431 (1992), DOI: 10.1080/07362999208809281

[14] J.G. Gaines, T.J. Lyons, *Random generation of stochastic area integrals*, SIAM J. Appl. Math., **54**, 1132 (1994), DOI: 10.1137/S0036139992235706

[15] T.A. Averina, S.M. Prigarin, *Calculation of stochastic integrals of Wiener processes* (Inst. Comp. Math. Math. Geophys. Siberian Branch Russ. Acad. Sci., Novosibirsk), Preprint 1048, 1–15 (1995)

[16] C.W. Li, X.Q. Liu, *Approximation of multiple stochastic integrals and its application to stochastic differential equations*, Nonlinear Anal. Theor. Meth. Appl., **30(2)**, 697 (1997), DOI: 10.1016/S0362-546X(96)00253-2

[17] S.M. Prigarin, S.M. Belov, *One application of series expansions of Wiener process* (Inst. Comp. Math. Math. Geophys. Siberian Branch Russ. Acad. Sci., Novosibirsk), Preprint 1107, 1–16 (1998)

[18] M. Wiktorsson, *Joint characteristic function and simultaneous simulation of iterated Ito integrals for multiple independent Brownian motions*, Ann. Appl. Prob., **11(2)**, 470 (2001), DOI: 10.1214/aoap/1015345301

[19] T. Ryden, M. Wiktorsson, *On the simulation of iterated Ito integrals*, Stoch. Proc. Appl., **91(1)**, 151 (2001), DOI: 10.1016/S0304-4149(00)00053-3

[20] E, Allen, *Approximation of triple stochastic integrals through region subdivision*, Commun. Appl. Analysis., **17**, 355 (2013)

[21] X. Tang, A. Xiao, *Asymptotically optimal approximation of some stochastic integrals and its applications to the strong second-order methods*, Adv. Comp. Math., **45**, 813 (2019), DOI: 10.1007/s10444-018-9638-0

[22] K.A. Rybakov, *Using spectral form of mathematical description to represent iterated Stratonovich stochastic integrals*, in Applied Mathematics and Computational Mechanics for Smart Applications. Smart Innovation, Systems and Technologies, vol. 217, ed. by L.C. Jain, M.N. Favorskaya, I.S. Nikitin, D.L. Reviznikov (Springer, Singapore, 2021), pp. 287-304, DOI: 10.1007/978-981-33-4826-4_20

[23] M.D. Kuznetsov, D.F. Kuznetsov, *SDE-MATH: A software package for the implementation of strong high-order numerical methods for Ito SDEs with multidimensional non-commutative noise based on multiple Fourier–Legendre*, Differential Equations and Control Processes, **1**, 93 (2021), URL `https://diffjournal.spbu.ru/EN/numbers/2021.1/article.1.5.html`

MATEC Web of Conferences **362**, 01015 (2022) https://doi.org/10.1051/matecconf/202236201015
CMMASS 2021

Attitude sensors relative angular misalignment estimation in integrated navigation systems

Ivan Kuznetsov[1,*], *Konstantin* Veremeenko[1], *Maksim* Zharkov[1], and *Andrey* Pronkin[1]

[1]Flight-Navigation and Information Measuring Systems Department, Moscow Aviation Institute (National Research University), 4, Volokolamskoe shosse, Moscow, 125993, Russia

Abstract. When using modern navigation systems as part of an on-board system, the navigation task can be solved in several ways: using positional, velocity and angular corrections, and systems using measurements of various physical nature—radio, such as short-range and long-range navigation radio systems, GLONASS/GPS satellite global navigation systems, optical—celestial navigation systems—can operate as correctors. The best performance of INS and its sensors (gyroscopes and accelerometers) errors estimation can be obtained when all types of correction are implemented simultaneously. At the same time, it is particularly difficult to implement correction according to attitude parameters due to the fact that the measuring axes of the INS and correctors may not match on board of moving objects. Such a mismatch is commonly called relative angular misalignment. The paper considers a possible approach to the attitude sensors relative angular misalignment estimation in integrated navigation systems (NS), carried out in motion when the NS is in operating mode.

1 Introduction

Most modern and prospective mobile objects navigation and attitude systems are a complex ones, the software and algorithmic support of which carries out complex processing of measuring information from onboard sensors and systems. Depending on the generation and measurements processing approach, the following classification of schemes for algorithms design has now developed [1, 2]: separate, loosely coupled, tightly coupled and deeply integrated. The last three of these approaches pretend the differences generation in the sensors and systems outputs (if necessary, after averaging, setting to a unified coordinate frame and dimensions by initial processing), combined as part of an integrated navigation system (NS), with the subsequent state vector errors estimation, usually using a Kalman filter.

In this case, the errors of the NS subsystems are included in the state vector, which are described by the corresponding mathematical models. In the vast majority of cases, the inertial navigation system (INS) acts as the NS information core, which is highly informative (the ability to calculate a complete list of motion parameters: coordinates, velocities, attitude parameters), autonomy, and a high frequency of data output. In this regard, the remaining NS sensors and systems are commonly referred to as INS correctors. Depending on the type

*e-mail: kuznetsovim@mai.ru

of correctors, the following types of INS correction can be considered: positional (by coordinates), by velocity and angular attitude parameters.

The best performance of INS and its sensors (gyroscopes and accelerometers) errors estimation can be obtained when all types of correction are implemented simultaneously. At the same time, it is particularly difficult to implement correction according to attitude parameters due to the fact that the measuring axes of the INS and correctors may not match on board of moving objects. Such a mismatch is commonly called relative angular misalignment. It is possible to minimize angular misalignment by increasing the accuracy of sensors and systems adjustment during their installation on board. However, the adjustment process also has finite accuracy and, in addition, is difficult or impossible in some cases. As examples of the INS correction by angular attitude parameters usage, the following can be mentioned: multi-antenna inertial-satellite navigation systems [3, 4], optical-inertial navigation systems [5–7], astroinertial navigation systems [8], INS initial alignment from the onboard navigation system of the carrier in motion [9–11], etc. The work describes a possible approach to the relative angular misalignment estimation of the NS attitude sensors, carried out in motion when the NS is in operating mode.

2 Mathematical models

The system state equation is presented in the following form:

$$\dot{\bar{X}} = F \cdot \bar{X} + G \cdot \bar{W}, \tag{1}$$

where F is the dynamics matrix, $\bar{X}$ is the system state vector, G is the noise matrix of the system, $\bar{W}$ is the system noise vector.

The state of the system is estimated using the measurement vector $\bar{Z}$:

$$\overline{Z} = H \cdot \overline{X} + \overline{V}, \tag{2}$$

where H is the measurements matrix, $\overline{V}$—the measurement noise vector.

It is proposed to include the following parameters to the state vector [1]:

$$\overline{X} = [X_{INS} \quad \Delta \quad A]^T. \tag{3}$$

Here $\Delta_{INS} = [x_1 \quad x_2 \quad x_3 \quad x_4 \quad x_5 \quad x_6 \quad \alpha \quad \beta \quad \gamma]$—coordinates, velocity projections and attitude angles errors determining by INS [11]; $\Delta = [\Delta\Omega_1 \quad \Delta\Omega_2 \quad \Delta\Omega_3 \quad \Delta n_1 \quad \Delta n_2 \quad \Delta n_3]$—gyros and accelerometers errors constant components (when extending the model of sensor errors, the state vector can also be extended by corresponding components) $A = [\psi^{ma} \quad \vartheta^{ma} \quad \gamma^{ma}]$—relative angular misalignment parameters between the INS and the attitude angles corrector frames.

The measurement vector in this case will have the following form:

$$\overline{Z} = [\delta\Phi \quad \delta U \quad \delta\Psi], \tag{4}$$

where $\delta\Phi = [\delta\varphi^{INS} - \delta\varphi^{C} \quad \delta\lambda^{INS} - \delta\lambda^{C} \quad \delta h^{INS} - \delta h^{C}]$—the difference in coordinate outputs of the INS and the corrector, $\delta U = [\delta U_E^{INS} - \delta U_E^{C} \quad \delta U_N^{INS} - \delta U_N^{C} \quad \delta U_Z^{INS} - \delta U_Z^{C}]$—the difference in velocities projections outputs of the INS and the corrector $\delta\Psi = [\delta\psi^{INS} - \delta\psi^{C} + \psi^{ma} \quad \delta\vartheta^{INS} - \delta\vartheta^{C} + \vartheta^{ma} \quad \delta\gamma^{INS} - \delta\gamma^{C} + \gamma^{ma}]$—the difference in heading, pitch and roll angles outputs of the INS and the corrector.

The dynamics matrix F is designed on the basis of the INS and its sensitive elements error model [11] and has the following form:

$$F = \begin{bmatrix} 0_{3\times3} & I_{3\times3} & 0_{3\times3} & 0_{3\times3} & 0_{3\times3} \\ F_1 & F_2 & F_3 & C_{3\times3} & 0_{3\times3} \\ 0_{3\times3} & 0_{3\times3} & F_4 & 0_{3\times3} & C_{3\times3} \\ & & 0_{6\times18} & & \end{bmatrix}, \quad (5)$$

$$F_1 = \begin{bmatrix} (\Omega_Y^2 + \Omega_Z^2 - \omega_0^2) & \dot{\Omega}_Z + \Omega_X\Omega_Y & -(\dot{\Omega}_Y + \Omega_X\Omega_Z) \\ -(\dot{\Omega}_Z + \Omega_X\Omega_Y) & (\Omega_X^2 + \Omega_Z^2 - \omega_0^2) & \dot{\Omega}_X - \Omega_Y\Omega_Z \\ \dot{\Omega}_Y + \Omega_X\Omega_Z & -(\dot{\Omega}_X - \Omega_Y\Omega_Z) & (\Omega_X^2 + \Omega_Y^2 + 2\omega_0^2) \end{bmatrix},$$

$$F_2 = \begin{bmatrix} 0 & 2\Omega_Z & -2\Omega_Y \\ -2\Omega_Z & 0 & 2\Omega_X \\ 2\Omega_Y & -2\Omega_X & 0 \end{bmatrix}, \quad F_3 = \begin{bmatrix} 0 & n_z & -n_y \\ -n_z & 0 & n_x \\ n_y & -n_x & 0 \end{bmatrix}, \quad F_4 = \begin{bmatrix} 0 & \Omega_Z & -\Omega_Y \\ -\Omega_Z & 0 & \Omega_X \\ \Omega_Y & -\Omega_X & 0 \end{bmatrix},$$

where ω_0—INS errors oscillation natural frequency, Shuler frequency $\omega_0 = 1.25 \cdot 10^{-3}\ \mathrm{s}^{-1}$, $\Omega_{X,Y,Z}$, $\dot{\Omega}_{X,Y,Z}$—absolute angular velocity vector of the base coordinate frame projections and its derivatives; $n_{X,Y,Z}$—specific force projections vector, $C_{3\times3}$—DCM between the body fixed frame (BFF) and local level frame (LLF). The parameters of relative angular misalignment are assumed to be constant. When forming the matrix H, a certain difficulty is presented by the fact that in the accepted model of the INS errors, attitude errors includes calculating the basic coordinate frame (CF) errors and attitude measurement frame relative to this basic CF errors. To solve this problem, when designing the matrix H, a link matrix between the INS attitude errors according to the model and the errors of the INS in determining heading, pitch and roll angles was formed. The structure of the system noise vector includes random components of the INS sensors errors, and the structure of the measurement noise vector includes random components of the correctors errors.

To estimate the state vector, an option of the discrete optimal Kalman filter is used, written in the Joseph form [10], which provides increased protection against computational divergence.

3 Simulation and car tests results

The study of operability and estimation of the proposed approach to determine the relative angular misalignment of attitude sensors achievable performance was carried out using simulation and post-processing of measurement records of a full-scale car test.

As the tested system a multiple antenna inertial-satellite integrated navigation system was used. Simulation scenarios assumed the use of INS sensors of different accuracy grades (from microelectromechanical (MEMS) up to navigation grade), a different number of GLONASS and GPS antennas, with different distances between them as long as the influence of trajectory during the car tests was studied.

Preliminary simulation results showed the potential possibility of using the proposed approach to estimate the relative angular mismatch of attitude parameters in motion when the system is in operating mode. The section will provide as well as the simulation and the car test results as the most illustrative ones to study the efficiency of a proposed approach to estimate the relative angular mismatch of attitude parameters in motion when the system is in operating mode under selected scenarios and experimental conditions.

As it was mentioned above car test was carried out with the use of NS designed at the department [12]. NS was mounted in the MAI Navigation System Car Testbed inside a car roof rack (figure 1) [13]. Reference attitude parameters were obtained from Novatel SPAN GNSS

Figure 1. MAI Navigation System Car Testbed

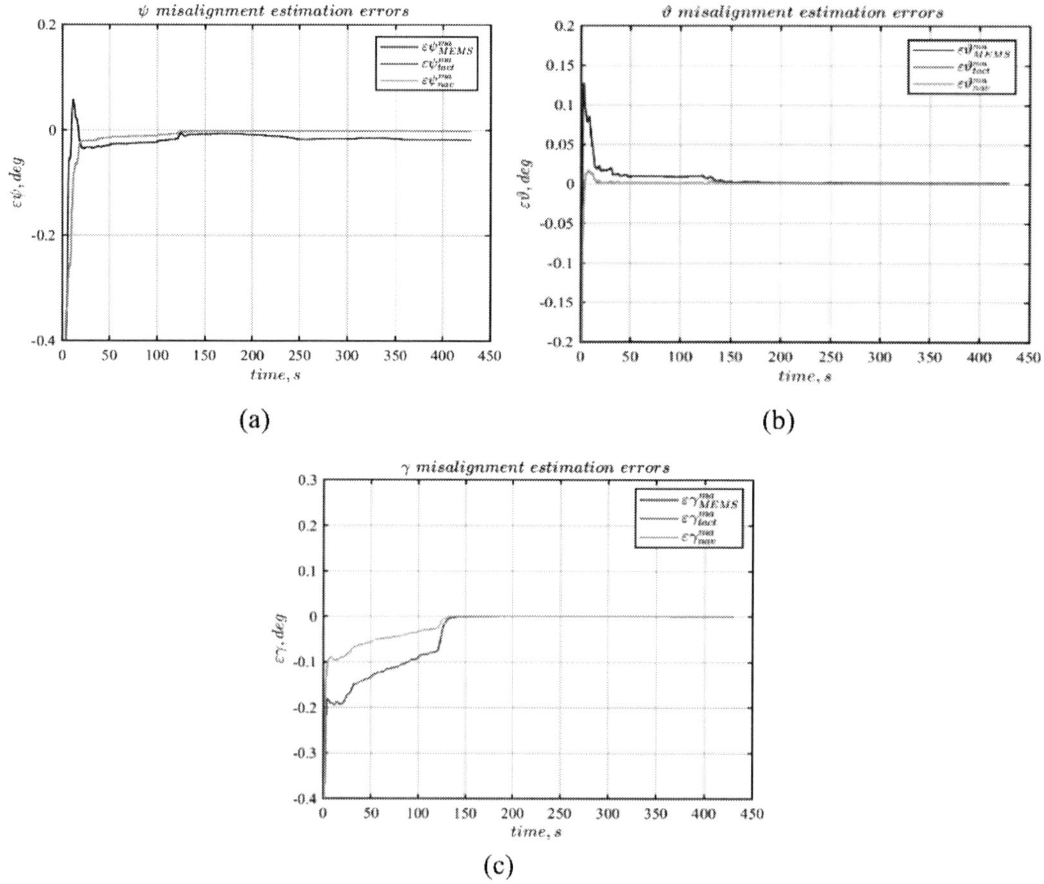

Figure 2. Simulation results (scenario 1)

Inertial Navigation System measurements and geodetic base GNSS receiver measurements postprocessed by NovAtel Inertial Explorer software.

Figures 2–4 show the estimation errors of angular misalignment $\varepsilon\psi^{ma}$, $\varepsilon\vartheta^{ma}$, $\varepsilon\gamma^{ma}$, $\varepsilon\psi^{ma}$, $\varepsilon\vartheta^{ma}$, $\varepsilon\gamma^{ma}$—in heading, pitch and roll angles measurements) in a multiple antenna inertial-satellite integrated navigation system, where in case of simulation an inertial navigation system from its structure was misaligned for a 1 degree to all three angles relative to a multiple antenna satellite navigation system and a multiple antenna inertial-satellite integrated navigation system and NS, designed by the department in case of car tests relatively misaligned the same way physically to a random angles.

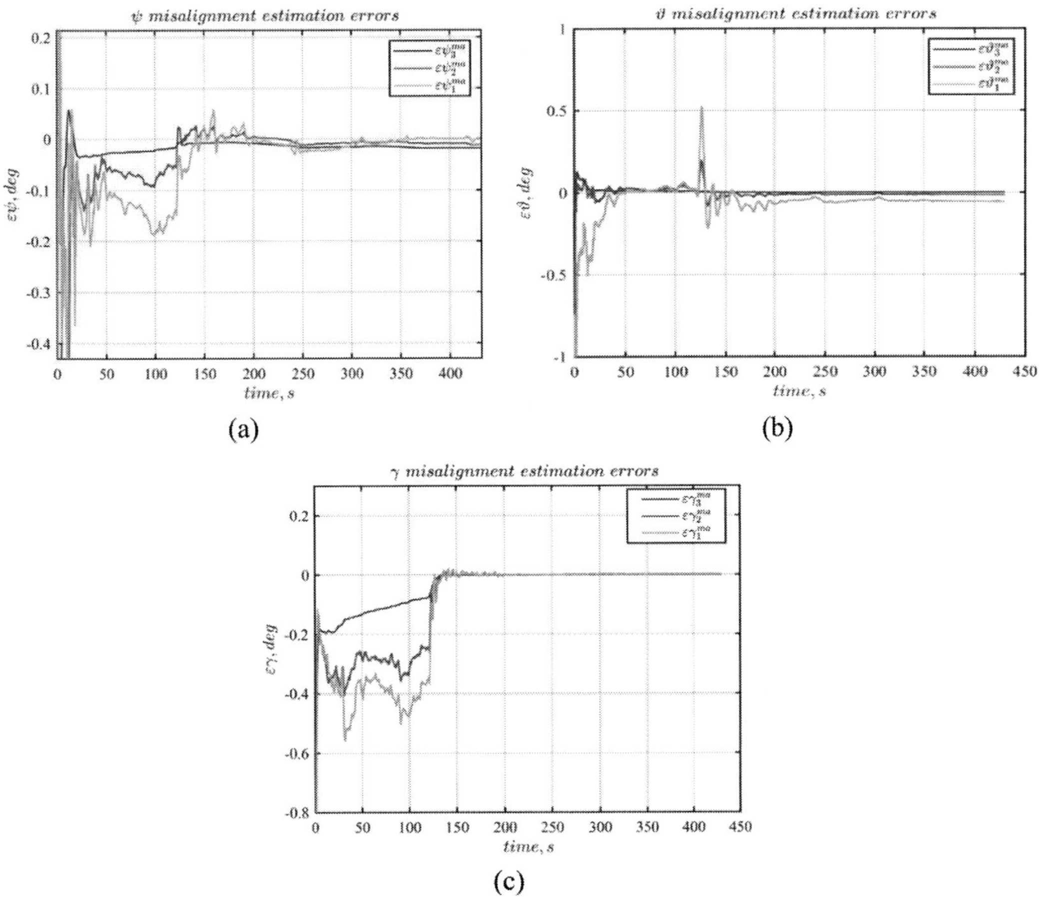

Figure 3. Car tests results (scenario 2)

To study the operability and estimation of the proposed approach to determine the relative angular misalignment of attitude sensors achievable performance three scenarios were designed:

- 1 (simulation)—in this scenario of simulation the output of a multiple antenna inertial-satellite integrated navigation system were generated using mathematical models, presented above. An inertial navigation system from its structure was misaligned for a 1 degree to all three attitude angles relative to a multiple antenna satellite navigation system. The trajectory included low and high maneuverable sections of object movement with different dynamic from a chosen range of navigation parameters—velocities and attitude angles. The feature of this scenario is the INS sensors of different accuracy grades (from micro-electromechanical (MEMS) up to navigation grade) influence study.
- 2 (car tests)—in this scenario of car tests an object—city car with a NS mounted in a roof rack was driving a complex trajectory with as well as low maneuverable with straight-forward movement sections and complex maneuverable with S-types movement sections trajectories (in the streets of Moscow city). The feature of this scenario is the difference of base length between onboard antennas influence study, which changed in the range of 1 to 3 meters for a similar trajectories of movement and the same experimental conditions.
- 3 (car tests)—this scenario studies the trajectory type influence on an approach to determine the relative angular misalignment of attitude sensors achievable performance.

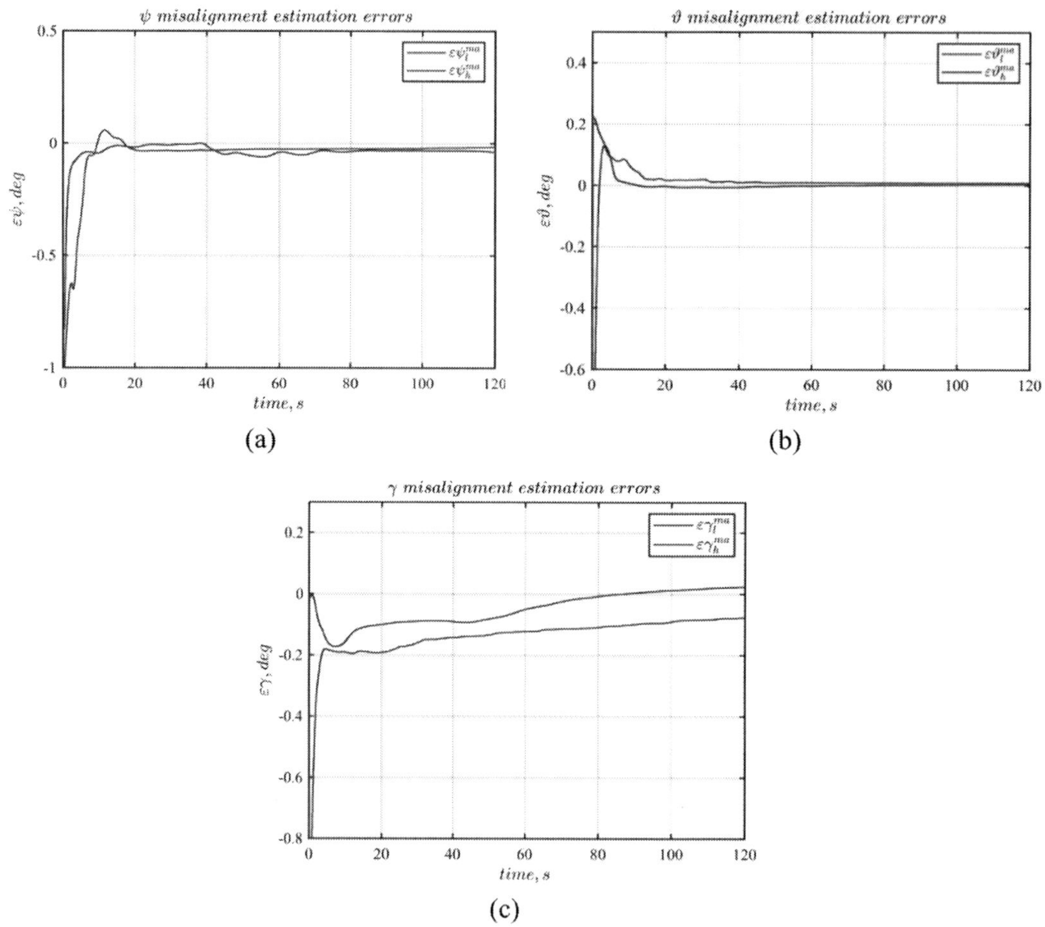

Figure 4. Car tests results (scenario 3)

Table 1 below shows the summary (simulation and car tests) levels of achieved accuracy at the certain timestamps (10, 60 and 160 s respectively), table 2 in the section below shows the summary (car tests) levels of achieved accuracy at the certain timestamps (10, 60 and 120 s respectively) for the scenarios implemented above.

4 Results analysis

The study of different grade IMUs in simulation (scenario 1) shows the absence of its use necessity better than tactical grade. In figure 1 it can be seen red and yellow lines are actually merged into one, and the absolute level of ultimate accuracy differs insignificantly (in the fourth decimal place in heading misalignment and in the fifth decimal place in two other angles).

The study of different base length between the 3 antennas of integrated navigation system (scenario 2) shows its influence on ultimate accuracy of estimation. The most widely spaced antennas relative location (3 meters) changes the qualitive nature of estimation process—more smoothed without sharp outliers during the initial transition process.

It is necessary to note the feature of trajectory influence on a performance of proposed approach to estimate the relative angular mismatch of attitude parameters in motion when

Table 1. Simulation and car test results (scenarios 1 and 2)

Scenario	Time, s	10			60			160		
1	c	*nav*	*tact*	*MEMS*	*nav*	*tact*	*MEMS*	*nav*	*tact*	*MEMS*
	$\varepsilon\psi^{ma}$, deg	0.0185	−0.182	−0.0184	−0.012	−0.012	−0.025	−0.001	−0.001	−0.007
	$\varepsilon\vartheta^{ma}$, deg	0.013	0.014	0.067	0.001	0.009	0.009	0.0009	0.0009	0.002
	$\varepsilon\gamma^{ma}$, deg	−0.09	−0.19	−0.19	−0.04	−0.04	−0.12	−0.0002	−0.0002	−0.008
2	l, m	3	2	1	3	2	1	3	2	1
	$\varepsilon\psi^{ma}$, deg	0.017	−0.72	−0.83	−0.02	−0.06	−0.11	−0.007	0.05	0.02
	$\varepsilon\vartheta^{ma}$, deg	0.06	−0.19	0.07	0.009	0.009	0.02	0.002	−0.003	−0.02
	$\varepsilon\gamma^{ma}$, deg	−0.19	−0.25	−0.28	−0.12	−0.35	−0.27	−0.0008	0.008	−0.0004

Table 2. Car test results (scenario 3)

Trajectory type	Time, s					
	10		60		120	
	low	*high*	*low*	*high*	*low*	*high*
$\varepsilon\psi^{ma}$, deg	0.02	−0.04	−0.025	−0.039	−0.01	−0.038
$\varepsilon\vartheta^{ma}$, deg	0.06	0.006	0.009	−0.001	0.01	0.004
$\varepsilon\gamma^{ma}$, deg	−0.19	−0.15	−0.12	−0.05	−0.07	0.02

the system is in operating mode under selected scenarios and experimental conditions. When comparing the results of simulation and car tests study it can be noted the level of achievable performance can be increased not only by maximizing the time of estimation, but also adding a dynamic maneuver after some time of estimation along low maneuverable movement. Moreover, the estimation accuracy can be increased many times, i.e. up to 10 times in heading misalignment estimation using navigation and tactical grade IMUs (see the results in table 1, scenario 1) and up to 100 times in roll misalignment estimation.

In table 1 "*c*" corresponds to a inertial measurement unit (IMU) class, where "*nav*" corresponds to a navigation grade (laser gyros, quartz accelerometers), "*tact*" corresponds to a tactical grade (FOG, high-precision silicon accelerometers), "*MEMS*"—microelectromechanical IMU grade respectively, "*l*" corresponds to a base length between onboard antennas.

Table 2 shows the results of car tests for the scenario which studies the trajectory type influence on an approach to determine the relative angular misalignment of attitude sensors achievable performance at a certain timestamps. The timestamps columns are splitted into two corresponding the types of the trajectory: a "*low*" maneuverable type—mostly with straightforward movement sections, "*high*" corresponds to a complex maneuverable with S-types movement sections trajectories.

5 Conclusions

The results of the study demonstrate the possibility of using the proposed approach to estimate the relative angular mismatch of attitude parameters in motion when the system is in operating mode under selected scenarios and experimental conditions. At the same time, in accordance with theoretical expectations, the angular misalignment is estimated jointly with the INS attitude parameters and the constant components of sensor errors. It is possible to increase the observability and improve the estimation parameters when implementing the movement of an object along a trajectory of a special type.
The research was carried out within the state assignment of Ministry of Science and Higher Education of the Russian Federation (theme No. FSFF-2020-0015).

References

[1] G. Schmidt, *INS/GPS Technology Trends*, in Draper Laboratory Report P-4036 (Cambridge, MA, October 2002)
[2] G. Schmidt, R. Phillips, *INS/GPS Integration Architecture Performance Comparisons*, in Draper Laboratory Report P-4105 (Cambridge, MA, February 2003)
[3] M.V. Zharkov, K.K. Veremeenko, D.A. Antonov, I.M. Kuznetsov, *Attitude Determination Using Ambiguous GNSS Phase Measurements and Absolute Angular Rate Measurements*, Gyroscopy and Navigation, **9(4)**, 277 (2018)
[4] G. Emel'yantsev, O. Stepanov, A. Stepanov, B. Blazhnov, E. Dranitsyna, M. Evstifeev, D. Eliseev, D. Volynskiy, *Integrated GNSS/IMU-Gyrocompass with Rotating IMU. Development and Test Results*, Remote Sens., **12**, 3736 (2020)
[5] A. Savkin, D. Antonov, L. Kolganov, M. Ryabinkin, E. Chekhov, *UAV navigation using opto-electronic and inertial means in GNSS-denied environment*, J. Phys.: Conf. Ser., **1925**, 012051 (2020)
[6] A. Savkin, D. Antonov, E. Chekhov, L. Kolganov, *Method for autonomous navigation based on integrated inertial and opto-electronic measurements*, in 28th Saint Petersburg International Conference on Integrated Navigation Systems, ICINS 2021, paper No. 9470426 (2021)
[7] V.N. Maksimov, A.I. Chernomorskii, *Integrated navigation and local mapping system for a single-axis wheeled module*, Gyroscopy and Navigation, **7(3)**, 285 (2016)
[8] G.A. Avanesov, R.V. Bessonov, A.N. Kurkina, M.B. Lyudomirskii, I.S. Kayutin, N.E. Yamshchikov, *Autonomous strapdown stellar-inertial navigation systems: Design principles, operating modes and operational experience*, Gyroscopy and Navigation, **4(4)**, 204 (2013)
[9] S.V. Sokolov, V.A. Pogorelov, A.B. Shatalov, *Solving the Autonomous Initial Navigation Task for Strapdown Inertial Navigation System on the Perturbed Basis Using Rodriguez–Hamilton Parameters*, Russian Aeronautics, **62**, 42 (2019)
[10] K.K. Veremeenko, M.V. Zharkov, I.M. Kuznetsov, A.N. Pron'kin, *Strapdown Inertial Navigation System Transfer Alignment: Algorithmic Features and Simulation Performance Analysis*, Russian Aeronautics, **63(4)**, 618 (2020)

[11] K.K. Veremeenko, V.M. Savel'Ev, *In-flight alignment of a strapdown inertial navigation system of an unmanned aerial vehicle*, J. Comput. Syst. Sci. Int., **52(1)**, 106 (2013)

[12] K.K. Veremeenko, D.A. Antonov, M.V. Zharkov, R.Yu. Zimin, I.M. Kuznetsov, A.N. Pron'kin, *Small-sized integrated navigation and landing complex of a variable structure*, in 19th Saint-Petersburg International Conference on Integrated Navigation Systems, ICINS 2012 (2012)

[13] D.A. Antonov, K.K. Veremeenko, M.V. Zharkov, R.Yu. Zimin, I.M. Kuznetsov, A.N. Pron'kin, *Test Complex for the Onboard Navigation System of an Airport Ground Vehicle*, J. Comput. Syst. Sci. Int., **55(5)**, 832 (2016)

Fuzzy controller synthesis algorithm with complete data on the object state

Svetlana Luneva[1,*] and *Anna* Luneva[1]

[1]Moscow Aviation Institute (National Research University), 4, Volokolamskoe shosse, Moscow, 125993, Russia

Abstract. The application of the classical control theory to nonlinear and linear objects control problems in some cases may not give effective results. Due to this, fuzzy controller synthesis methods for objects controlling have been developed based on fuzzy sets and fuzzy logic methods. A suggested method makes it possible to construct a fuzzy system without the help from human experts. The algorithm for this method consists of three steps: on step 1 the known input data ranges are determined and divided into fuzzy regions to form membership functions; on step 2 the data pairs are formed and then used to generate a fuzzy rules table; on step 3 the membership functions are scaled for specific values, so the fuzzy system works more correct. The method has shown it's good performance in the underwater vehicle stabilization problem with complete data on the object state.

1 Introductions

Fuzzy systems synthesis method consists in fuzzification of input and output data, formation of fuzzy rules table by human experts in the field of the problem being solved and its subsequent use, fuzzy data processing with the resulting fuzzy system and its defuzzification [1, 2, 7, 8, 11, 12].

This paper is devoted to the solving of fuzzy controller synthesis problem with complete data on the object state in the stabilization of an underwater vehicle problem. The solution is based on the algorithm of the generating rules from the known numerical data method, which makes it possible to construct a set of fuzzy rules [3].

2 Statement of the problem

An underwater vehicle moving in a vertical plane at a constant longitudinal travel speed [4, 5] is considered as a control object (see figure 1).

The mathematical model of the control object is described by a system of differential equations

$$\begin{cases} \dot{\omega} = a_{11}\omega + a_{12}\psi + b\delta, \\ \dot{\psi} = \omega, \\ \dot{\delta} = u \end{cases}$$

with the given initial conditions: $\omega(0) = \omega_0$, $\psi(0) = \psi_0$, $\delta(0) = \delta_0$.

*e-mail: LunevaSY@mai.ru

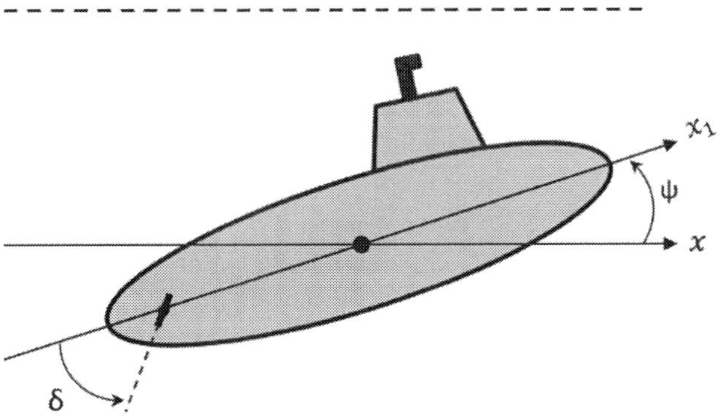

Figure 1. Control object

Here ψ—trim angle, δ—aft horizontal rudders deviation angle from the neutral position, ω—angular velocity, u—control.

The position of the underwater vehicle is characterized by a state vector consisting of three coordinates $x(t) = (\omega(t), \psi(t), \delta(t))^T$. The control is carried out through one channel.

The base initial values are considered as $\omega(0) = 0$, $\psi(0) = 4$, $\delta(0) = 0$ [4, 5].

The task here is to find the control $u(t, x)$ and its corresponding trajectories of motion $x(t)$, that ensure the object bringing to a horizontal position for any initial conditions from a given set of initial states.

One of the possible solutions to this problem is to minimize the quadratic criterion $I = \int_0^{\infty} [B_1 \cdot \omega(t)^2 + B_2 \cdot \psi(t)^2 + B_3 \cdot u(t)^2]\, dt$ using an optimal linear controller $u(t, x) = K_1\omega(t) + K_2\psi(t) + K_3\delta(t)$. The procedure of finding the controller is reduced to finding a positive definite solution to the algebraic Riccati equation [4]. As a result, the coefficients of the controller K_1, K_2, K_3 are found.

Let's consider the problem solution using fuzzy logic: we have to synthesize a fuzzy controller with three input variables and one output variable ψ, δ, ω using the method of generating fuzzy rules based on known data.

3 The algorithm of generating fuzzy rules from known data method

Step 1. Setting fuzzy variables and their fuzzy values

For each variable, we define the domain intervals in which its values will probably lie during the control system functioning. Each domain interval is then divided into N generally intersecting regions for input variables ω, ψ, δ and into M regions for the output variable u, and each region is assigned with a triangular or trapezoidal fuzzy membership function (figures 2(a)–(d)). The membership function determines the corresponding linguistic meaning [10]: P1 (Psy 1), ..., PN (Psy N), D1 (Delta 1), ..., DN (Delta N), O1 (Omega 1), ..., ON (Omega N) for the input variables C1 (Control 1), ..., CM (Control M) for the output variable.

We assign linguistic variables $\tilde{\psi}$, $\tilde{\delta}$, $\tilde{\omega}$ to each input variable and a linguistic variable $\tilde{u}$ to the output variable. Their linguistic meanings are defined as: $\tilde{A}_i = \{\tilde{A}_i^j, j = 1, \ldots, 2N + 1\}$, $\tilde{B} = \{\tilde{B}^j, j = 1, \ldots, 2M + 1\}$.

Step 2. Generating fuzzy rules from given data pairs

To apply the generating fuzzy rules method, the known data must be represented as pairs, where the first element of the pair is the state vector, and the second is the value of control [3, 6, 9]. The values corresponding to the vertices of the membership functions and determining the linguistic values of the coordinates are selected as possible values for each coordinate of the state vector. To form fuzzy rules, all possible combinations of these values are fed to the input of the fuzzy system, and the control value is calculated for each of these combinations. Then the values of the corresponding membership functions are found, and the largest of them is selected. The found highest value corresponds to a certain linguistic meaning of control. If the values of two neighboring membership functions are equal, if the corresponding membership functions lie to the left of zero, the value corresponding to the right membership function and the left one in the opposite case is selected. Thus, one linguistic rule is assigned to each data pair.

2.1. Forming data pairs

2.1.1. To generate data pairs, the values of the state vector variables corresponding to the vertices of the membership functions are taken and all possible combinations are composed.

2.1.2. The values of the control u corresponding to each combination are calculated. Thus, data pairs, where the first element of the pair is the state vector, and the second element of the pair is the control value u are formed.

2.1.3. The obtained values of the variables are written into separate ordered sets, where the ordinal numbers of the elements of the set correspond to the ordinal number of the data pairs obtained on step 2.1.2.

2.2. Computing the values of membership functions for each data pair
The calculation of the values of the membership functions is iterative. At each iteration:

2.2.1. The values of each membership function are calculated for each value of the state vector and control from the corresponding sets created on step 2.1, and then written into separate ordered sets. The values of the membership functions are calculated sequentially, in accordance with their ordinal numbers j. Writing to a set occurs in the same order.

2.2.2. The maximum value among the values of the control membership functions from the obtained set is calculated. Then the ordinal number of the membership function with the maximum value is written into the set to generate the rules. After that, the set with already calculated the membership functions values is cleared to write new values for the next iteration.

2.3. Forming the rule tables according to the principle: IF $\tilde{\psi} = \tilde{A}_1^j$ AND $\tilde{\delta} = \tilde{A}_2^j$ AND $\tilde{\omega} = \tilde{A}_3^j$, THEN $\tilde{u} = \tilde{B}^i$.

The j numbers of linguistic values correspond to the ordinal numbers of the membership functions. When compiling the rules, the sets obtained on step 2.2 are used, where the ordinal numbers of the elements of each set coincide with the number of the rule.

Since there are usually lots of data pairs, and each data pair generates one rule, it is highly possible that there will be some conflicting rules, i.e., rules that have the same "if" part but a different "then" parts [3]. One way to resolve this conflict is to assign a degree to each of the conflicting rules. In practice the additional information about the data pair utility is known, i.e., it is known which data pairs are very useful and generate advice rules and which data pairs are unlikely to be used for generating rules. In this case we can assign a weight coefficient to every conflicting data pair, representing our confidence in generated

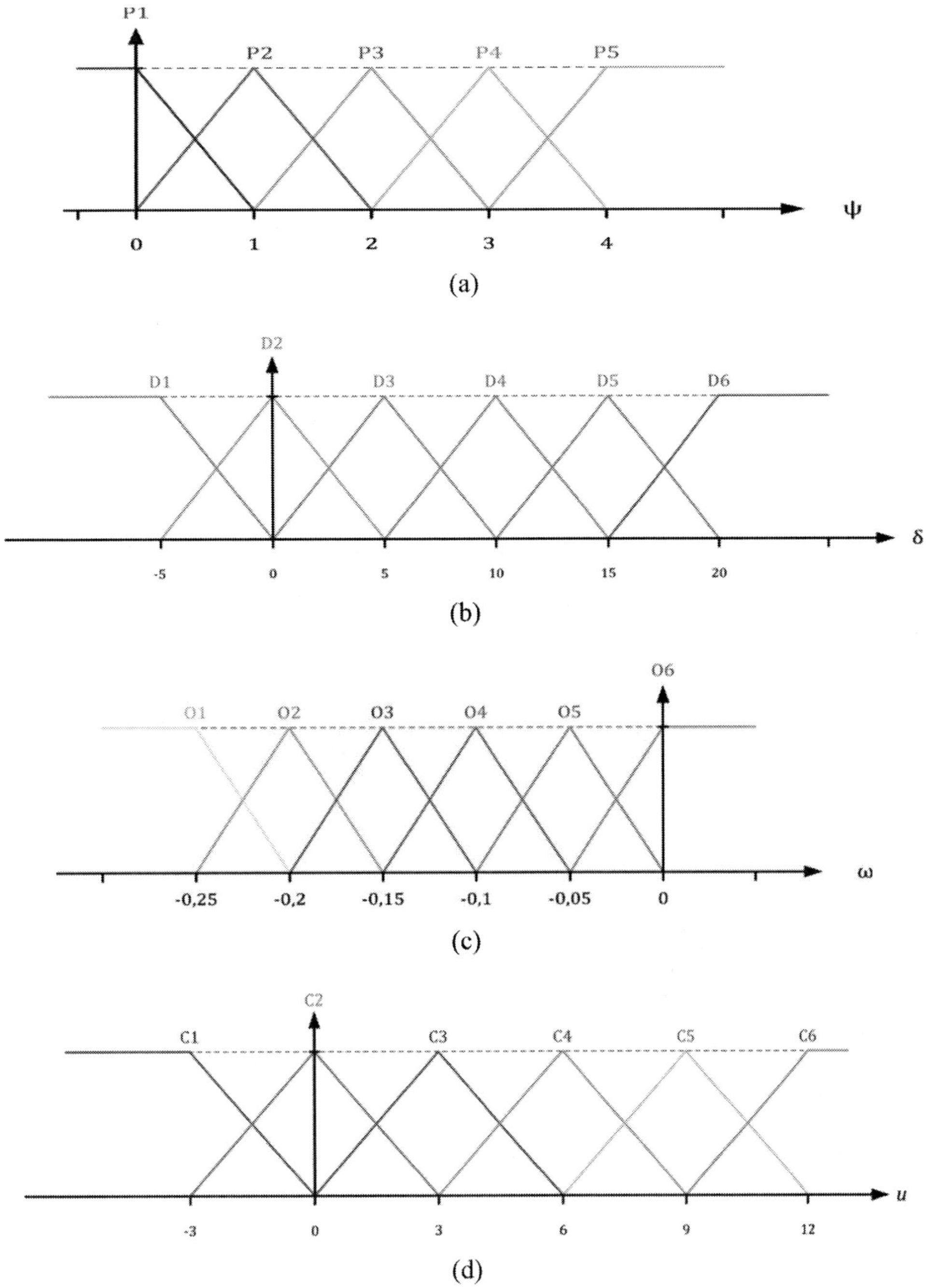

Figure 2. Membership functions for ψ ($N_1 = 5$) (a), δ ($N_2 = 6$) (b), ω ($N_3 = 6$) (c), u ($M = 6$) (d)

rule correctness. In this way, we'll get a fuzzy set, where every data pair belongs to this set with a chosen weight coefficient. Thus in case in case of conflict a rule that has a maximum degree is accepted, and the other ones are deleted.

Since there are three input variables, the rule table can be presented as a parallelepiped. For ease of use, it is convenient to present it in the form of flat sections sequence, i.e. the obtained linguistic rules are ordered in the form of a set of two-dimensional tables with fixed

$\tilde{\omega}$ = «O1»							
		$\tilde{\delta}$					
		D1	**D2**	**D3**	**D4**	**D5**	**D6**
$\tilde{\psi}$	**P1**	C1	C1	C1	C1	C1	C1
	P2	C1	C1	C1	C1	C1	C1
	P3	C2	C2	C1	C1	C1	C1
	P4	C3	C3	C2	C1	C1	C1
	P5	C4	C3	C3	C2	C2	C1

$\tilde{\omega}$ = «O2»							
		$\tilde{\delta}$					
		D1	**D2**	**D3**	**D4**	**D5**	**D6**
$\tilde{\psi}$	**P1**	C1	C1	C1	C1	C1	C1
	P2	C2	C1	C1	C1	C1	C1
	P3	C3	C2	C2	C1	C1	C1
	P4	C4	C3	C2	C2	C1	C1
	P5	C4	C4	C3	C3	C2	C2

$\tilde{\omega}$ = «O3»							
		$\tilde{\delta}$					
		D1	**D2**	**D3**	**D4**	**D5**	**D6**
$\tilde{\psi}$	**P1**	C1	C1	C1	C1	C1	C1
	P2	C2	C2	C1	C1	C1	C1
	P3	C3	C3	C2	C1	C1	C1
	P4	C4	C3	C3	C2	C2	C1
	P5	C5	C4	C4	C3	C3	C2

$\tilde{\omega}$ = «O4»							
		$\tilde{\delta}$					
		D1	**D2**	**D3**	**D4**	**D5**	**D6**
$\tilde{\psi}$	**P1**	C2	C1	C1	C1	C1	C1
	P2	C3	C2	C1	C1	C1	C1
	P3	C4	C3	C2	C2	C1	C1
	P4	C4	C4	C3	C3	C2	C2
	P5	C5	C5	C4	C4	C3	C3

$\tilde{\omega}$ = «O5»							
		$\tilde{\delta}$					
		D1	**D2**	**D3**	**D4**	**D5**	**D6**
$\tilde{\psi}$	**P1**	C2	C2	C1	C1	C1	C1
	P2	C3	C2	C2	C1	C1	C1
	P3	C4	C3	C3	C2	C2	C1
	P4	C5	C4	C4	C3	C3	C2
	P5	C6	C5	C5	C4	C4	C3

$\tilde{\omega}$ = «O6»							
		$\tilde{\delta}$					
		D1	**D2**	**D3**	**D4**	**D5**	**D6**
$\tilde{\psi}$	**P1**	C3	C2	C1	C1	C1	C1
	P2	C3	C3	C2	C2	C1	C1
	P3	C4	C4	C3	C3	C2	C2
	P4	C5	C5	C4	C4	C3	C2
	P5	C6	C6	C5	C5	C4	C3

Figure 3. Rule table for $\tilde{\omega}$

linguistic values of one of the three input linguistic variables. The obtained rule table for is presented on figure 3.

Step 3. Scaling membership functions

After the table of linguistic rules is compiled, it is saved and remains unchanged for any combination of initial conditions belonging to the specified neighborhoods around the base values. However, the parameters of the membership functions constructed in step 1 of the algorithm must be scaled for specific values taken by the input and output variables during the operation of the system (figures 4(a)–(d)).

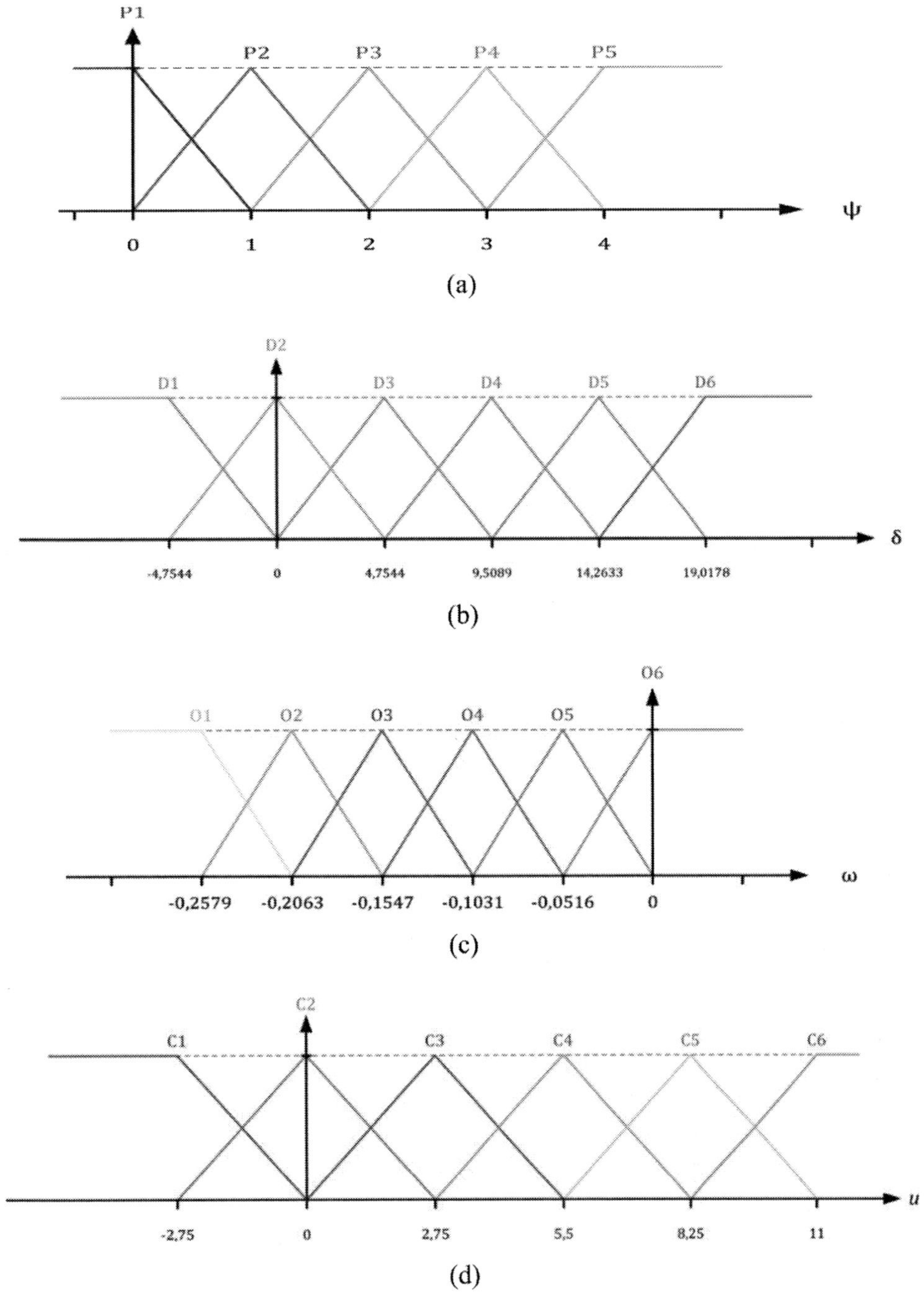

Figure 4. Scaled membership functions for ψ ($N_1 = 5$) (a), δ ($N_2 = 6$) (b), ω ($N_3 = 6$) (c), u ($M = 6$) (d)

The problem was solved numerically using an optimal linear controller and using a synthesized fuzzy controller (figure 5). The fuzzy inference system was implemented using FuzzyLogicToolbox in the MATLAB computer mathematics system, the "min" rule was used to find the intersection of the sets, the "max" rule was used for combining, the "min" rule was used for the implication, and the "centroid" method was used for defuzzification.

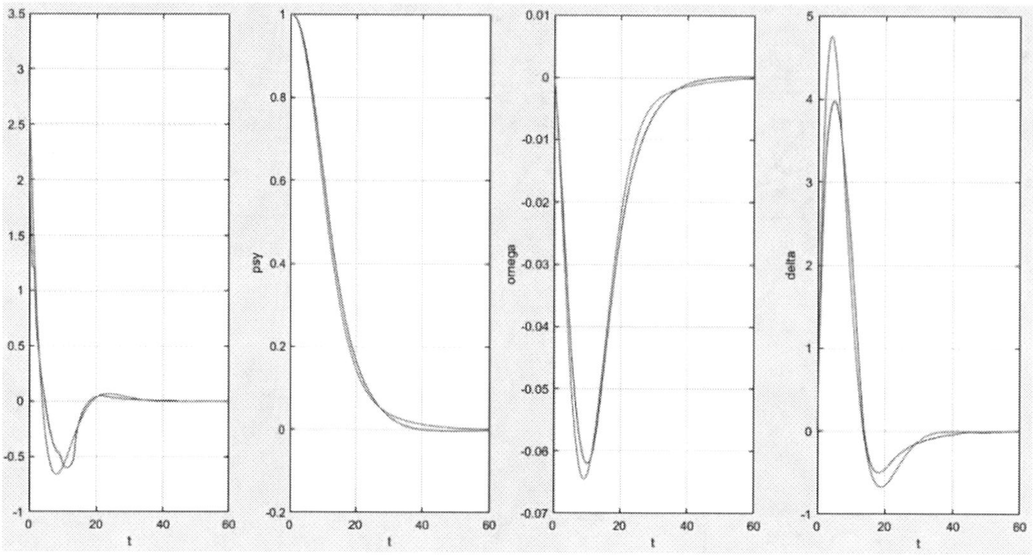

Figure 5. Simulation results for $\omega(0) = 0$, $\psi(0) = 1$, $\delta(0) = 0$

The following parameters were used in the simulation: $a_{11} = -0.1253$; $a_{12} = -0.004637$; $b = -0.002195$; $B_1 = 1.97$; $B_2 = 0.03$; $B_3 = 0.05$; $K_1 = 26.1$; $K_2 = 2.75$; $K_3 = -0.399$.

4 Conclusion

The simulation results comparison showed that the proposed algorithm for the synthesis of a fuzzy controller gives acceptable results and can be used to solve similar problems in conditions of uncertainty of the plant parameters.

References

[1] K.M. Passino, S. Yurkovich, *Fuzzy control* (Addison Wesley Longman, Menlo Park, CA, 1998)
[2] A.G. Badalova, A.V. Panteleev, *Enterprise Risk Management* (Vuzovskaya kniga, Moscow, 2017)
[3] L.X. Wang, J.M. Mendel, *Generating Fuzzy Rules by Learning from Examples*, IEEE Transactions on Systems, Man, and Cybernetics, **22(6)**, 1414 (1992)
[4] E.I. Veremej, *Line systems with feedback* (Lan', Saint-Petersburg, 2013)
[5] A.V. Panteleev, A.S. Bortakovskij, *Control Theory in Examples and Tasks* (Vysshaya shkola, Moscow, 2003)
[6] D. Meng, Z. Pei, *Extracting Linguistic Rules from Data Sets Using Fuzzy Logic and Genetic Algorithms*, Neurocomputing, **78**, 48 (2012)
[7] E.H. Mamdani, S. Assilian, *An experiment in linguistic synthesis with a fuzzy logic controller*, International Journal of Man-Machine Studies, **7(1)**, 1 (1975)
[8] J.M. Alonso, C. Castiello, C. Mencar, *Interpretability of fuzzy systems: Current research trends and prospects*, in Springer Handbook of Computational Intelligence (Springer, Berlin, Heidelberg, 2015)

[9] S. Guillaume, *Designing fuzzy inference systems from data: an interpretability-oriented review*, IEEE Trans. Fuzzy Syst., **9(3)**, 426 (2001)
[10] U. Bodenhofer, P. Bauer, *A formal model of interpretability of linguistic variables*, in Interpretability Issues in Fuzzy Modeling (Springer, 2003)
[11] W. Siler. J.J. Buckley, *Fuzzy expert systems and fuzzy reasoning* (John Wiley & Sons, Hoboken, New Jersey, 2005)
[12] G. Feng, *A survey on analysis and design of model-based fuzzy control systems*, IEEE Trans. Fuzzy Syst., **14(5)**, 676 (2006)

Methods and algorithms for determining the main quasi-homogeneous forms of polynomials and power series

Viktor Nefedov[1,*]

[1]Moscow Aviation Institute (National Research University), 4, Volokolamskoe shosse, Moscow, 125993, Russia

Abstract. Methods are proposed that allow one to determine the special forms of polynomials and power series used in solving a number of practical problems. The most important of them are the construction of necessary and sufficient conditions for an extremum for polynomials and power series, as well as checking matrices for D-stability arising in the study of ecosystem stability. This special forms (the so-called main quasi-homogeneous polynomial forms) are generalizations of the concept of a homogeneous polynomial form. They correspond to the sum of the terms of the polynomial belonging to some face of the Newton polytope of this polynomial. In some cases, the main quasi-homogeneous polynomial forms necessary for research can also be determined for power series (in particular, when constructing necessary and sufficient conditions for an extremum). In the case of polynomials, two cases are investigated separately: the selection of all the main forms of the polynomial and the selection of the main forms corresponding to the faces of the Newton polytope in its "southwestern" part (such forms also can be distinguished for an arbitrary power series), since both cases have their practical applications. Practically applicable methods are described for each of these cases. Several methods are considered sequentially (starting with a simple enumeration and ending with a method with a significant reduction in the number of options in the enumeration). The last (most economical) method is described as a practically realizable algorithm. A practically realizable rather economical algorithm for solving an auxiliary problem is described—finding the set of corner points of the Newton polytope.

1 Introduction

Let us briefly recall the main stages of the study of the polynomial $p(x)$, where $x=(x_1,\ldots,x_n)$, to an extremum, for definiteness—to a minimum. Let $x^{(0)}$—some stationary point for which $p'(x^{(0)})=0_{(n)}\in\mathbb{R}^n$. The next step is to check the sufficient minimum condition. For this, the quadratic form $\varphi(h)=\langle p''(x^{(0)})h,h\rangle$, where $h\in\mathbb{R}^n$, is investigated. Then, if this form is positive definite, i.e. $\varphi(h)>0$ for all $h\neq 0_{(n)}$ (which in fact means that $p(x)$ is strongly convex in a neighborhood of $x^{(0)}$) then $x^{(0)}$ is a local minimum point of the polynomial $p(x)$. If $\exists h^{(0)}\in\mathbb{R}^n$: $\varphi(h^{(0)})<0$, then it is not a local minimum point. And if $\varphi(h)\geq 0$ for all $h\in\mathbb{R}^n$, and at the same time $\exists h^{(0)}\in\mathbb{R}^n$: $h^{(0)}\neq 0_{(n)}$, $\varphi(h^{(0)})=0$, then it says: "more subtle research methods are required", which are usually not given. Thus, even in the case of a weak convexity in the neighborhood of $x^{(0)}$, the proposed schemes "do not work". Mean-

*e-mail: nefedovvn54@yandex.ru

while, cases with $\varphi(h) \equiv 0$, when $p''(x^{(0)})$ is a zero matrix are possible and for these cases sufficient conditions for a minimum are also described [1, 2]. In the case $\varphi(h) \equiv 0$, we can only assert that there is no quadratic form, but other forms can exist, i.e. a generalization of the notion of a homogeneous form of a polynomial is required. By such a generalization we mean the sum of the terms of the polynomial corresponding to some face of the Newton polytope [3] of the polynomial $p(x)$. It should be noted that Newton's polytopes are a means for studying a wide class of problems (see, for example, [3–6]). In particular, Newton's theory of polyhedra connects the geometry of polyhedra with algebraic geometry [6]. Let us introduce some notation. For a vector $k \in \mathbb{Z}^n_{\geq}$, where $\mathbb{Z}_{\geq} = \mathbb{N} \cup \{0\}$, a polynomial $p(x)$, and a set $N \subseteq \mathbb{Z}^n_{\geq}$ we denote: $x^k = x_1^{k_1} x_2^{k_2} \cdots x_n^{k_n}$; $coef(p, k) = a$—coefficient at the term ax^k in $p(x)$; $N_p = \{k \in \mathbb{Z}^n_{\geq} \mid coef(p, k) \neq 0\}$—is the carrier of the polynomial $p(x)$; CoN_p—is the convex hull of the set N_p (Newton's polytope of the polynomial $p(x)$); $p_N(x) = \sum_{k \in N} coef(p, k)x^k$. We will call a polynomial $\varphi(x)$ an A-quasi-homogeneous (polynomial) form, where $A \in \mathbb{N}^n$, if $\varphi \not\equiv 0$, $\exists B \in \mathbb{Z}_{\geq}$: $\forall k \in N_\varphi$ $\langle A, k\rangle = B$; we will call $\varphi(x)$ a quasi-homogeneous form, if $\exists A \in \mathbb{N}^n$, such that $\varphi(x)$ is an A-quasi-homogeneous form. Polynomial $p(x)$ will be called: nonnegative, if $\forall x \in \mathbb{R}^n$ $p(x) \geq 0$; nondegenerate in the weak sense, if $x_{i_1} \neq 0, \ldots, x_{i_r} \neq 0 \Rightarrow p(x) \neq 0$, where $\{x_{i_1}, \ldots, x_{i_r}\}$—is the set of variables from which $p(x)$ essentially depends. We will write for the polynomials $q(x), p(x)$: $q \preccurlyeq p$, if (a) $N_q \subseteq N_p$; (b) $\forall k \in N_q$ $coef(q, k) = coef(p, k)$. For the polynomials $p(x), \varphi(x)$ and the vector $A \in \mathbb{N}^n$ we say that $\varphi(x)$ is the main A-quasi-homogeneous form of $p(x)$ if (a) $\varphi \preccurlyeq p$; (b) $N_\varphi = \{k \in N_p \mid \langle A, k\rangle = B_A\}$, $B_A = \min\{\langle A, k\rangle \mid k \in N_p\}$. We say that a polynomial $\varphi(x)$ is a main quasi-homogeneous form of $p(x)$ if $\exists A \in \mathbb{N}^n$ such that $\varphi(x)$ is a main A-quasi-homogeneous form of $p(x)$.

The use of quasi-homogeneous forms gives the following necessary and sufficient conditions for an extremum.

Statement 1 [1]. Let $p(x)$ be a polynomial, $p(0_{(n)}) = 0$, $p'(0_{(n)}) = 0_{(n)}$, and all main quasi-homogeneous forms of $p(x)$ are non-negative and non-degenerate in the weak sense. Then $0_{(n)}$ is a point of a local minimum.

Statement 2 [1]. Let $0_{(n)}$ be a point of local minimum of the polynomial $p(x)$, $p(0_{(n)}) = 0$, $p'(0_{(n)}) = 0_{(n)}$. Then all main quasi-homogeneous forms of $p(x)$ are non-negative.

Example 1. (1) Let $p_1(x_1, x_2) = x_1^2x_2^6 - 1.98x_1^4x_2^5 + x_1^6x_2^4 + x_2^{10} - 10x_1x_2^9 - 20x_1^8x_2^4$. Then the main quasi-homogeneous forms of this polynomial are: (a) $\varphi(x_1, x_2) = x_1^2x_2^6 - 1.98x_1^4x_2^5 + x_1^6x_2^4 = 0.99x_1^2x_2^4(x_2 - x_1^2)^2 + 0.01x_1^2x_2^4(x_2^2 + x_1^4)$—non-negative and non-degenerate in the weak sense (here $A = \langle 1, 2\rangle$; if $x_1 \neq 0$, $x_2 \neq 0$, then $\varphi(x_1, x_2) > 0$); (b) x_2^{10}, $x_2^{10} + x_1^2x_2^6$, $x_1^2x_2^6$, $x_1^6x_2^4$ are non-negative and non-degenerate in the weak sense. Then, by virtue of statement 1, $0_{(2)}$ is the point of the local minimum of the polynomial $p_1(x_1, x_2)$. (2) Let $p_2(x_1, x_2) = x_1^2x_2^6 - 2.01x_1^4x_2^5 + x_1^6x_2^4 + x_2^{10} - 10x_1x_2^9 - 20x_1^8x_2^4$. Then the main quasi-homogeneous form $\eta(x_1, x_2) = x_1^2x_2^6 - 2.01x_1^4x_2^5 + x_1^6x_2^4$ of the polynomial $p_2(x_1, x_2)$ is not non-negative ($\eta(1, 1) = -0.01$), and, by virtue of statement 2, $0_{(2)}$ is not a local minimum point.

Remark 1. The definition of the main quasi-homogeneous forms of the polynomial was given in relation to the problem of studying stationary points for local minimality. In this regard, a vector $A \in \mathbb{N}^n$ was used, and with this choice of A, the quasi-homogeneous forms corresponded to the faces of the Newton polyhedron in its "southwestern" part. But when solving more general problems (for example, when studying a polynomial for definiteness in a positive orthant [7, 8]), quasi-homogeneous forms corresponding to all faces of the Newton polytope may be required, and in this case we will use $A \in \mathbb{Z}^n \backslash \{0_{(n)}\}$ instead of $A \in \mathbb{N}^n$ (see [9]).

Remark 2. In the case $A \in \mathbb{N}^n$, we can pose the problem of finding the set of main quasi-homogeneous forms not only for the polynomial, but also for the power series, which is the result of the expansion of the analytic function in powers of the variables in the vicinity of the stationary point, since in this case their number will be finite and it will be required a finite number of "first" terms from this expansion. Moreover, statements 1, 2 remain valid for the power series $p(x)$ (see [2]).

2 Methods for finding the main quasi-homogeneous forms of a polynomial

The simplest is

Method 1. Let us first consider the case when it is required to select all main A-quasi-homogeneous forms of the polynomial $p(x) \not\equiv 0$ for $A \in \mathbb{Z}^n \setminus \{0_{(n)}\}$ (see remark 1). We iterate over all non-empty subsets $N \subseteq N_p$ (i.e. $N \in 2^{N_p} \setminus \{\emptyset\}$). Then, to check whether $p_N(x)$ is the main quasi-homogeneous form of the polynomial $p(x)$, it is enough to verify whether there exists an integer solution to the system of equalities and inequalities (with respect to the unknown vector $e \in \mathbb{Z}^n$):

$$\langle e, k \rangle = \langle e, k^{(0)} \rangle, \quad k \in N \setminus \{k^{(0)}\}; \quad \langle e, k \rangle \geq \langle e, k^{(0)} \rangle + 1, \quad k \in N_p \setminus N, \tag{1}$$

where here and below, $k^{(0)}$ is any fixed element from N (if $|N| = 1$, then there are no equalities in (1)). Then, if there is an integer solution $e = A$ of system (1), then $A \in \mathbb{Z}^n \setminus \{0_{(n)}\}$ (if $A = 0_{(n)}$, then inequalities in (1) do not hold), and $p_N(x)$ is the main A-quasi-homogeneous form of the polynomial $p(x)$; otherwise, $p_N(x)$ is not the main quasi-homogeneous form of $p(x)$. This follows directly from the definition of the main quasi-homogeneous form.

To solve the problem of finding an integer solution to system (1), one can consider a problem $\max\{|e_1|, \dots, |e_n|\} \to \min$ with constraints (1) or an equivalent LPP (add a new variable u)

$$u \to \min; \quad e_1 \leq u, \ -e_1 \leq u, \ \dots, \ e_n \leq u, \ -e_n \leq u \quad (\Leftrightarrow \max\{|e_1|, \dots, |e_n|\} \leq u), \tag{2}$$

under additional constraints (1). To solve the LPP (1), (2), it is better to use the simplex method, since any solution obtained by this method will be an extreme point of a convex polyhedral set given by a system of equalities and inequalities with integer coefficients, and, therefore, the components of this solution are rational numbers. Note further that if e is a solution to system (1), then $\forall \lambda \geq 1$ λe is also a solution to this system. But then, taking any rational solution e of system (1), and multiplying it by the least common multiple of the denominators of the nonzero components of this solution, we obtain an integer solution to system (1).

As already noted above, for solving some problems one is interested in main A-quasi-homogeneous forms not for all $A \in \mathbb{Z}^n \setminus \{0_{(n)}\}$, but only for $A \in \mathbb{N}^n$. In this case, instead of (1), one should consider the system (1),

$$e_i \geq 1, \quad i = 1, 2, \dots, n, \tag{3}$$

to find some solution of which, instead of LPP (1), (2), one can consider the LPP of the form (1),

$$e_1 \to \min; \quad e_i \geq 1, \quad i = 1, 2, \dots, n. \tag{4}$$

Remark 3. The simplest examples show that the same quasi-homogeneous form can be the main A-quasi-homogeneous for different $A \in \mathbb{Z}^n \setminus 0_{(n)}$, including for $A \in \mathbb{N}^n$ and for $A \in \mathbb{Z}^n \setminus \mathbb{N}^n$. For instance, consider a polynomial $p(x_1, x_2) = x_1^2 + x_2^2 + x_1^2 x_2^2$, for which x_2^2 is the main A-quasi-homogeneous form for $A = \langle 2, 1 \rangle$, and also for $A = \langle 1, -1 \rangle$ and for $A = \langle 1, 0 \rangle$. Thus, for this example, solving LPP (1), (2), we may obtain $A = \langle 1, -1 \rangle$ and the question of the existence of $A \in \mathbb{N}^n$ remains open. When using LPP (1), (4), this situation does not arise.

Acting in this way, we can define all main A-quasi-homogeneous polynomial forms of the polynomial $p(x)$ (either for $A \in \mathbb{Z}^n \backslash \{0_{(n)}\}$, or, in accordance with remark 1, only for $A \in \mathbb{N}^n$). However, in the case when the quantity $|N_p|$ is large enough, the number of elements in 2^{N_p} is extremely large, which makes it impossible to apply this approach. In this regard, the question arises of reducing the number of subsets that should be checked by solving the problem of finding at least one integer solution in LPP (1), (2) (or, respectively, in (1), (4)). This will require some definitions and notation. We will use the following binary relations for $a, b \in \mathbb{R}^n$: $a \leq b \Leftrightarrow a_i \leq b_i$, $i = 1, 2, \ldots, n$; $a < b \Leftrightarrow a_i < b_i$, $i = 1, 2, \ldots, n$. Let Y be a finite set from $\mathbb{R}^n$. We denote: $P(Y) = \{y \in Y \mid \forall y' \in Y\ [y' \leq y \Rightarrow y' = y]\}$—the set of Pareto-optimal points from Y; $S(Y) = \{y \in Y \mid \forall y' \in Y\ [y' < y \Rightarrow y' = y]\}$—the set of Slater-optimal points from Y. We also introduce generalizations of these sets. Let $z = (z_1, \ldots, z_n) \in \{0, 1\}^n$. Let us denote for $a, b \in \mathbb{R}^n$: $a \leq_z b \Leftrightarrow (-1)^{z_i} a_i \leq (-1)^{z_i} b_i$, $i = 1, 2, \ldots, n$; $a <_z b \Leftrightarrow (-1)^{z_i} a_i < (-1)^{z_i} b_i$, $i = 1, 2, \ldots, n$. We further denote $P_z(Y) = \{y \in Y \mid \forall y' \in Y\ [y' \leq_z y \Rightarrow y' = y]\}$, $S_z(Y) = \{y \in Y \mid \forall y' \in Y\ [y' <_z y \Rightarrow y' = y]\}$, $P^*(Y) = \bigcup_{z \in \{0,1\}^n} P_z Y)$, $S^*(Y) = \bigcup_{z \in \{0,1\}^n} S_z(Y)$. In addition, for $A \in \mathbb{R}^n$ denote $z^A = (z_1^A, \ldots, z_n^A) \in \{0, 1\}^n$, where $z_i^A = \begin{cases} 1, & if \quad A_i < 0, \\ 0, & if \quad A_i \geq 0. \end{cases}$ Obviously, $\forall a, b, A \in \mathbb{R}^n$: $a \leq_{z^A} b \Rightarrow \langle A, a \rangle \leq \langle A, b \rangle$, and if $A_i \neq 0, i = 1, 2, \ldots, n$, then $a \leq_{z^A} b$, $a \neq b \Rightarrow \langle A, a \rangle < \langle A, b \rangle$.

We present the following almost obvious

Statement 3. Let Y be a finite set from $\mathbb{R}^n$, $A \in \mathbb{R}^n$, $Y^* = Arg \min\{\langle A, y \rangle \mid y \in Y\}$. Then

(a) if $A_i > 0$, $i = 1, 2, \ldots, n$, then $Y^* \subseteq P(Y)$;

(b) if $A_i \neq 0$, $i = 1, 2, \ldots, n$, then $Y^* \subseteq P_{z^A}(Y)$;

(c) if $A_i \geq 0$, $i = 1, 2, \ldots, n$, $A \neq 0_{(n)}$, then $Y^* \subseteq S(Y)$;

(d) if $A \neq 0_{(n)}$, then $Y^* \subseteq S_{z^A}(Y)$.

A consequence of statement 3 is

Statement 4. Let some polynomial $p(x) \not\equiv 0$. Then

(a) $\forall A \in \mathbb{N}^n$ for the main A-quasi-homogeneous form $\varphi_A(x)$ of this polynomial, $N_{\varphi_A} \subseteq P(N_p)$ holds;

(b) $\forall A \in \mathbb{Z}^n \backslash \{0_{(n)}\}$ for the main A-quasi-homogeneous form $\varphi_A(x)$ of this polynomial, $N_{\varphi_A} \subseteq S_{z^A}(N_p) \subseteq S^*(N_p)$ holds;

(c) $\forall A \in \mathbb{Z}^n \backslash \{0_{(n)}\}$, if $A_i \neq 0$, $i = 1, 2, \ldots, n$, then for the main A-quasi-homogeneous form $\varphi_A(x)$ of this polynomial, $N_{\varphi_A} \subseteq P_{z^A}(N_p) \subseteq P^*(N_p)$ holds.

Statement 5. Let Y be a finite set from $\mathbb{R}^n$, $A \in \mathbb{R}^n$. Then

(a) $\forall y \in Y, \forall z \in \{0, 1\}^n\ \exists y^* \in P_z(Y) \subseteq S_z(Y)$: $y^* \leq_z y$;

(b) $\forall y \in Y\ \exists y^* \in P_{z^A}(Y) \subseteq S_{z^A}(Y)$: $y^* \leq_{z^A} y$, $\langle A, y^* \rangle \leq \langle A, y \rangle$.

Using statements 4, 5, we obtain the possibility of reducing the number of options $N \in 2^{N_p} \backslash \{\emptyset\}$ for enumerating all main quasi-homogeneous forms of the polynomial $p(x)$. In the case of finding all main A-quasi-homogeneous forms for $A \in \mathbb{N}^n$, it is sufficient to consider LPP (1), (4) for $N \in 2^{N_p} \backslash \{\emptyset\}$, and in (1) we can replace N_p by $P(N_p)$ (by virtue of statement 5(b)). Accordingly, in the case of finding all main A-quasi-homogeneous forms for $A \in \mathbb{Z}^n \backslash 0_{(n)}$, it is sufficient to consider LPP (1), (2) for $N \in 2^{S^*(N_p)} \backslash \{\emptyset\}$, and in (1) we can replace N_p with $S^*(N_p)$ (by virtue of statement 5(b)). This alone can result in a significant reduction in the total amount of computation.

To further reduce the number of calls to the solution of LPP (1), (2) (or (1), (4)) one more concept is needed. Let Y be a convex set from $\mathbb{R}^n$. A point $y \in Y$ is called a corner point of Y if condition $y = \alpha y^{(1)} + (1 - \alpha) y^{(2)}$, where $y^{(1)}, y^{(2)} \in Y$, $\alpha \in (0, 1)$, implies $y^{(1)} = y^{(2)} = y$. For a finite set $Y \subset \mathbb{R}^n$ we denote: $\Psi(Y)$—the set of corner points of the CoY (CoY is the convex hull of the set Y), $\Omega(Y) = \Psi(Y) \cap P(Y)$.

Statement 6. Let Y be a finite set from $\mathbb{R}^n$. Then $\Psi(Y) \subseteq P^*(Y)$. Statement 6 is a consequence of the fact that for any corner point $v \in \Psi(Y)$ there is a vector $A \in \mathbb{R}^n$ such that $A_i \neq 0$, $i = 1, 2, \ldots, n$, and $\forall y \in \Psi(Y)\backslash\{v\}$ $\langle A, y\rangle < \langle A, v\rangle$ (hence $v \in Arg\min\{\langle A, y\rangle \mid y \in Y\} \subseteq$ $\subseteq P_{z^A}(Y) \subseteq P^*(Y)$; see statement 3(b)).

Method 2. Let us first consider the case when it is required to select all main A-quasi-homogeneous forms of the polynomial $p(x) \not\equiv 0$ for $A \in \mathbb{N}^n$. Let us show that in this case it will be sufficient to consider LPP (1), (4) for all nonempty subsets N of the set $\Omega(N_p) = \Psi(N_p) \cap P(N_p)$, that is, for $N \in 2^{\Omega(N_p)}\backslash\{\emptyset\}$. Moreover, the set N_p in (1) can be replaced by $\Omega(N_p)$. Then in the case when for LPP $e_1 \to \min$ under constraints

$$\langle e, k\rangle = \langle e, k^{(0)}\rangle, \quad k \in N\backslash\{k^{(0)}\}; \quad \langle e, k\rangle \geq \langle e, k^{(0)}\rangle + 1, \quad k \in \Omega(N_p)\backslash N; \quad e_i \geq 1, \quad i = 1, 2, \ldots, n, \tag{5}$$

where $N \in 2^{\Omega(N_p)}\backslash\{\emptyset\}$, there is an integer solution $e = A$, for the set

$$\tilde{N} = \{k \in P(N_p) \mid \langle A, k\rangle = \langle A, k^{(0)}\rangle\} \tag{6}$$

the following holds: $p_{\tilde{N}}(x)$ is the main A-quasi-homogeneous form of the polynomial $p(x)$. Indeed, suppose it is not. Then $\exists \tilde{k} \in \Omega(N_p)\backslash N$: $\langle A, \tilde{k}\rangle < \langle A, k^{(0)}\rangle$ (here we used the fact that the minimum value of a linear function on a bounded convex polyhedral set is attained at the corner points of this set (in this case, due to $A \in \mathbb{N}^n$ belonging to $P(N_p)$), and this contradicts the fact that $e = A$ is a solution to system (5).

Let us show that the main$A$-quasi-homogeneous forms of the polynomial $p(x)$ for all $A \in \mathbb{N}^n$ will be distinguished in this way. We need

Statement 7. Let $N \subset \mathbb{R}^n$ be a finite set,$A \in \mathbb{R}^n$, $N_1 = Arg\min\{\langle A, x\rangle \mid x \in N\}$. Further let $k \in N_1$. Then $\min\{\langle A, x\rangle \mid x \in N\} = \min\{\langle A, x\rangle \mid x \in \Psi(N)\}$ and $\exists k^{(1)}, \ldots, k^{(r)} \in N_1 \cap \Psi(N)$, $\alpha_1, \ldots, \alpha_r \in \mathbb{R}$:

$$\sum_{i=1}^{r} \alpha_i k^{(i)} = k, \quad \sum_{i=1}^{r} \alpha_i = 1, \quad \alpha_i > 0, \quad i = 1, 2, \ldots, r. \tag{7}$$

Evidence. It is known (see, for example, [10, 11]) that there are $k^{(1)}, \ldots, k^{(r)} \in \Psi(N)$, $\alpha_1, \ldots, \alpha_r \in \mathbb{R}$, for which (7) is true. Note that due to (7) we have: $B = \langle A, k\rangle = \sum_{i=1}^{r} \alpha_i \langle A, k^{(i)}\rangle$, whence, by virtue of $\langle A, k^{(i)}\rangle \geq B$, $\alpha_i > 0$, $i = 1, 2, \ldots, r$, $\sum_{i=1}^{r} \alpha_i = 1$, we obtain $\langle A, k^{(i)}\rangle = B$, $i = 1, 2, \ldots, r$. Thus, $k^{(1)}, \ldots, k^{(r)} \in N_1 \cap \Psi(N)$.

Corollary 1. Let the conditions of statement 7 be satisfied. Then $Arg\min\{\langle A, y\rangle \mid y \in \Psi(N)\} =$ $= \Psi(N) \cap N_1$.

Remark 4. Let the conditions of statement 7 be satisfied and $A_i \neq 0$, $i = 1, 2, \ldots, n$. Then due to Statement 3 the condition $\exists k^{(1)}, \ldots, k^{(r)} \in N_1 \cap \Psi(N)$ can be replaced by $\exists k^{(1)}, \ldots, k^{(r)} \in$ $\in N_1 \cap P_{z^A}(N) \cap \Psi(N)$, and in the case $A_i > 0$, $i = 1, 2, \ldots, n$,—by $\exists k^{(1)}, \ldots, k^{(r)} \in N_1 \cap \Omega(N)$.

Let for some $\breve{N} \subseteq P(N_p)$, $\breve{A} \in \mathbb{N}^n$ (where $\breve{N} \neq \emptyset$) $p_{\breve{N}}(x)$ is the main $\breve{A}$-quasi-homogeneous form of the polynomial $p(x)$, that is, $\breve{N} = Arg\min\{\langle \breve{A}, y\rangle \mid y \in N_p\}$. Let us show that it will be isolated by the method 2. We denote $N = \breve{N} \cap \Omega(N_p)$. Obviously (see statement 7 and remark 4) that $N \neq \emptyset$. Let $k^{(0)} \in N$. Then system (5) has an integer solution (in particular, the $e = \breve{A}$). Let us show that for any solution $e = A$ of system (5) (not even necessarily integer) for the set $\tilde{N}$ satisfying (6), $\tilde{N} = \breve{N}$ holds (and thus $p_{\breve{N}}(x)$ is distinguished by the method under consideration). (a) Let us show that $k \in \breve{N} \Rightarrow k \in \tilde{N}$. Since $\breve{N} = Arg\min\{\langle \breve{A}, k\rangle \mid k \in N_p\}$, by virtue of statement 7 and remark 4 $\exists k^{(1)}, \ldots, k^{(r)} \in \Omega(N_p) \cap \breve{N} = N$, $\alpha_1, \ldots, \alpha_r \in \mathbb{R}$, which holds (7). Moreover, according to the choice of A, we have: $k^{(i)} \in N \Rightarrow \langle A, k^{(i)}\rangle = \langle A, k^{(0)}\rangle$, $i = 1, 2, \ldots, r$, whence, by virtue of (7), we obtain $\langle A, k\rangle = \langle A, k^{(0)}\rangle$, i.e. $k \in \tilde{N}$. (b) Now let us show that $k \in \tilde{N} \Rightarrow k \in \breve{N}$. Note that from (5) it follows: $\langle A, k^{(0)}\rangle = \min\{\langle A, k\rangle \mid k \in \Omega(N_p)\} =$ $= \min\{\langle A, k\rangle \mid k \in N_p\}$ (since the minimum of the linear function is attained on $\Psi(N_p)$ and by virtue of $A_i = e_i \geq 1$, $i = 1, 2, \ldots, n$, simultaneously on $P(N_p)$), $N = Arg\min\{\langle A, k\rangle \mid k \in \Omega(N_p)\} = \Omega(N_p) \cap Arg\min\{\langle A, k\rangle \mid k \in N_p\}$, $Arg\min\{\langle A, k\rangle \mid k \in N_p\} \subseteq P(N_p)$ and therefore $\tilde{N} = Arg\min\{\langle A, k\rangle \mid k \in N_p\}$, $\Omega(N_p) \cap \tilde{N} = N$, whence, by virtue of statement 7 and remark 4, $\exists k^{(1)}, \ldots, k^{(r)} \in \Omega(N_p) \cap \tilde{N} = N$, $\alpha_1, \ldots, \alpha_r \in \mathbb{R}$, which holds (7). Moreover, $\langle \breve{A}, k\rangle = \sum_{i=1}^{r} \alpha_i < \langle \breve{A}, k^{(i)}\rangle = \langle \breve{A}, k^{(1)}\rangle$, due to $k^{(1)}, \ldots, k^{(r)} \in N = \breve{N} \cap \Omega(N_p) \subseteq \breve{N} = Arg\min\{\langle \breve{A}, y\rangle \mid y \in N_p\}$, that is, $k \in \breve{N}$.

Let us now consider the case when it is required to select all main A-quasi-homogeneous forms for $A \in \mathbb{Z}^n \backslash \{0_{(n)}\}$. Let us show that in this case it will be sufficient to consider system LPP (1), (2) for all nonempty subsets N of the set $\Psi(N_p)$, that is, for $N \in 2^{\Psi(N_p)} \backslash \{\emptyset\}$. Moreover, the set N_p in (1) can be replaced by $\Psi(N_p)$. Then in the case when for $N \in 2^{\Psi(N_p)} \backslash \{\emptyset\}$ for the system considered in this way

$$\langle e, k\rangle = \langle e, k^{(0)}\rangle, \quad k \in N \backslash \{k^{(0)}\}; \quad \langle e, k\rangle \geq \langle e, k^{(0)}\rangle + 1, \quad k \in \Psi(N_p) \backslash N, \tag{8}$$

there is an integer solution $e = A$, which is found by a solution of the LPP (2), (8) for the set

$$\tilde{N} = \{k \in S^*(N_p) \mid \langle A, k\rangle = \langle A, k^{(0)}\rangle\} \tag{9}$$

the following holds: $p_{\tilde{N}}(x)$ is the main A-quasi-homogeneous form of the polynomial $p(x)$. Indeed, suppose it is not. Then either $\exists \tilde{k} \in \Psi(N_p)$: $\langle A, \tilde{k}\rangle < \langle A, k^{(0)}\rangle$ (here we used the fact that the minimum value of a linear function on a bounded convex polyhedral set is attained at the corner points of this set), which contradicts the fact that $e = A$ is a solution to system (8), or $\langle A, k^{(0)}\rangle = \min\{\langle A, k\rangle \mid k \in N_p\}$. In the latter case, by virtue of statement 4(b), $Arg\min\{\langle A, k\rangle \mid k \in N_p\} \subseteq S^*(N_p)$, whence it follows that $Arg\min\{\langle A, k\rangle \mid k \in N_p\} = \tilde{N}$, and $p_{\tilde{N}}(x)$ is the main A-quasi-homogeneous form of the polynomial $p(x)$.

Let us show that the main A-quasi-homogeneous forms of the polynomial $p(x)$ for all $A \in \mathbb{Z}^n \backslash \{0_{(n)}\}$ will be distinguished in this way. Let for $\breve{N} \subseteq P(N_p)$ (where $\breve{N} \neq \emptyset$) $p_{\breve{N}}(x)$ is the main $\breve{A}$-quasi-homogeneous form of the polynomial $p(x)$ for some $\breve{A} \in \mathbb{Z}^n \backslash \{0_{(n)}\}$, i.e. $Arg\min\{\langle \breve{A}, k\rangle \mid k \in N_p\} = \breve{N}$. Let us show that it will be distinguished by the described method. We denote $N = \breve{N} \cap \Psi(N_p)$. Obviously, $N \neq \emptyset$ (by virtue of statement 7, $\breve{N} \neq \emptyset \Rightarrow N \neq \emptyset$), that is, $N \in 2^{\Psi(N_p)} \backslash \{\emptyset\}$. Let $k^{(0)} \in N$. Then, using the fact that $\Psi(N_p) \backslash N = \Psi(N_p) \backslash \breve{N} \subseteq N_p \backslash \breve{N}$, we obtain that system (8) has an integer solution (in particular, the vector $\breve{A}$). Let us show that for any solution $e = A$ of system (8) (even not necessarily integer) for the set $\tilde{N}$ satisfying (9), $\tilde{N} = \breve{N}$ holds (and thus $p_{\breve{N}}(x)$ is distinguished by the method under consideration). (a) Let us show that $k \in \breve{N} \Rightarrow k \in \tilde{N}$. Since $\breve{N} = Arg\min\{\langle \breve{A}, k\rangle \mid k \in N_p\}$, by virtue of statement 7, $\exists k^{(1)}, \ldots, k^{(r)} \in \Psi(N_p) \cap \breve{N} = N$, $\alpha_1, \ldots, \alpha_r \in \mathbb{R}$, which holds (7). Moreover, according to the choice of A, we have: $\langle A, k^{(i)}\rangle = \langle A, k^{(0)}\rangle$, $i = 1, 2, \ldots, r$, whence, by virtue of (7), we obtain $\langle A, k\rangle = \langle A, k^{(0)}\rangle$, i.e. $k \in \tilde{N}$ (by statement 4(b), $k \in \breve{N} \Rightarrow k \in S^*(N_p)$). (b) Let us show that $k \in \tilde{N} \Rightarrow k \in \breve{N}$. Note that from (8) it follows: $\langle A, k^{(0)}\rangle = \min\{\langle A, k\rangle \mid k \in \Psi(N_p)\} = \min\{\langle A, k\rangle \mid k \in N_p\}$. Using (8) and corollary 1, we find that $N = Arg\min\{\langle A, k\rangle \mid k \in \Psi(N_p)\} = \Psi(N_p) \cap Arg\min\{\langle A, k\rangle \mid k \in N_p\}$, and hence, by virtue of $k^{(0)} \in N$, as well as of statement 4(b), $\tilde{N} = Arg\min\{\langle A, k\rangle \mid k \in N_p\}$, $\Psi(N_p) \cap \tilde{N} = N$. Since $k \in \tilde{N}$, by virtue of statement 7 $\exists k^{(1)}, \ldots, k^{(r)} \in \Psi(N_p) \cap \tilde{N} = N$, $\alpha_1, \ldots, \alpha_r \in \mathbb{R}$, which holds (7). Moreover, $\langle \breve{A}, k\rangle = \sum_{i=1}^{r} \alpha_i \langle \breve{A}, k^{(i)}\rangle = \langle \breve{A}, k^{(1)}\rangle$, due to $k^{(1)}, \ldots, k^{(r)} \in N = \breve{N} \cap \Psi(N_p) \subseteq \breve{N}$, that is, $k \in \breve{N}$.

Method 3. Let us first consider the case when it is required to select all main A-quasi-homogeneous forms for $A \in \mathbb{Z}^n \backslash \{0_{(n)}\}$.

Statement 8. Let $N_p \neq \emptyset$ (that is, $p(x) \not\equiv 0$) hold for some polynomial $p(x)$. Let $k^{(1)}, \ldots, k^{(r)} \in \Psi(N_p)$, $k^{(0)} = \frac{1}{r}(k^{(1)} + \ldots + k^{(r)}) \in riCoN_p$, $r \in \mathbb{N}$. Then any set of points from N_p, including the points $k^{(1)}, \ldots, k^{(r)}$, cannot enter any main quasi-homogeneous polynomial form of $p(x)$ other than the polynomial $p(x)$ itself.

Evidence. Suppose that such a form has been found, i.e. $\exists A \in \mathbb{Z}^n \backslash 0_{(n)}$, $B \in \mathbb{Z}$: $\langle A, k^{(1)} \rangle = B, \ldots, \langle A, k^{(r)} \rangle = B$, $\forall k \in N_p$ $\langle A, k \rangle \geq B$, and $\langle A, k^* \rangle > B$ holds for some $k^* \in N_p$. We can assume that $k^* \in \Psi(N_p)$ (since the linear function reaches its extreme values at the corner points of the bounded polytope). Then $\langle A, k^{(0)} \rangle = B$. On the other hand, since $k^{(0)} \in riCoN_p$, then [12] $\exists \alpha_k > 0$, $k \in \Psi(N_p)$: $\sum_{k \in \Psi(N_p)} \alpha_k = 1$, $k^{(0)} = \sum_{k \in \Psi(N_p)} \alpha_k k$, and by virtue of $\alpha_{k^*} > 0$ we obtain $\langle A, k^{(0)} \rangle > B$, that is, came to a contradiction.

Remark 5. When $A \in \mathbb{N}^n$ statement 6 can be strengthened. Let $\bar{k} \in \mathbb{Z}^n_{\geq}$, where $\bar{k}_i = \max\{k_i \mid k \in N_p\}$, $i = 1, 2, \ldots, n$. Then the condition $k^{(0)} \in riCoN_p$ can be replaced by a weaker one: $k^{(0)} \in riCo\{N_p \cup \{\bar{k}\}\}$. The proof is similar, taking into account the fact that $\bar{k} \in \Psi(N_p \cup \{\bar{k}\})$.

Statement 8 points to the following possible organization of the process (algorithm) for the selection of all main quasi-homogeneous polynomial forms of the polynomial $p(x)$ different from this polynomial (for example, in the case of $intCoN_p \neq \emptyset$ any main quasi-homogeneous polynomial form of the polynomial $p(x)$ is different from it).

Algorithm 1 (finding all main quasi-homogeneous polynomial form of the polynomial $p(x)$)

Step 1. Select $\Psi(N_p), S^*(N_p)$. Let's number the points from $\Psi(N_p)$, i.e. let $\Psi(N_p) = \{k^{(1)}, \ldots, k^{(m)}\}$.

Step 2. Further work of the algorithm is divided into m stages. At the 1st stage, we list all main quasi-homogeneous polynomial forms with the occurrence of $k^{(1)}$ in them. At the 2nd stage—with the occurrence of $k^{(2)}$, but without $k^{(1)}$. At the 3rd stage—with the entry of $k^{(3)}$, but already without $k^{(1)}$, $k^{(2)}$, etc.

Step 3. The described multistage process is conveniently represented as a set of growing trees (forest). Each point $k^{(i)} \in \Psi(N_p)$ is the root of the tree along which all main quasi-homogeneous polynomial forms will be located with a survey entry of the point $k^{(i)}$ in them, but in the absence of points $k^{(1)}, \ldots, k^{(i-1)}$. As a result, sets of pairwise disjoint sets of main quasi-homogeneous polynomial forms will be obtained.

Let us describe a tree with the root $k^{(1)}$ (the rest are described similarly). We will consider $k^{(1)}$ as the top of the 0th level of this tree. Level 1 vertices adjacent to $k^{(1)}$ are all $k \in \Psi(N_p) \backslash \{k^{(1)}\}$ for which $\frac{1}{2}(k + k^{(1)}) \notin riCoN_p$ (in this case, by virtue of statement 8, all obviously inappropriate cases are cut off), which we arrange in ascending order of numbers. Each vertex $\bar{k}$ of the 1st level is adjacent to the vertices of the 2nd level ("sons" of the vertex $\bar{k}$). Moreover, any vertex $\tilde{k}$ that is a "son" of a vertex $\bar{k}$ is taken from the set of level 1 vertices lying to the right of $\bar{k}$ (to avoid repetitions) and such that $\frac{1}{3}(k^{(1)} + \bar{k} + \tilde{k}) \notin riCoN_p$ (in this case, by virtue of statement 8, all obviously inappropriate cases are cut off), which we again arrange in the order of increasing numbers. We continue this process as long as possible. During this process, for each current vertex $\check{k}$ of the tree under construction, select the set of vertices included in the chain connecting $\check{k}$ with $k^{(1)}$ (it is the only one in the tree) and check for the set of vertices $N \subseteq \Psi(N_p)$, included into this chain, the existence of an integer solution in LPP (2), (8). In the case of the existence of a solution, we single out the next main quasi-homogeneous polynomial form $\tilde{N}$ by formula (9), where $k^{(0)}$ is any representative from N.

Let us now consider the case when it is required to select all main A-quasi-homogeneous forms for $A \in \mathbb{N}^n$. In this case, we modify algorithm 1. At step 1, we will now use $\Omega(N_p) \backslash riCo\{N_p \cup \{\bar{k}\}\}$, where $\bar{k} \in \mathbb{Z}^n_{\geq}$, $\bar{k}_i = \max\{k_i \mid k \in N_p\}$, $i = 1, 2, \ldots, n$, instead of $\Psi(N_p)$. Accordingly, at step 3, instead of LPP (2), (8), we use LPP $e_1 \to \min$ under constraints (5), and find $\tilde{N}$ by formula (6) (instead of (9)). In addition, we replace everywhere in the algorithm $riCoN_p$ on the $riCo\{N_p \cup \{\bar{k}\}\}$.

3 Method for finding the set of corner points $\Psi(Y)$

Consider the following problem. Let $Y \subset \mathbb{R}^n$ be a finite set. It is required to select $\Psi(Y)$—the set of corner points of the convex polytope CoY. We will use

Statement 9. Let $Y \subset \mathbb{R}^n$ be a finite set. For a point $v \in Y$ to have $v \in \Psi(Y)$ it is necessary and sufficient that

$$\exists e \in \mathbb{R}^n: \quad e \neq 0_{(n)}, \quad \forall y \in Y \backslash \{v\} \langle y - v, e \rangle < 0. \tag{10}$$

Corollary 2. Using the fact that a linear function attains its extremal values on a convex polytope at the corner points, in condition (10) we can replace Y with any finite set V such that $\Psi(Y) \subseteq V \subseteq Y$, and since $\Psi(Y) \subseteq P^*(Y)$, then in (10) one can, for example, change Y to $P^*(Y)$.

Corollary 3. Using the fact that if the inequalities in (10) hold for some $e \in \mathbb{R}^n$, these inequalities will also hold for λe, where $\lambda > 0$, condition (10) can be reformulated in terms of LPP. Suppose we are in the conditions of statement 9 and a finite set V is such that $\Psi(Y) \subseteq V \subseteq Y$ (for example, $V = P^*(Y)$). Then, for the point $v \in V$ to hold $v \in \Psi(Y)$, it is necessary and sufficient that for the LPP (11): $t(v) < 0$, where

$$u \to \min(= t(v)); \quad \langle y - v, e \rangle \leq u, \quad y \in V \backslash \{v\}, \quad -1 \leq e_i \leq 1, \quad i = 1, 2, \ldots, n. \tag{11}$$

The following simple technique allows significantly reduce the number of calls to the solution of the LPP (11) to find $\Psi(Y)$. Let $Y \subset \mathbb{R}^n$ be a finite set, $g_i(x)$, $i = 1, 2, \ldots, k$, be real functions defined on Y. Introduce the sets $Y_0 = Y$, $Y_i = \{x \in Y_{i-1} \mid g_i(x) = \min\{g_i(x') \mid x' \in Y_{i-1}\}\}$, $i = 1, 2, \ldots, k$. Let us denote $\text{lexArg}[Y, g_1, \ldots, g_k] = Y_k$.

Statement 10. Let $Y \subseteq \mathbb{R}^n$ be a finite set, $A \in \mathbb{R}^n$, $Y_A = \text{lexArg}[Y, \langle A, x \rangle, x_1, \ldots, x_n]$. Then Y_A consists of a single point $y^A \in \Psi(Y)$.

Remark 5. Statement 10 remains valid in the case $Y_A = \text{lexArg}[Y, \langle A, x \rangle, (-1)^{i_1} x_1, \ldots$ $\ldots, (-1)^{i_n} x_n]$, where $i_1, \ldots, i_n \in \{0, 1\}$.

Consider the point $y^\Sigma = (1 \backslash |Y|) \sum_{y \in Y} y \in \text{ri}CoY$ (instead of Y we can use $P^*(Y)$ or any set V, satisfying condition $\Psi(Y) \subseteq V \subseteq Y$).

Algorithm 2 (finding the set $\Psi(Y)$)

Step 1. Put $U = P^*(Y)$, $W = \emptyset$, $V = U \cup W$. At each stage of the algorithm we have

$$W \subseteq \Psi(Y) \subseteq V = U \cup W \subseteq P^*(Y). \tag{12}$$

Step 2. Consider any point $v \in U$. Let $A = v - y^\Sigma$ (obviously, $A \neq 0_{(n)}$). Define $\{y^A\} =$ $= Y_A = \text{lexArg}[Y, \langle A, x \rangle, x_1, \ldots, x_n]$ (see statement 10) and set $W := W \cup \{y^A\}$. If $y^A = v$, then assign $U := U \backslash \{v\}$ and go to step 3. If $y^A \neq v$, then solve the LPP (11) and assign $U := U \backslash \{v\}$. If $t(v) < 0$, then $v \in \Psi(Y)$ (see corollary 3), and assign $W := W \cup \{v\}$, otherwise $v \notin \Psi(Y)$. Finally, assign $V := U \cup W$ and go to step 3.

Step 3. If $U = \emptyset$, then the process ends and in this case $W = \Psi(Y)$ (according to (12)). Otherwise, go to step 2.

4 Results

Thus, three practical methods for finding all main quasi-homogeneous forms of a polynomial are described (1st is the easiest and 3rd is the most difficult). However, with a large number of terms in the polynomial, the 3rd one is most preferable. Let us show by examples the advantage of the 2nd method in comparison with the 1st, and also the 3rd in comparison with the 2nd.

Example 2. (1) Consider the polynomial $p(x, y, z) = 3xy + 5xz - yz + 7xyz + 4x^2yz-$ $-6xy^2z + 2xyz^2$, where $N_p = \{(1, 1, 0), (1, 0, 1), (0, 1, 1), (1, 1, 1), (2, 1, 1), (1, 2, 1), (1, 1, 2)\}$.

Let us consider the case when it is required to select all main A-quasi-homogeneous forms of $p(x, y, z)$ for $A \in \mathbb{N}^n$. Then, in the case of applying the method 1, one would have to enumerate $2^{|N_p|} - 1 = 2^7 - 1 = 127$ possible cases, in each of which the corresponding LPP would be solved. In the case of using method 2, $\Omega(N_p) = \{(1, 1, 0), (1, 0, 1), (0, 1, 1)\}$, and therefore it suffices to consider $2^{|\Omega(N_p)|} - 1 = 2^3 - 1 = 7$ cases.

(2) Consider the polynomial $p(x, y) = 3y^{10} + 5x^2y^9 - x^3y^8 + 7x^4y^7 + 4x^5y^5 - 6x^6y^3 + 2x^7$, where $N_p = \{(0, 10), (2, 9), (3, 8), (4, 7), (5, 5), (6, 3), (7, 0)\}$. Let us consider the case when it is required to select all main A-quasi-homogeneous forms of $p(x, y)$ for $A \in \mathbb{N}^n$. Then, in the case of applying the method 2, $\Omega(N_p) = N_p$, and therefore one would have toenumerate $2^{|\Omega(N_p)|} - 1 = 2^7 - 1 = 127$ possible cases. In the case of using method 3, we have: $\bar{k} = (7, 10)$, $\Omega(N_p)\backslash riCo\{N_p \cup \{\bar{k}\}\} = \{(0, 10), (7, 0)\}$, and therefore consideration of no more than 3 cases will be required, in each of which we obtain one of the appropriate quasi-homogeneous forms: $3y^{10}$ for $A = (2, 1)$, $2x^7$ for $A = (1, 4)$ and $3y^{10} + 2x^7$ for $A = (10, 7)$.

5 Conclusion

We consider the problem of identifying the main quasi-homogeneous forms of a polynomial (quasi-homogeneous form of a polynomial is a generalization of the concept of its homogeneous form), which are the sums of the terms of the polynomial belonging to some face of the Newton polytope of this polynomial.

Finding these forms is necessary for solving many practical problems, in particular, for construction of necessary and sufficient conditions of an extremum for polynomials and power series [1, 2], as well as checking matrices for D-stability arising in the study of ecosystem stability [6, 7].

For various practical applications, it is sometimes required to select all the main quasi-homogeneous forms. Meanwhile, when solving some problems, it is enough to use some part of them (for example, "southwest" forms—in the case of two variables). In the latter case, it is possible to consider not only polynomials, but also power series (i.e., analytic functions). In accordance with this, two variants of methods (algorithms) are considered.

On the basis of the proposed methods, practically realizable algorithms are described that use a significant reduction in the number of options to be enumerated, in comparison with the methods of simple enumeration.

A practically realizable rather economical algorithm for solving an auxiliary problem is described—finding the set of corner points of the polytope CoY, where Y is a finite set from $\mathbb{R}^n$.

References

[1] V.N. Nefedov, *Estimation of the error in convex polynomial optimization problems*, Comput. Math. and Math. Phys., **30(1)** (1990)

[2] V.N. Nefedov, *Necessary and sufficient conditions for an extremum in analytical optimization problems*, Electronic journal "Trudy MAI", **33** (2009)

[3] S.G. Gindikin, *Energy estimates associated with Newton's polyhedron*, Trudy Moskov. mat. soc., **101** (1974)

[4] A.D. Bruno, *Power geometry in algebraic and differential equations* (Nauka, Fizmatlit, Moscow, 1998)

[5] L.R. Volevich, S.G. Gindikin, *Newton polytope method in the theory of partial differential equations* (URSS, Moscow, 2002)

[6] A.G. Khovanskiy, *Polyhedra and algebra*, Proceedings of ISA RAS, **38** (2008)

[7] G.V. Kanovei, D.O. Logofet, *Relations, properties and invariant transformations of D- and aD-stable matrices*, Vestn. Moscow Un-ty. Ser. 1. Mat., Mekh., **6** (2001)
[8] D.O. Logofet, *Stronger-than-Lyapunov notions of matrix stability, or how "flowers" help solve problems in mathematical ecology*, Linear Algebra Appl., **398** (2005)
[9] V.N. Nefedov, *On one method for studying a polynomial for definiteness in positive orthant*, Electronic journal "Trudy MAI", **22** (2006)
[10] E.G. Belousov, *Introduction to convex analysis and integer programming* (Publishing house of Moscow University, Moscow, 1977)
[11] I.I. Eremin, N.N. Astafiev, *Introduction to the theory of linear and convex programming* (Nauka, Moscow, 1976)
[12] A. Brondsted, *An introduction to convex polytopes* (Springer-Verlag, New York, 1983)

Calculation of fatigue fracture in structural elements using a multimode damage model

Ilia Nikitin[1], *Alexander* Nikitin[1,*], and *Chong* Wang[2]

[1]Institute for Computer Aided Design of Russian Academy of Sciences, 2nd Brestskaya st, Moscow, 123056, Russia
[2]Failure Mechanics and Engineering Disaster Prevention and Mitigation Key Laboratory of Sichuan Province, Sichuan University, Chengdu, 610207, China

Abstract. The problem of nucleation and development for the fatigue fracture process in element of aircraft structure is solved using previously developed cyclic damage model. The model contains the evolution equation for the damage function depending on stress state and amplitudes of the cyclic loads. On this basis, calculations of the initiation and development of crack-like fracture zones are carried out for an aircraft structure element of a disk with a blade in a flight loading cycle (HCF mode). The damage function was calculated based on the elastic solution for the element of structure loaded by centrifugal forces in disk and blade and aerodynamical pressure on blade. An alternative cyclic loading regime associated with high-frequency vibrations of the blades was investigated. Fatigue fracture in this process corresponds to the VHCF mode. Simulation of the combined cyclic loading mode was also carried out. In all these modes, the real time of the nucleation and development of quasi-cracks was estimated up to reaching the observed disk surface.

1 Introductions

The developing of a new actual models and effective numerical methods for prediction the fatigue failure of structural elements is very important problem for different industries. In the case of aircraft industry, it was reported that more than 90% in service fractures are due to fatigue. It is important to note that the stress state of the structural elements are often combines different loading modes and regimes. In the case of engine applications, it is usually considering the low amplitude high frequency vibrations and high amplitude low frequency loadings or its combination. The high frequency vibrations are due to instability in air flow, not-perfect coaxially of shafts, design errors and other reasons. The low frequency loadings are usually associated with centrifugal force, changes in engine operating regime. For a long time, it was assumed that low amplitude vibrations are not significantly effects on fatigue life of the element if its natural frequency far from the resonant frequency of structure. However, modern advances in fatigue science have shown the destructive potential of such loadings in the very high cycle range. It was reported that material degradation and fatigue failure in structural materials can be observed well below the classical fatigue limit. In the recent years a series of in-service accidents in aircraft industry due to high frequency vibrations have

*e-mail: nikitin_alex@bk.ru

shown the importance of the problem and the limitation of used design protocols. To improve the situation and increase a safety of fatigue designs it is necessary to develop models that are considering modern achievements in material science and fatigue.

Most of the actual models of fatigue life prediction are based on hypothesis of static stress field during the loading process. But the real material behavior and in service conditions are different. During the cyclic loading the mechanical properties of material are permanently degrading. Depending on the load amplitude this degradation can be volumetric or localized. Nonetheless, this degradation is affecting the stress state and can lead to premature crack initiation compared to predictions of model that are based on static stress state hypothesis. The developing of progressive fatigue damage accumulation theory and corresponding models is very perspective scientific field with good industrial application perspectives. This paper is focused on developing of a such model with progressive degradation of material and its application to real structures. The example of spatial damage zone calculation in the joint area of turbojet engine disk and blade under cyclic loading are presented. The lifetime of structure was estimated with taken considering as high amplitude low frequency loadings, as well low amplitude high frequency vibrations. The outstanding results of numerical simulations shows a good agreement with the in-service fatigue fractures.

2 Multimode two-criterion model of cyclic damage

Let us formulate a multimode two-criterion model of cyclic damage, first proposed, and developed in [1–3]. A fatigue fracture criterion corresponding to the left branch of the bimodal fatigue curve in classical LCF–HCF mode has the following form:

$$\sigma_{eq} = \sigma_u + \sigma_L N^{-\beta_L}, \quad \sigma_L = 10^{3\beta_L}(\sigma_B - \sigma_u). \tag{1}$$

In these formulas σ_B is the ultimate tensile strength of the material, σ_u is the classic fatigue limit of the material during a fully reverse cycle (stress ratio $R = -1$, β_L is power index of the left branch of the bimodal fatigue curve.

The fatigue fracture criterion corresponding to the right branch of the bimodal fatigue curve (VHCF mode) is assumed in the following form:

$$\sigma_{eq} = \tilde{\sigma}_u + \sigma_V N^{-\beta_V}, \quad \sigma_V = 10^{8\beta_V}(\sigma_u - \tilde{\sigma}_u). \tag{2}$$

Here $\tilde{\sigma}_u$—fatigue limit of the material under symmetric loading cycle in the VHCF regime, β_V is power index of the right branch of the bimodal fatigue curve.

Expressions for equivalent stresses σ_{eq} are determined by the microdamage mechanism and are associated with two well-known criteria of multiaxial fatigue fracture [4–6]. The stress-based Smith–Watson–Topper (SWT) criterion is associated with the development of microcracks of normal opening [4, 5]. The Carpinteri–Spagnoli–Vantadori (CSV) criterion is associated with the development of shear microcracks [6].

The process of gradual cyclic material failure is described by a damage function $0 \le \psi(N) \le 1$ [6, 7]. When $\psi = 1$ a material particle (physically small volume) is completely destroyed. Its Lamé modules become equal to zero. The damage function ψ is described by the kinetic equation:

$$\frac{\partial \psi}{\partial N} = \frac{B\psi^{\gamma}}{(1 - \psi^{1-\gamma})}, \tag{3}$$

where $0 < \gamma < 1$ is the model parameter that determines the rate of fatigue damage development.

The coefficient B is determined by the procedure that is clearly associated with the selected criterion for multiaxial fatigue failure of one type or another. The expressions for the coefficient B in various modes were obtained in [1, 2].

For the LCF–HCF at $\sigma_u + \Delta\sigma_u < \sigma_{eq} < \sigma_B$, $\Delta\sigma = 10^{-5\beta}(\sigma_B - \sigma_u)$

$$B = B_L = 10^{-3}[\langle\sigma_{eq} - \sigma_u\rangle/(\sigma_B - \sigma_u)]^{1/\beta_L}/(1-\gamma)/2 \tag{4}$$

and for VHCF at $\tilde{\sigma}_u < \sigma_{eq} \leq \sigma_u + \Delta\sigma_u$

$$B = B_V = 10^{-8}[\langle\sigma_{eq} - \tilde{\sigma}_u\rangle/(\sigma_B - \tilde{\sigma}_u)]^{1/\beta_V}/(1-\gamma)/2. \tag{5}$$

When $\sigma_{eq} < \tilde{\sigma}_u$ the fatigue failure does not occur, when $\sigma_{eq} > \sigma_B$ the failure comes instantly.

The degradation of a material is described by a decrease in its elastic Lamé moduli λ and μ to zero with an increase of damage ψ to the value $\psi = 1$. The following nonlinear law of decreasing Lamé moduli is adopted:

- for $\psi \leq \psi_*$, $\lambda(\psi) = \lambda_0(1 - k\psi)$, $\mu(\psi) = \mu_0(1 - k\psi)$;
- for $\psi_* \leq \psi \leq 1$, $\lambda = 0, \mu = 0$.

The parameter $0 < k < 1/2$ characterizes a relatively weak change in the moduli until the critical level $\psi_* \lesssim 1$ of damage is reached, after which a state of complete fracture occurs with the loss of the ability of the elastic resistance of the material to any loads.

Let us describe the selected criteria for two mechanisms of fatigue failure of a material under cyclic loading.

The Smith–Watson–Topper criterion of multiaxial fatigue failure in the LCF–HCF mode with the development of normal opening micro-cracks (stress-based SWT) [4, 5] is corresponding to the left branch of the bimodal fatigue curve and has the following form:

$$\sqrt{\langle\sigma_{1\,\max}\rangle\Delta\sigma_1/2} = \sigma_u + \sigma_L N^{-\beta_L}, \tag{6}$$

where σ_1 is the largest principal stress, $\Delta\sigma_1$ is the spread of the largest principal stress per cycle, $\Delta\sigma_1/2$ is its amplitude. According to the chosen criterion only tensile stresses lead to failure, so it has the value $\langle\sigma_{1\,\max}\rangle = \sigma_{1\,\max}$. We assume that the right branch of SN-curve for the VHCF regime is also describing by the similar equation (2).

We suppose that the equivalent stress has the following expression for both LCF–HCF and VHCF modes:

$$\sigma_{eq} = \sigma^n = \sqrt{\langle\sigma_{1\,\max}\rangle\Delta\sigma_1/2}. \tag{7}$$

The Carpinteri–Spagnoli–Vantadori criterion of multiaxial fatigue failure in the LCF–HCF mode, including the concept of a critical plane (stress-based CSV) [6], corresponding to the left branch of the bimodal fatigue curve, in a simplified formulation has the form:

$$\sqrt{(\langle\Delta\sigma_n\rangle/2)^2 + 3(\Delta\tau_n/2)^2} = \sigma_u + \sigma_L N^{-\beta_L}, \tag{8}$$

where $\Delta\tau_n/2$ is the amplitude of the tangential stress on the plane (critical), where it reaches its maximum value, $\Delta\sigma_n/2$ is the amplitude of the normal (tensile) stress on the critical plane, $\langle\Delta\sigma_n\rangle = \Delta\sigma_n H(\sigma_{n_{\max}})$. This criterion includes the mechanism of fatigue fracture with the formation of shear micro-cracks.

We also suppose that the equivalent stress has the following expression for both LCF–HCF and VHCF modes:

$$\sigma_{eq} = \sigma^\tau = \sqrt{(\langle\Delta\sigma_n\rangle/2)^2 + 3(\Delta\tau_n/2)^2}. \tag{9}$$

3 Numerical algorithm for fatigue damage development calculation

To integrate the equation $\partial\psi/\partial N = B\psi^\gamma/(1 - \psi^{1-\gamma})$, the damage function approximation was applied at the k-node of the computational grid for given discrete values ψ_k^t at moments N^t and sought ψ_k^{t+1} at moments N^{t+1}.

After integrating the damage equation, the new value of damage function depends on the increment of the number of cycles ΔN^t as in [2]:

$$\psi_k^{t+1} = \left(1 - \sqrt{[1-(\psi_k^t)^{1-\gamma}]^2 - 2(1-\gamma)B\Delta N^t}\right)^{1/(1-\gamma)}, \tag{10}$$

$$\Delta N^t = \min_k \frac{1}{2}\Delta\tilde{N}_k^t, \tag{11}$$

$$\Delta\tilde{N}_k^t = \left[\psi^{1-\gamma}/(1-\gamma) - \psi^{2(1-\gamma)}/2/(1-\gamma)\right]\Big|_{\psi_k^t}^{1} \Big/ B. \tag{12}$$

All elements are sorted out, for each of them the most damaged node is searched and according to its damage the mechanical properties of the element are adjusted for $\psi_k^t < \psi_*$:

$$\lambda(\psi_k^t) = \lambda_0(1 - k\psi_k^t), \quad \mu(\psi_k^t) = \mu_0(1 - k\psi_k^t). \tag{13}$$

Those elements that belong to nodes with damage $\psi_* \leq \psi_k^t \leq 1$ are removed from the calculation area and form a localized zones (quasi-cracks) of completely destroyed material.

4 Calculation results for real structural element

Mathematical modelling of the experiments [8–12] using specimens of various shapes, subjected to cyclic loading in modes HCF and VHCF showed good qualitative and quantitative agreement with the geometry of real cracks and the values of durability obtained during tests.

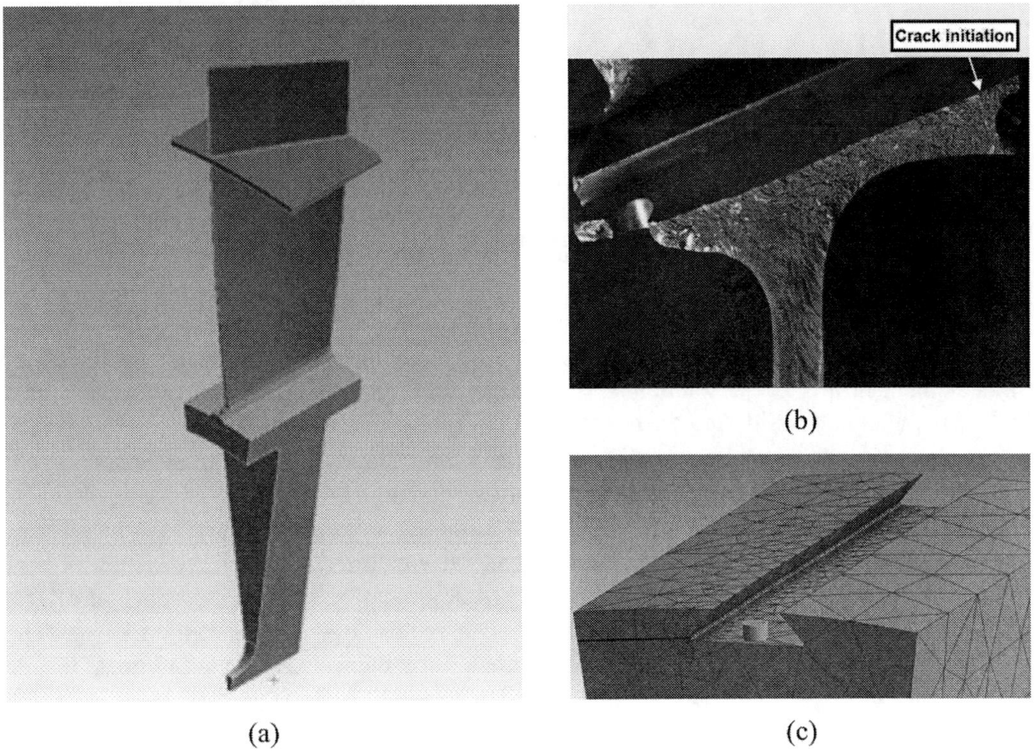

Figure 1. General view of the disk sector with blade (a), fracted disk (b) and the finite element mesh of the disk (c)

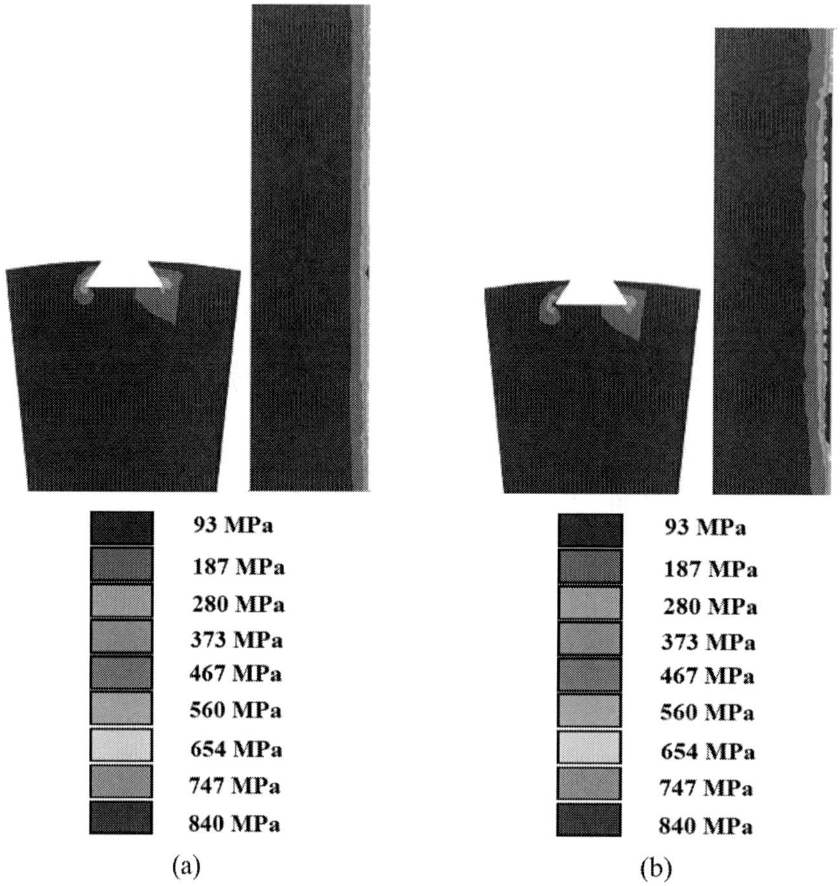

Figure 2. View in cross-section and longitudinal section at the quasi-crack initiation, $N = 5.2 \cdot 10^4$ (a); view in cross-section and longitudinal section with further growth of the quasi-crack, $N = 2.5 \cdot 10^5$ (b)

Here numerical experiments are carried out to calculate the nucleation and development of quasi-cracks in the constituent elements of aircraft structures. The turbojet engine (TJE) compressor disk with blades is loaded in the flight loading regime (HCF mode) and by high frequency vibrations of blades (VHCF mode). Previously the stress-strain state of the structural element in this cycle loadings and the time of its primary damage were determined in [13, 14] without considering the development of quasi-cracks before macro fracture.

The Ansys software was used to solve the quasi static and dynamic solid mechanics problems supplemented by a code to calculate the damage equation and changes of elastic modulus.

All numerical experiments were conducted on an element of structure such a sector of the disk with blade. On figure 1(a) shows a general view of the model, consisting of disk sector and one blade. The fracture surface of turbojet engine is presented on figure 1(b). The finite element mesh of the corresponding disk sector is presented in figure 1(c). The material used for calculation of the disk is titanium alloy with the following ultimate tensile strength and fatigue parameters $\sigma_B = 1135$ MPa, $\sigma_u = 330$ MPa, $\tilde{\sigma}_u = 330$ MPa, $\beta_L = 0.3$, $\beta_V = 0.25$. Elastic moduli are $\lambda_0 = 77$ GPa, $\mu_0 = 44$ GPa. Parameters of damage equation are $k = 0.5$, $\gamma = 0.5$, $\psi_* = 0.98$.

In the first series of numerical experiments, calculations of the flight loading cycle were carried out (stress ratio is $R = 0$) The damage function was calculated based on the elastic

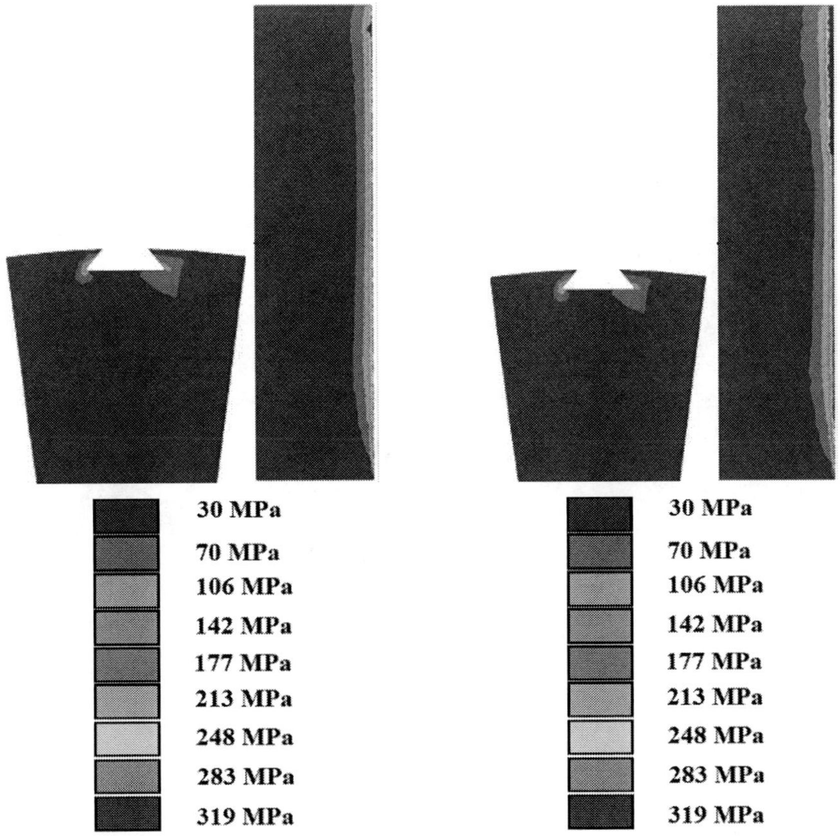

(a) (b)

Figure 3. View in cross-section and longitudinal section at the moment of quasi-crack nucleation, $N = 7.3 \cdot 10^{10}$ (a); view in cross-section and longitudinal section with further growth of the quasi-crack, $N = 1.5 \cdot 10^{11}$ (b)

solution for the element of structure loaded by centrifugal forces in disk and blade and aerodynamical pressure on blade on the hypothesis of an isolated profile. The values of the external load were chosen in the range that allows us to study crack nucleation and development for high-cycle fatigue. The angular speed of rotation of the disk has a value of$\omega = 425$ rad/s, flow speed $V = 200$ m/s.

The estimation of the distributed aerodynamic pressure on the blade, leading to its bending and additional stresses in the contact zone with the disk, was carried out by the method [15]. The contact interaction of the disk and the blade assumes the possibility of normal separation of the contact surfaces along the normal and shear with friction, the coefficient of friction $q = 0.1$.

In the second series of numerical experiments the calculation of the observed high-frequency bending-torsional vibrations of the blade caused by axial displacements of the ends of the blade shroud ring with an amplitude of ~ 0.5 mm and frequency 1000 Hz (reverse cycle, stress asymmetry ratio is $R = -1$ VHCF mode) are presented. On figure 3(a) shows the levels of equivalent stresses in cross-section and longitudinal section one for quasi-crack nucleation, number of cycles $N = 7.3{\cdot}10^{10}$, on figure 3(b) shows the levels of equivalent stresses in cross-section and longitudinal section with further growth of the quasi-crack, $N = 1.5{\cdot}10^{11}$.

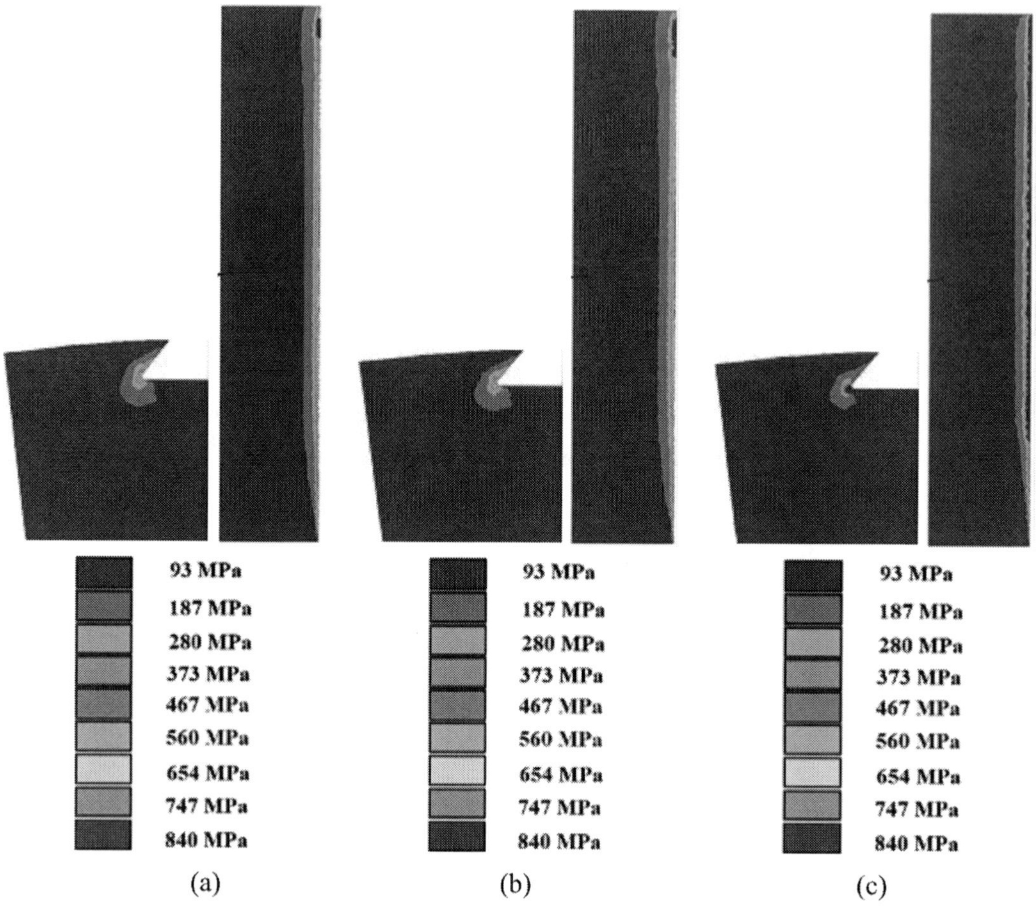

Figure 4. View in cross-section and longitudinal section (nucleation), VHCF mode, $N = 7.3 \cdot 10^{10}$ (a); view in these sections with growth of the quasi-crack, HCF mode, $N = 9.3 \cdot 10^{3}$ after switching of the modes (b); development of a quasi-crack to the lateral surface of the disk, HCF mode, $N = 4.2 \cdot 10^{4}$ cycles after switching (c)

The nucleation zone is located at the far zone of the "dovetail", arises at the number of cycles $N = 7.3 \cdot 10^{10}$ (dark spot in rectangle in figure 3(a). The quasi-crack grows, reaches the end of the disk, and becomes observable from the outside at $N = 1.5 \cdot 10^{11}$ (figure 3(b)).

In both loading modes, the growth of a quasi-crack from its nucleation to reaching the observed surface takes a significant number of cycles, which can be estimated using the proposed damage model.

In the third series of numerical experiments the combined cyclic loading of the disk is considered. At the beginning of the cyclic process, a quasi-crack nucleates according to the VHCF mode from the oscillations of the blades. Then, because of the appearance of a concentrator, it develops in flight loading cycles according to the HCF mode. Such a sequence of damage development in a real disk during exploitation is described in [16].

In figure 4(a) shows the levels of equivalent stresses in cross-section and longitudinal section at the moment of quasi-crack nucleation in VHCF mode, number of cycles $N = 7.3 \cdot 10^{10}$, in figure 4(b) shows the levels of equivalent stresses in cross-section and longitudinal section with further growth of the quasi-crack in HCF mode, $N = 9.3 \cdot 10^{3}$ after mode switching. In figure 4(c) shows the levels of equivalent stresses in cross-section and longitudinal section at moment of a quasi-crack reaching the lateral surface of the disk $N = 4.2 \cdot 10^4$ after mode switching.

It is possible to assess the real time of the nucleation and development of a quasi-crack in each of the modes.

Assuming that the average flight cycle is approximately 3 hours, in the HCF mode this time is $t_1 = 5.2 \cdot 10^4 \cdot 3\ \text{h} \sim 1.6 \cdot 10^5$ h. Time to reach the surface is $t_2 = 2.5 \cdot 10^5 \cdot 3\ \text{h} \sim 7.5 \cdot 10^5$ h. In the VHCF mode at an oscillation frequency of 1000 Hz, the time for crack initiation is $t_3 = 7.3 \cdot 10^{10} \cdot 0.001\ \text{s} \sim 2 \cdot 10^4$ h. Thus, these estimates show that the microcrack initiation is more probable in the VHCF mode. However, the minimum time before the emergence of a quasi-crack on the surface is achieved with the combined mechanism and mode switching after the initiation of the VHCF damage. This time is $t_4 = 2 \cdot 10^4\ \text{h} + 4.2 \cdot 10^4 \cdot 3\ \text{h} \sim 1.5 \cdot 10^5 < t_1$. These estimates confirm the high probability of the combined mechanism realization for fatigue fracture of the disk.

5 Conclusion

The problem of nucleation and development for the fatigue fracture process in element of aircraft structure is solved using previously developed cyclic damage model. The model contains the evolution equation for the damage function depending on stress state and amplitudes of the cyclic loads. The calculation algorithm is based on a special post proceeding of a numerical solution for solid mechanics problem. The normal crack opening and shear crack mechanisms are taken into account in this model.

On this basis, calculations of the initiation and development of crack-like fracture zones are carried out for an aircraft structure element of a disk with a blade in a flight loading cycle (HCF mode). The damage function was calculated based on the elastic solution for the element of structure loaded by centrifugal forces in disk and blade and aerodynamical pressure on blade.

An alternative cyclic loading regime associated with high-frequency vibrations of the blades was investigated. Fatigue fracture in this process corresponds to the VHCF mode. Simulation of the combined cyclic loading mode was also carried out. In all these modes, the real time of the nucleation and development of quasi-cracks was estimated up to reaching the observed disk surface. The geometry and shape of quasi-cracks in the vicinity of the contact zone of the disk and the blade are similar to those observed in real operation.

This study is supported by the grant of the Russian Science Foundation No. 19-19-00705.

References

[1] I.S. Nikitin, N.G. Burago, A.B. Zhuravlev, A.D. Nikitin, Mechanics of Solids, **55(8)**, 298 (2020)
[2] I.S. Nikitin, N.G. Burago, A.D. Nikitin, B.A. Stratula, IOP Conf. Ser.: Mater. Sci. Eng., **927**, 012019 (2020)
[3] I.S. Nikitin, N.G. Burago, A.D. Nikitin, B.A. Stratula, Procedia Structural Integrity, **28**, 2032 (2020)
[4] R.N. Smith, P. Watson, T.H. Topper, J. Mater., **5**, 767 (1970)
[5] N. Gates, A. Fatemi, Int. J. Fatigue, **91**, 337 (2016)
[6] A. Carpinteri, A. Spagnoli, S. Vantadori, Int. J. of Fatigue, **33**, 969 (2011)
[7] J. Lemaitre, J.L. Chaboche, *Mechanics of solid materials* (Cambridge University Press, 1994)
[8] A.K. Marmi, A.M. Habraken, L. Duchene, Int. J. Fatigue, **31**, 2031 (2009)

[9] A. Nikitin, T. Palin-Luc, A. Shanyavskiy, C. Bathias, Eng. Fract. Mech., **167**, 1125 (2016)
[10] H.Q. Xue, H. Tao, F. Montembault, Q.Y. Wang, C. Bathias, Int. J. Fatigue, **29**, 2085 (2007)
[11] I.S. Nikitin, A.D. Nikitin, B.A. Stratula, J. Phys.: Conf. Ser., **1945**, 012043 (2021)
[12] I.S. Nikitin, N.G. Burago, A.D. Nikitin, B.A. Stratula B., J. Phys.: Conf. Ser., **1945**, 012042 (2021)
[13] N.G. Bourago, A.B. Zhuravlev, I.S. Nikitin, Mechanics of Solids, **46**, 828 (2011)
[14] N.G. Burago, A.B. Zhuravlev, I.S. Nikitin, V.L. Yakushev, Mathematical Models and Computer Simulations, **8(5)**, 523 (2016)
[15] N.G. Burago, I.S. Nikitin, P.A. Yushkovskiy, Mechanics of Solids, **50**, 546 (2015)
[16] A.A. Shanyavskiy, *Modeling of Metal Fatigue fracture* (Monografia, Ufa, 2007)

Explosive technologies for testing composite constructions of aircraft for strength to heat-power non-stationary loading

Afanasy Ostrik[1,*], *Dmitry* Nikolaev[1], *Vladimir* Bakulin[2], and *Irina* Bugay[3]

[1]Institute of Problems of Chemical Physics of the Russian Academy of Sciences, 1, Academician Semenov prospekt, Chernogolovka, Moscow Region, 142432, Russia
[2]Institute of Applied Mechanics of Russian Academy of Sciences, 7, Leningradsky prospekt, Moscow, 125040, Russia
[3]Leonov Moscow Region University of Technology, 42, Gagarina st., Korolev, Moscow Region, 141070, Russia

Abstract. Sets of explosive and heating devices for experimental study of strength of composite constructions of aircraft against complex thermal and mechanical action of radiation and particles fluxes of different physical nature are considered. Two new explosive devices for generating low-pulse loads of microsecond duration are proposed. Devices based on thermoelectric heating elements, conductive plates and pyrotechnic compositions applied to the surface of the tested structure have been developed to simulate thermal action of radiation and particles fluxes. A universal test bench using these sets of devices has been developed. This test bench allows testing the strength of composite constructions to the joint action of thermal and mechanical loads having given space-time distributions.

1 Introduction

The intensive development of sources of radiations and particles having various physical nature requires the design of flight vehicle (FV) operating under the conditions of high energy density fluxes. Constructive means of protection of FV based on heterogeneous porous materials and new generation of protecting coatings are developed no less intensively. The use of modern protective heterogeneous packages for FV constructions requires experimental confirmation of their effectiveness.

An experimental study of the aftereffects of the thermal and mechanical actions of radiation and particle flows (RPF) on constructive FV elements by direct irradiation of them is not possible because there are no powerful laboratory sources of this flows capable of generating the necessary energy densities on surfaces with dimensions of the order of several meters. Therefore, at the present stage, the main method of studying the RPF effects is testing of full-scale VF constructions for non-stationary loading by explosive and heating devices to simulate thermal and mechanical loads modeling RPF actions [1, 2].

The development of a set of such devices is relevant and of great practical importance [2–5]. However, there is still no set of devices that fully satisfy the requirements of practice. In particular, low pulse submicrosecond load generators are required to provide si-

*e-mail: ostrik@ficp.ac.ru

multaneous pressure loading the surface of the tested structure. In addition, there are only few devices for modeling the joint thermal and mechanical RPF action. But there are many cases realized during flight, when the FV is subjected to joint multiple thermal and mechanical RPF action, in which its complex character is impotent [5, 6]. The essence of the complexity [6] is that the result of a joint (it's not necessarily simultaneous) action (temperature increase, thermal ablation of material, accumulation of plastic deformations and microdamage, cracking and stratification) is not determined by simply summation of the result of each factor and the principle of superposition of damage is not true. The mutual influence of heating and mechanical destruction is one of the cases of realization of complex RPF action.

An approach based on the joint application of heating devices and explosive generation of mechanical loads is used for strength tests for complex thermal and mechanical RPF action. In turn, the joint use of devices requires the development of universal test benches. The universality of the test bench is understood as: dual purpose (suitable for verification of individual devices and for testing); multifunctionality (applicable for different combinations of RPF actions); wide region of use (in addition to RPF actions, it can be used for any transient effects of heat and power loads that have a complex space-time profiles). Versatility is achieved by modularity of individual modeling devices and their interchangeability in the test bench.

The development of a new set of RPF modeling devices and a universal bench for testing FV constructions for strength to the complex RPF action is the main goal of this work.

2 Thermal RPF action

Gas fluxes (for example, aerodynamic flows or convective and radiant flows from jet flowing from combustion chambers of reactive engines) are an operational sources of thermal action on FV constructions [7]. As a rule, flight heat flows are close to quasi-stationary, and a slowly changing temperature profile is established in the constructions. In addition, aircraft can be subjected to impulse actions of various types of RPF [6].

Some practically important (these are important for the case of thermal action) types of RPF actions and their corresponding energy absorption regimes [6] are given in table 1. It should be noted that, of course, not all possible combinations of the considered types of RPF actions are realized in practice for FV. However, a number of practically important combinations are known when the complex RPF action takes place. Influences of volumetric absorbed radiation (VAR) is the most dangerous for FV construction [2, 5, 6].

The complex action of the VAR, having a hard and soft parts in its spectrum, turns out to be the most noticeable: heating with hard radiation changes the thermodynamic and mechanical characteristics of the material, which affects the formation of a thermomechanical impulse, and the soft part leads to ablation of the material (thermal action) and the formation of the evaporative part of the mechanical impulse. Therefore, even the conditional separation of thermal and mechanical RPF actions is of little use in this case.

Thermal action leads to different temperature profiles depending on RPF parameters and properties of composite materials (CM) (in particular, it depends on free path length of particles in the absorbing material and its temperature conductivity), as well as the thickness of the construction.

At least, when studying the complex thermal and mechanical action, it becomes necessary to create the following set of its thermal states in the thin-walled FV construction before generating a mechanical impulse:

- near-surface heating when the material is warmed to a certain thickness and the rear side of the structure remains substantially cold;

Table 1. RPF influence and corresponding absorption regime

Type of RPF	Spatial regime	Temporary regime
radio radiation of high frequency	volume absorption	gradual absorption
monochromatic radiation of optical range	superficial absorption	gradual absorption
light radiation of a powerful explosion	superficial absorption	gradual absorption
soft X-rays of a powerful explosion or laser fusion reactor	superficial absorption	instantaneous absorption
hard X-rays of a powerful explosion	volume absorption	instantaneous absorption
γ-radiation of powerful explosion or laser fusion reactor	volume absorption	instantaneous absorption
electrons of radiation belts	volume absorption	gradual absorption
neutrons of a powerful explosion or laser fusion reactor	volume absorption	instantaneous absorption
nuclear reactor neutrons	volume absorption	gradual absorption
heavy accelerator ions	volume absorption	instantaneous absorption

- volumetric non-uniform heating when the CM is warmed over the entire thickness of the structure, but with a significant difference in temperatures of the irradiated and rear surfaces;
- uniform heating when the CM temperature over the thickness of the construction is constant.

Near-surface heating is characteristic when CM is irradiated to monochromatic radiation, light pulse of powerful explosion or aerodynamic flows in developed evaporation regime [7, 8]. In this case, the temperature near the ablating (evaporating or sublimating) surface is 1000–2000°C.

Volumetric non-uniform heating occurs when irradiated VAR has photon energy of more than 20 keV and low levels of surface density of energy (mechanical action prevails at high levels of energy, that accompanied by CM destruction). Non-uniform heating can be caused by influence of monochromatic radiation, light impulse, gamma-neutron radiation and aerodynamic flows in thermal conductivity regime. In this case, the temperature varies in the range 100–1000°C. Non-uniform heating is also realized for nozzles made of carbon-carbon CM of operating rocket engines [9]. The temperatures of the nozzle materials are 1000–3000°C.

3 Mechanical RPF action

An approach based on the joint application of heating devices and explosive generation of mechanical loads is used for strength tests for complex thermal and mechanical RPF action. Explosive technologies for reproducing the mechanical RPF action are quite developed and are widely used for testing of construction from CM [1–5, 10]. Therefore, we consider only two relatively new explosive devices from the used set.

As already noted, it is necessary to develop low-pulse devices for modeling the mechanical RPF action. Tape charge [11] and guided initiation charge [12] are the most promising of explosive devices for generating low-pulse loads of microsecond duration.

The tape charge is made of explosive (EX) tapes placed on round tubes (see figure 1). The tubes are equally spaced from the surface of the object to be loaded. The multi-point beam

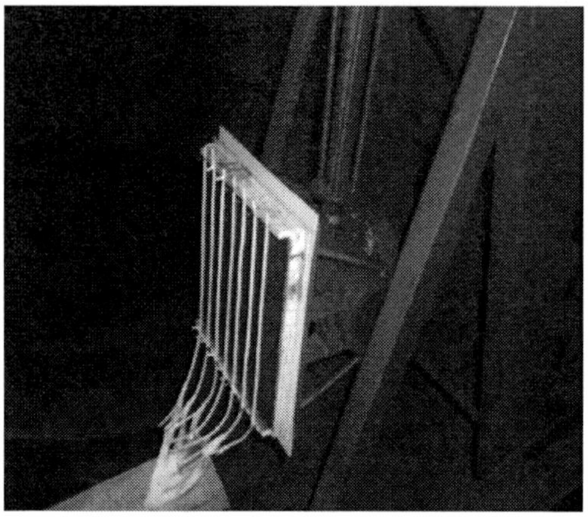

Figure 1. Tape explosive charge

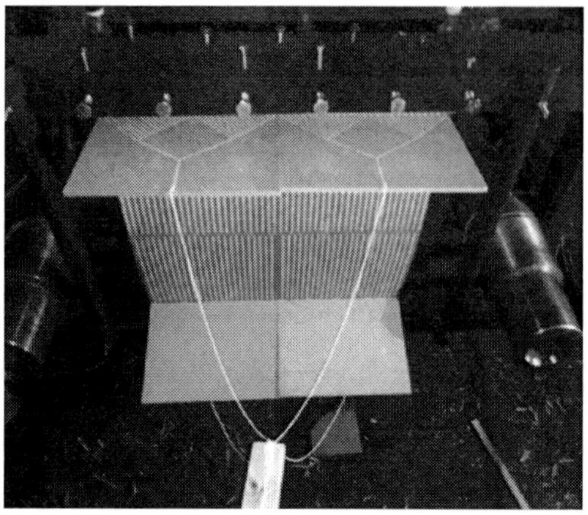

Figure 2. Controlled initiation explosive charge

initiation system from the end of the EX tapes is used to provide the simultaneous applying of the load.

The charge having controlled initiation is made in the form of a thin (≤ 3 mm) shell of hard cardboard with milled channels filled with plastic EX (see figure 2). Placement of EX in channels gives a number of advantages compared to other methods of localizing EX (for example, a continuous layer or sectors on a porous substrate). Controlled initiation is implemented by method adopted from technology of generation of megabar level pressures [13].

The set [2], completed with these new devices, is more capable and is used for testing thin-walled FV constructions for strength to RPF of different physical nature.

4 Complex RPF action on flat packages from CM

Several methods of creating the thermal state of the barrier (using a contact conductive plate, sheet pyrotechnic charges, microwave radiation, etc.) in combination with various types of explosive generators of mechanical loads are used in modeling the complex action on elements of the aircraft structure. The choice of the set of modeling devices depends on the types and combinations of the acting RPF and the characteristics of the test objects.

The temperature distribution in the thin-walled composite plate must be measured to verify the heating devices reproducing the thermal RPF action. Contactless method of temperature profile measurement in structural materials [14] is used. Method consists in formation of channels of different depth in investigated sample. Temperature scanning is carried out through prepared channels by a system of remote infrared pyrometers. Temperature profile in sample is determined by measuring pyrometric information. In this work, temperature measurement is carried out by a non-contact portable pyrometer Raytek Raynger MX.

The non-uniform temperature profile in the flat sample is created by surface heating with the help of a contact plate with current, which varies according to a given law to provide the required temperature distribution in the near-surface layer of the plate. The test scheme is shown in figure 3.

Reproduction of preset flux $q(t)$ of absorbed RPF energy is provided by control of current change in electric circuit. Preliminary estimation of dependence $I = I(t)$ can be made by relation

$$I(t) = \sqrt{\frac{q(t) \cdot S}{\eta R}},$$

 https://doi.org/10.1051/matecconf/202236201019

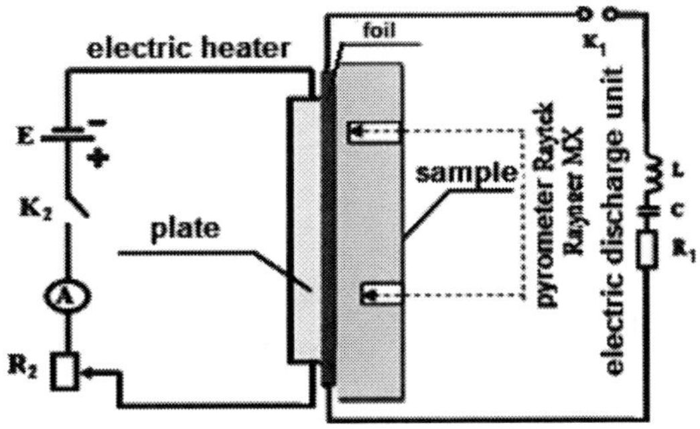

Figure 3. Scheme of testing plate

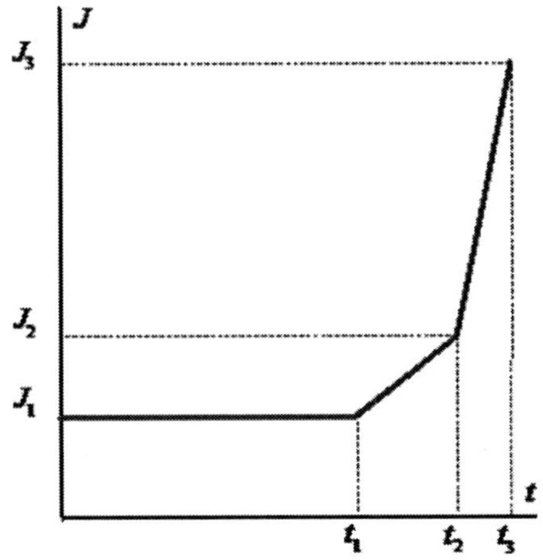

Figure 4. Current-Time Variant

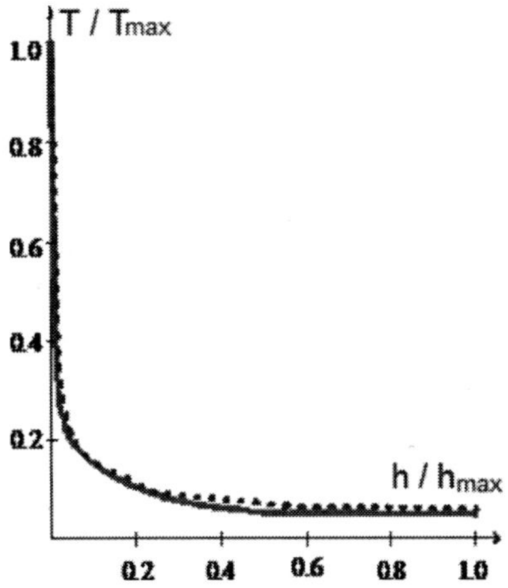

Figure 5. Temperature profile (dashed line is calculation profile at VAR absorption)

where S is heating area (cm^2); R—resistance of current-conducting layer (Ohm); $q(t)$ is density of RPF flux (W/cm^2); η is experimentally determined coefficient of energy transfer from current-conducting layer to sample ($\eta < 1$). Then, the time dependence required to form a given spatial temperature profile is experimentally corrected.

One variation of current versus time is shown in figure 4. Tests showed that the proposed method allows reproducing the required temperature profiles with an error of not more than 15%. Figure 5 shows the comparison of the obtained temperature profile with the one given in the case of simulated RPF absorption.

The considered heating method is applicable for creating monotonically descending temperature profiles from the irradiation surface. In general, combined heating from various energy sources can be used to reproduce more complex temperature distributions in multilayer composite packets.

A version of the combined heating schema from the conductive layer and high-frequency radiation (extremely high-frequency radiation with a wavelength of 1–10 mm) is presented in figure 6.

The result of heating the double-layer sample from the CM according to this scheme is shown in figure 7. It can be seen that the total profile is non-monotonic and experiences a jump at the boundary of the layers, that is characteristic for influence of the a rigid VAR

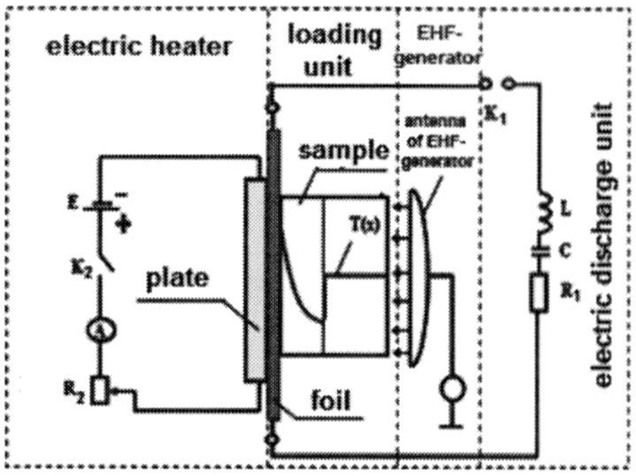

Figure 6. Sample combined heating experiment scheme

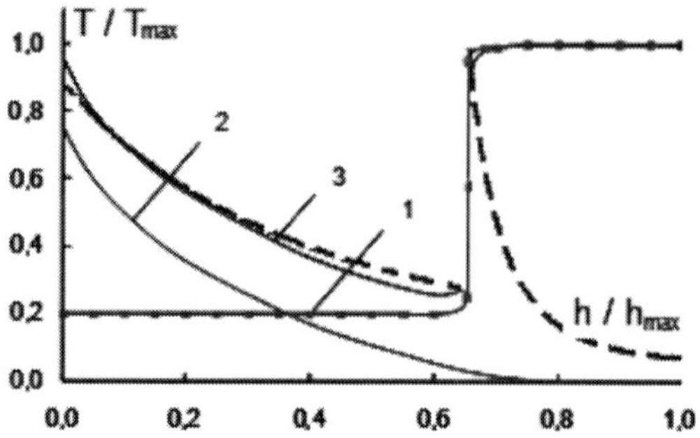

Figure 7. Temperature profiles in the case of combined heating of a two-layer composite plate: 1 is EHF-heating; 2 is electric heating; 3 is total heating; dashed line is simulated profile

(a jump of energy release from the VAR occurs due to a sharp change in the cross sections of the interaction of radiation quants with matter [6]). The simulated temperature profile corresponding to VAR is less than the total profile from heating for the second layer. This means that the thermal action is overestimated compared to what occurs when RPF influence, but it goes to the safety margin of the designed construction.

When required temperature distribution is reached, samples are loaded with pressure pulse formed by the discharge of capacitor bank through metal foils shown in figures 3 and 6.

5 Complex RPF action on FV construction: universal test bench

Tests of full-scale constructive elements of the FV suggest the presence of powerful heating sources and explosive generators of mechanical loading. Equidistant-surface charge (ESC) [2] and charge with controlled initiation (CCI) [12] are the most convenient and suitable devices for modeling the complex RPF action. Thermoelectric heaters (TEH) and pyrotechnic compositions applied to the surface of the tested structure are used to simulate thermal action. Devices that simulate RPF actions are located for joint operation on a universal test bench.

A variant of the universal test bench equipment is shown in figure 8 (it's side view). The tested thin-walled composite cylindrical construction (3) is suspended on a ballistic pendulum (1, 2). It is hermetically closed by two strong steel lids. The construction is heated from one side as a result of combustion of the pyrotechnic composition (5). Change of construction temperature is measured by industrial optical pyrometer (11) and is continuously recorded. Internal pressure (the internal pressure value is measured by the piezoelectric sensor, and is also calculated by changing the readings of the strain gauges) is created simultaneously with heating by burning a bundle of a pyrotechnic composition suspended inside the construction. The stopper (6) opens at the moment of combustion completion (5), and the spring rotary mechanism (7) rotates the construction on the 180° to the position for which the heated part is opposite the explosive loading device of the CCI (4) located stationary. Detonation of device (4) occurs as soon as new position of structure is fixed by stopper (8). Circumferential and longitudinal non-stationary deformations at various points of the inner surface of the tested shell are measured using strain gauges (10) to diagnose the state of the loaded structure. Electric terminals of strain gauges are connected by means of sealed feed-through (9) which is also used for electric ignition of pyrotechnic composition inside construction.

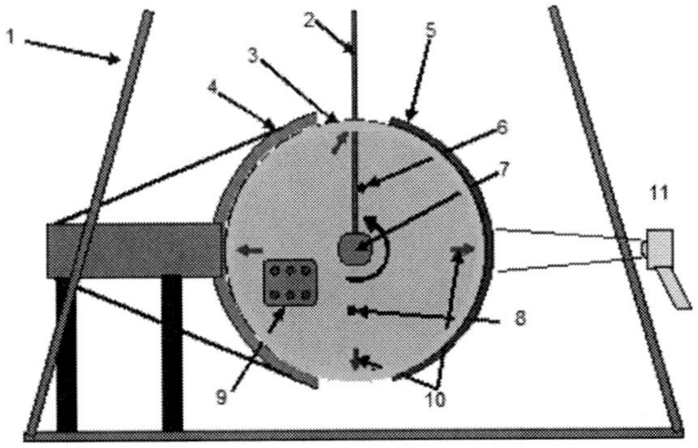

Figure 8. Bench scheme (side view): 1 is pendulum post (top hinge is not shown); 2 is pendulum suspension; 3 is test cylindrical shell with lids; 4 is explosive device (CCI or ESC) located on the stand; 5 is pyrotechnic composition or TEH; 6 is detachable stopper; 7 is spring rotary mechanism; 8 is non-detachable stopper; 9 is sealed feed-through; 10 is locations of strain gauges; 11 is optical pyrometer

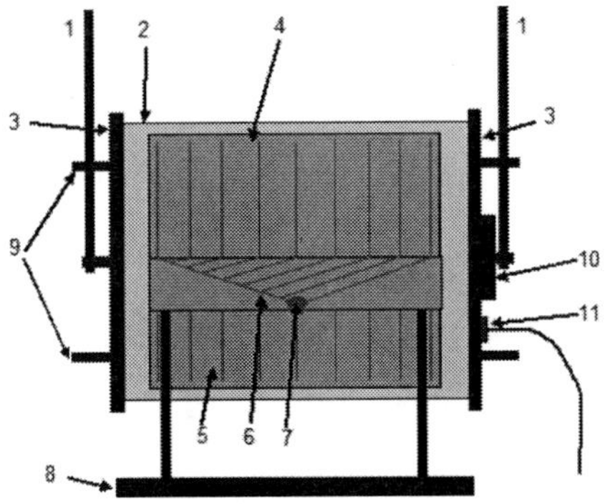

Figure 9. Bench scheme (front view): 1 is pendulum suspension; 2 is test cylindrical shell; 3 hermetic lids; 4 is explosive device of distributed impulse loading, upper part; 5 is explosive device, lower part; 6 is detonation wiring; 7 is initiation point; 8 is explosive loading device stand; 9 are stoppers; 10 is spring rotary mechanism; 11 is sealed feed-through

A view of the test bench from the front is shown in figure 9. Device of impulse loading (it is CCI in the described variant of equipment) is installed on post (8) and consists of upper (4) and lower (5) parts in the form of curved plates having milled channels filled with plastic explosive. Parts are united by detonation wiring (6) transmitting detonation from initiation point (7) to channels on parts (4), (5). Parts (4, 5, 6) are made by milling from hard cardboard.

The bench is original and significantly expands the possibilities for testing FV construction for joint thermal and mechanical RPF action. A distinctive feature and advantage of the bench is the control of reproducible parameters of thermal and mechanical actions during the tests. This is essential, since the stability of these parameters when using explosive and pyrotechnic devices in some cases is insufficient. In particular, control of mechanical action parameters is provided by using the pendulum method to measure the momentum of the force transmitted to the tested construction.

Previous developments were taken into account when creating a universal bench. A previously developed bench rocket engine with a rotation housing [2] was taken as the basis. But a fundamental change that expanded the possibilities of universalization of the bench was made. Previously, modeling devices moved over the tested construction when they were replaced. In the new variant, the tested construction is turned when the heating device simulating the thermal RPF action and flight heat flows is changed to an explosive device for modeling the mechanical RPF action. Rotation of the shell on the 180° provide combustion of the pyrotechnic composition layer (5, figure 8) away from the explosive loading device (4, figure 8), ensuring its preservation.

6 Conclusion

Methods and devices for modeling complex thermal and mechanical RPF action on flat packages from CM and full-scale (with loading area from 2 to 5 m^2) cylindrical composite constructions of FV are proposed and verified.

A universal test bench has been developed to study the strength of cylindrical composite constructions to the complex RPF action.

The work is carried out with the financial support of the Russian National Fund (Agreement No. 21-72-20023). The research was carried out within the state assignment AAA-A19-119071190040-5 (thematic program No. 0089-2019-0001) and 121112200126-5.

References

[1] V. Loborev, A. Ostrik, V. Petrovsky, A. Cheprunov, *Methods of Modeling the Mechanical Action of Radiation on Materials and Constructions* (CPTI, S. Posad, 1997)
[2] V. Bakulin, V. Gribanov, A. Ostrik, E. Romadinova, A. Cheprunov, *Mechanical Action of X-ray Radiation on Thin-walled Composite Constructions* (Fizmatlit, Moscow, 2008)
[3] Y. Mao, H. Deng, Y. Li et al., Appl. Mech. Mater., **72**, 29 (2010)
[4] S. Gerasimov, V. Kuzmin, *Transactions of Nizhni Novgorod state technical university*, **3(105)**, 202 (2014)
[5] V. Loborev, V. Fortov, A. Ostrik et al., *Physics of Nuclear Explosion. Vol. 3: Reproduction of Factors of Explosion*, ed. by V. Loborev, B. Shilobreev, V. Fortov et al. (Fizmatlit, Moscow, 2013)
[6] V. Bakulin, A. Ostrik, *Complex Action of Radiation and Particles on Thin-walled Constructions with Heterogeneous Sheetings* (Fizmatlit, Moscow, 2015)
[7] Yu. Polezhaev, F. Yurevich, *Thermal Protection* (Energy Publ., Moscow, 1976)
[8] J. Ready, *Effects of High-power Laser Radiation* (Academic Press, New York, 1971)
[9] R. Ellis, J. Lee, F. Payne et. al., American Institute of Aeronautics and Astronautics Papers, **2672**, 1 (1997)
[10] A. Ostrik, V. Bakulin, D. Nikolaev, IOP Conf. Ser.: Mater. Sci. Eng., **927** 012068 (2020)
[11] Yu. Girin, A. Martynov, A. Cheprunov, et al., Patent for invention RUS 2557298 (2015)
[12] A. Ostrik, D. Nikolaev, I. Bugay, *Book of Abstracts of XXXIV Inter. Conf. on Interaction of Intense Energy Fluxes with Matter* (Moscow, JIHT RAS, 2019)
[13] D. Nikolaev, V. Ternovoi, V. Kim, A. Shutov, J. Phys.: Conf. Ser., **500**, 142026 (2014)
[14] Yu. Girin, S. Slobodchikov, A. Cheprunov et al., Patent for invention RUS 2521217 (2012)

Application of the modified self-organizing migration algorithm MSOMA in optimal open-loop control problems of switching systems

Andrei Panteleev[1] and *Vladislav* Rakitianskii[1,*]

[1]Moscow Aviation Institute (National Research University), 4, Volokolamskoe shosse, Moscow, 125993, Russia

Abstract. The article discusses the application of the metaheuristic algorithm of global constrained optimization for solving the problem of finding the optimal open-loop control for nonlinear switching deterministic dynamical systems. The quality of control is assessed by the value of the functional defined on individual trajectories. The optimal control problem is reduced to a parametric optimization problem, which is solved using the MSOMA algorithm, which belongs to the evolutionary group. The MSOMA algorithm is a new algorithm based on the SOMA self-organizing migration algorithm. The modification consists in identifying three leaders among the individuals forming the current population. For each member of the population, two clones are generated with the same position. Thus, in fact, three populations are generated, each of which then realizes a migration cycle (evolutionary process) relative to one of the three selected leaders. A step-by-step algorithm for piecewise-constant, piecewise-linear, quadratic spline and cubic spline methods of control laws approximation is proposed. The effectiveness of the proposed method is demonstrated by the example of solving optimal control problems for switching systems with two and three subsystems. The influence of the parameters of the MSOMA algorithm on the quality of the obtained result is investigated. Comparison of the operation of the algorithm with the known solution is carried out.

1 Introduction

The solution of the problem of finding the open-loop optimal control of deterministic dynamical switching systems is considered. The object model is described by a system of ordinary differential equations. The controls are constrained by a parallelepiped type. The quality of control is assessed by the value of the quality functional.

For the numerical solution of the problem of finding the optimal open-loop control of deterministic switching systems, as a rule, the necessary optimality conditions in the form of the maximum principle are applied together with methods for solving two-point boundary value problems (shooting method, residual minimization, grid method, differential sweep method, finite element method, etc.) [1]. The collocation method, methods of approximation of control by piecewise-constant, piecewise-linear functions, splines, expansions in various systems of basis functions, spectral and quasi-spectral methods still remain popular [2–6].

*e-mail: VMRakityanskij@mai.education

In most approaches, the problem of finding the optimal control is reduced to the problem of parametric optimization, for the solution of which both classical methods of zero, first and second orders are used, as well as new metaheuristic algorithms [7, 8].

In this article, to solve optimal control problems, the transition to a discrete problem is carried out, and then the solution to the original problem is constructed by interpolating the values at the grid nodes. When searching for a control in the form of a piecewise constant, piecewise linear function or splines of different orders, the calculation formulas of the Runge–Kutta method of the fourth order are used, into which expressions for the control law are substituted corresponding to the used type of approximation. Next, a single-criterion optimization algorithm is used to select the optimal values of the parameters that define the desired control.

As an optimization method, it is proposed to use a modified self-organizing migration algorithm MSOMA [9]. The MSOMA algorithm is a new algorithm based on the SOMA self-organizing migration algorithm [10]. A computational procedure is proposed, the effectiveness of which is demonstrated by the example of optimal control problems for a chemical process in a mixing reactor and a singular optimal control problem [7].

2 Statement of the problem

The behavior of the control object model is described by a system of differential equations:

$$\dot{x}_{i(t)} = f_{i(t)}(t, x_{i(t)}(t), x_{i(t)}(t-h), u_{i(t)}(t)), \tag{1}$$

where $x_{i(t)}$ is system state vector, $x_{i(t)} = (x_{1,i(t)}, \dots, x_{n,i(t)})^T \in \mathbb{R}^n$; t is time, $t \in T = [t_0, t_f]$ is the time interval of the system operation; the start t_0 and end points t_f of the process are set; h is time lag; $u_{i(t)}$ is control vector, $u_{i(t)} = (u_{1,i(t)}, \dots, u_{q,i(t)}) \in U_{i(t)}(t) \subseteq \mathbb{R}^q$; $U_{i(t)}(t)$ is the set of possible control values in the mode $i(t)$, described by the direct product of segments $[a_{j,i(t)}(t), b_{j,i(t)}(t)]$, $j = 1, \dots, q$; $f_{i(t)}(t, x, y, u)$ is continuously differentiable vector function of the right parts of the subsystem equation with number i, $f_{i(t)}(t, x, y, u)$: $T \times \mathbb{R}^n \times \mathbb{R}^n \times U_{i(t)}(t) \to \mathbb{R}^n$; $i(t)$ is piecewise constant function of switching modes, $i(t) = i_k$ for $t \in [t_k, t_{k+1})$, $k = 0, 1, \dots, K$, where K is number of switches, $t_{K+1} = t_f$; $i_k \in I = \{1, 2, \dots, M\}$ is multiple switching modes, M is the number of subsystems; $i(t_0) = i_0$, $\{t_k\}_{k=1}^K$ is many switching points $t_k \in T$.

The change of one subsystem to another is determined by a switching sequence σ over a period of time $[t_0, t_f]$:

$$\sigma = ((t_0, i_0), (t_1, i_1), \dots, (t_K, i_K)), \tag{2}$$

where $0 \le K < \infty$, $t_0 \le t_1 \le \dots \le t_K \le t_f$, $i_k \in I$, $k = 0, \dots, K$. Each element of the sequence contains the moment of switching from one subsystem to another and the number of the next (active) subsystem.

Initial conditions:

$$x_{i(t_0)}(t) = \varphi_{i(t_0)}(t), \quad t \in [t_0 - h, t_0), \quad x_{i(t_0)}(t_0) = x_{0,i(t_0)}, \tag{3}$$

where $x_{0,i(t_0)}$ is initial state of the system, $\varphi_{i(t_0)}(t) \colon \mathbb{R}^1 \to \mathbb{R}^n$ is a piecewise continuous function of the system's prehistory.

At the moments of switching, the state vector of the system generally changes according to the transition equation:

$$x(t_k) = M_{i(t_k)i(t_k-0)} x(t_k - 0) + T_{i(t_k)i(t_k-0)}, \tag{4}$$

where M_{ij} is transition matrix of size $(n \times n)$, $T_{ij} \in \mathbb{R}^n$ is the transition vector from the subsystem with the number j to a subsystem with a number i. If $T_{ij} \equiv \theta$, where θ is zero vector, then linear jumps of the form $x(t_k) = M_{i(t_k)i(t_k-0)}x(t_k - 0)$ are observed in the system, and if $M_{ij} \equiv E$ is a single matrix of order n, then transitions between subsystems occur while maintaining continuity of trajectories. Since the system (1) is a system with a delay, after the transition to a new subsystem, the background of the system with the number is used $i(t_k - 0)$: $x_{i(t_k)}(t) = x_{i(t_k-0)}(t)$, $t \in [t_k - h, t_k)$.

The set of acceptable controls U_0 form multi-link piecewise continuous functions $u(t) = \bigcup_{i(t)=i_0}^{i_K} u_{i(t)}(t)$, satisfying the constraint $u_{i(t)}(t) \in U_{i(t)}(t)$, $\forall t \in [t_0, t_f]$, $i(t) = i_k$, $i = 0, 1, \ldots, K$.

The switching system is a corte $S = \{D, \bar{F}\}$, where $D(I, \Gamma)$ is a directed graph with a set of vertices $I = \{1, 2, \ldots, M\}$ and lots of arcs Γ, where $e_{i,j} \in \Gamma, e_{i,j} = \langle i, j \rangle$ is the arc coming out of the vertex. $i \in I$. and entering the vertex $j \in I$, as well as many vector functions $f_{i(t)} \in \bar{F}$, where $\bar{F} = \{f_1(t, x, y, u), \ldots, f_M(t, x, y, u)\}$.

Set of vertex heirs i this is a set of nodes reachable from a vertex i: $succ(i) \in I$, $i_{k+1} \in succ(i_k)$, $k = 0, \ldots, K$. The graph is given by the adjacency matrix.

The performance index to be minimized is given by

$$I = \int_{t_0}^{t_f} f^0_{i(t)}(t, x_{i(t)}(t), x_{i(t)}(t-h), u_{i(t)}(t))\, dt + F(x_{i_K}(t_f)) + \sum_{k=0}^{K} s(x_{i(t_k)}(t_k), i(t_k), i(t_k - 0)), \quad (5)$$

where $f^0_{i(t)}(t, x, y, u)$, $F(x)$ is given continuous functions, $s(x_{i_k}(t_k), i(t_k), i(t_k - 0))$ is penalty function for switching from one subsystem with a number $i(t_k - 0)$ to another subsystem with a number $i(t_k)$.

In order to calculate the value of the functional I it is necessary to find the trajectory of the system $x(t) = \bigcup_{i(t)=i_0}^{i_K} x_{i(t)}(t)$, formed by the trajectories of subsystems (links) determined by the switching sequence (2) corresponding to the permissible control $u(t)$, from the equation of the system (1) taking into account the initial condition (3) and the transition equation (4).

It is required to find such a switching sequence $\sigma^* = ((t_0, i_0), (t_1, i_1), \ldots, (t_K, i_K))$ and control $u^*(t) = \bigcup_{i(t)=i_0}^{i_K} u^*_{i(t)}(t)$ from the set of permissible controls on which the minimum value of the functional I is achieved.

3 Solution search strategy

The basis of the strategy is the decomposition of control in the interval $[t_k, t_{k+1})$ according to the basis system of functions [11] using the saturation function, taking into account the specified control constraints:

$$u_{j,i(t)}(t) = \text{sat}\left\{\sum_{m=0}^{N} u_{j,i(t),m} \cdot C(m,t)\right\} =$$

$$= \begin{cases} a_{j,i(t)}(t), & \text{if } \left\{\sum_{m=0}^{N} u_{j,i(t),m} \cdot C(m,t)\right\} < a_{j,i(t)}(t), \\ \left\{\sum_{m=0}^{N} u_{j,i(t),m} \cdot C(m,t)\right\}, & \text{if } a_{j,i(t)}(t) \le \left\{\sum_{m=0}^{N} u_{j,i(t),m} \cdot C(m,t)\right\} \le b_{j,i(t)}(t), \\ b_{j,i(t)}(t), & \text{if } \left\{\sum_{m=0}^{N} u_{j,i(t),m} \cdot C(m,t)\right\} > b_{j,i(t)}(t). \end{cases} \quad (6)$$

where $\text{sat}\left\{\sum_{m=0}^{N} u_{j,i(t),m} \cdot C(m,t)\right\}$ is saturation control function, $t \in [t_k, t_{k+1})$, $j = 1, \ldots, q$, N is the number of unknown decomposition coefficients set by the user.

At the same time, the structure of the block matrix is a column of optimized control parameters

$$\mathfrak{A} = \Bigg[\Bigg[\underbrace{[K]}_{\substack{\text{number}\\\text{of switches}}} \quad \underbrace{[t_1 t_2 \cdots t_K]}_{\substack{\text{moments of switching}\\\text{subsystems}}} \quad \underbrace{[i_0 i_1 \cdots i_K]}_{\substack{\text{numbers of sequentialy}\\\text{active subsystems}}} \Bigg]$$

$$\underbrace{\Bigg[\underbrace{[u_{1,0,0} u_{1,0,1} \cdots u_{1,0,N}]}_{\text{control decomposition coefficients } u_{1,i_0}} \quad \cdots \quad \underbrace{[u_{q,0,0} u_{q,0,1} \cdots u_{q,0,N}]}_{\text{control decomposition coefficients } u_{q,i_0}} \Bigg]}_{\text{control } u_{i_0} \text{ subsystem } i_0} \tag{7}$$

$$\underbrace{\Bigg[\underbrace{[u_{1,K,0} u_{1,K,1} \cdots u_{1,K,N}]}_{\text{control decomposition coefficients } u_{1,i_K}} \quad \cdots \quad \underbrace{[u_{q,K,0} u_{q,K,1} \cdots u_{q,K,N}]}_{\text{control decomposition coefficients } u_{q,i_K}} \Bigg]}_{\text{control } u_{i_K} \text{ subsystem } i_K} \Bigg]^T.$$

The optimization problem is to choose the vector coefficients by minimization of the cost functional value (5).

4 Algorithm

Step 1. Set:

$K_{\max}$ is a maximum number of switches from one subsystem to another;

$N_{\max}$ is a maximum number of intervals of the sign of the control of each subsystem;

N_K is a maximum number of attempts to improve the number of switches;

N_{sw} is a maximum number of attempts to improve the switching moments from one subsystem to another with a fixed number of switches;

N_{mode} is a maximum number of attempts to search for sequences of changing mode numbers with a fixed number of switches and switching moments;

N_{contr} is a maximum number of attempts to search for subsystem control values.

Step 2. Start the iteration counter to improve the number of switches.

Step 3. (The first level of the search procedure). Generate a random number of switches in a switching sequence $K \in \{0, 1, \ldots, K_{\max}\}$ and fix it. One can generate a number K, using a uniform distribution law on a segment $[0, K_{\max}]$, rounding with a disadvantage, removing the value already used in the search from the set of acceptable values. One can also start with $K = 0$, by increasing the number of switches by one each time Check the iteration counter to improve the number of switches. If the number N_K is reached, then complete the procedure and proceed to step 9. If not, then enter the resulting solution into the *Pool* set and continue.

Start the counter of iterations of the search for switching moments.

Step 4. (Second level of the search procedure.) The time interval of the system operation $[t_0, t_f]$ should be divided into segments $[t_0, t_1], [t_1, t_2], \ldots, [t_K, t_f]$, where $\{t_k\}_{k=1}^{K}$ is a sequence of switching moments generated randomly with a condition check $t_0 \le t_1 \le \ldots \le t_K \le t_f$; fix the resulting sequence. Check the iteration counter of the search for switching moments. If the number N_{sw} is reached, then go to step 3.

Start the counter of attempts to search for a sequence of subsystem shift numbers.

Step 5. (The third level of the search procedure.) Randomly generate a sequence based on a given adjacency matrix $\sigma = ((t_0, i_0), (t_1, i_1), \ldots, (t_K, i_K))$, containing information about the change of one subsystem to another. From the set of heirs of each of the vertices of the graph, one of the possible heirs is selected: $i_0 \in I$, $i_{k+1} \in succ(i_k)$, $i_{k+1} \in I$, $k = 0, 1, \ldots, K-1$. Fix the resulting sequence. Check the iteration counter of searching for a sequence of subsystem shift numbers. If the number N_{mode} is reached, then go to step 4.

Start the counter for the number of attempts to search for subsystem control values.

Step 6. (The fourth level of the search procedure).

Set the control on the interval $[t_k, t_{k+1})$ in the form of a decomposition of the desired control law according to a basic system of functions using a saturation function that takes into account the specified control constraints (6).

Thus, the control of the switching system can be represented by a block matrix-column (7)

Step 7. Using the control $u(t) = \bigcup_{i(t)=i_0}^{i_K} u_{i(t)}(t)$, set by the matrix $\mathbb{A}$, find the trajectory of the switching system $x(t) = \bigcup_{i(t)=i_0}^{i_K} x_{i(t)}(t)$ and calculate the value of the control quality functional (5): $I = \int_{t_0}^{t_f} f_{i(t)}^0(t, x_{i(t)}(t), x_{i(t)}(t-h), u_{i(t)}(t))\, dt + F(x_{i_K}(t_f)) + \sum_{k=0}^{K} s(x_{i(t_k)}(t_k), i(t_k), i(t_k-0))$. To calculate the functional, it is required to integrate the system (1) together with the equation for the auxiliary variable: $\dot{x}_{i(t),n+1} = f_{i(\tau)}^0(\tau, x_{i(\tau)}(\tau), x_{i(\tau)}(\tau-h), u_{i(\tau)}(\tau))$, $x_{i_0,n+1}(t_0) = 0$, $x_{i(t_k),n+1}(t_k) = x_{i(t_{k+1}),n+1}(t_k)$ using the switching sequence (2), the initial conditions (3) and the transition equation (4) using the MSOMA [12] method.

Step 8. Go to the next matrix-column to find the minimum functional I by one of the metaheuristic methods, generating a new piecewise constant control for each of the active subsystems (step 6). Check the counter of the number of attempts to search for subsystem control values. If the number N_{contr} is reached, then go to step 5.

Step 9. Choose the best solution from the *Pool* set.

5 Numerical examples

Example 1

Consider a switching system consisting of three nonlinear subsystems in table 1.

Table 1. Nonlinear subsystems

Subsystem 1	Subsystem 2	Subsystem 3
$\begin{cases} \dot{x}_1 = x_1 + u \sin x_1 \\ \dot{x}_2 = -x_2 - u \cos x_2 \end{cases}$	$\begin{cases} \dot{x}_1 = x_2 + u \sin x_2 \\ \dot{x}_2 = -x_1 - u \cos x_1 \end{cases}$	$\begin{cases} \dot{x}_1 = -x_1 - u \sin x_1 \\ \dot{x}_2 = x_2 + u \cos x_2 \end{cases}$

Let $t_0 = 0$, $t_f = 3$. The system switches at a moment in time $t = t_1$ from subsystem 1 to subsystem 2 and at a time $t = t_2$ from subsystem 2 to subsystem 3 $(0 \le t_1 \le t_2 \le 3)$. Initial conditions: $x_1(0) = 2$, $x_2(0) = 3$. Initial switching time values: $t_1 = 1$, $t_2 = 2$.

It is required to find optimal switching moments t_1, t_2 and optimal control $u(t)$, providing a minimum of the cost functional:

$$I = \frac{1}{2} \int_0^3 \left((x_1(t) - 1)^2 + (x_2(t) + 1)^2 + u^2(t) \right) dt + \frac{1}{2}(x_1(3) - 1)^2 + \frac{1}{2}(x_2(3) + 1)^2 \to \min .$$

Table 2 shows the results of the influence analysis of the MSOMA parameters for various types of control law approximations.

Table 2. Influence of MSOMA parameters for different types of approximation

Parameters					min I	min I	min I	min I
$NStep$	PRT	NP	$Migration$	$MinDist$	piecewise constant	piecewise linear	quadratic spline	cubic spline
20	0.6	30	50	10^{-4}	5.457996	5.43687	5.436123	5.438055
40	0.6	60	50	10^{-6}	5.461033	5.43429	5.434128	5.434687
30	0.5	40	50	10^{-6}	5.456356	5.43828	5.441293	5.437901
50	0.4	70	60	10^{-7}	5.458779	5.43400	5.434498	5.435280

Decomposition coefficients in control at the best value (with quadratic spline control approximation):

$$u(\cdot) = \{-1.80061465;\ -1.16927298;\ -0.94401819;\ -0.1357218;\ 0;\ 0;\ -0.07919648;\ -0.04200354;\ -0.07259585;\ -0.14019546\}.$$

Switching moments (with the quadratic spline approximation of control law): $t_1 = 0.22366355$, $t_2 = 1.03408362$. The obtained control law and the phase trajectory are shown in figure 1.

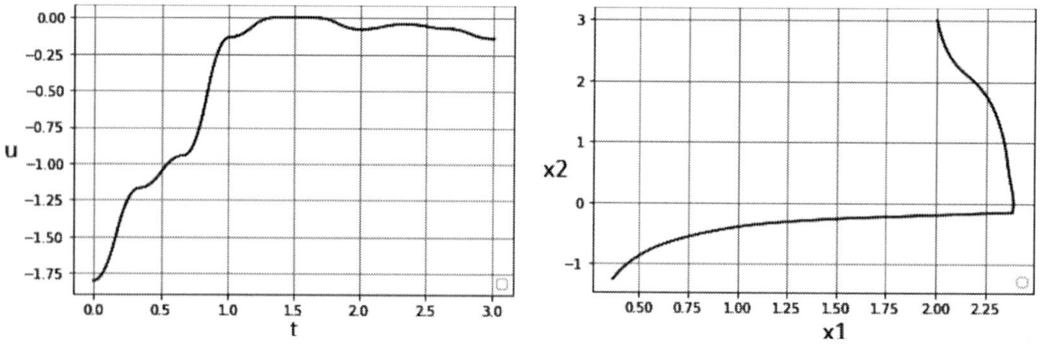

(a) (b)

Figure 1. Control law using the quadratic spline (a) and phase trajectory (b)

The known solution was obtained in [13]: $t_1^{opt} = 0.2262$, $t_2^{opt} = 1.0176$, $I_{opt} = 5.4399$. So, the best solution obtained by the presented algorithm is better by about 0.005.

Example 2

Consider a switching system consisting of two linear subsystems in table 3.

Let $t_0 = 0$, $t_f = 2$. The system switches once at a time t_1 from subsystem 1 to subsystem 2 ($0 \le t_1 \le 2$). In this case, switching occurs with a discontinuous jump:

$$\begin{cases} x_1(t_1+) = x_1(t_1-) + 0.2, \\ x_2(t_1+) = x_2(t_1-) + 0.2. \end{cases}$$

Table 3. Nonlinear subsystems

Subsystem 1	Subsystem 2
$\begin{cases} \dot{x}_1 = 2x_1 + u \\ \dot{x}_2 = -x_2 \end{cases}$	$\begin{cases} \dot{x}_1 = -x_1 \\ \dot{x}_2 = 2x_2 + u \end{cases}$

Initial conditions: $x_1(0) = 4$, $x_2(0) = 4$. It is required to find the optimal switching moment t_1 and optimal control $u(t)$, providing a minimum of the cost functional:

$$I = \frac{1}{2}\int_0^2 u^2(t)\,dt + \frac{1}{2}x_2^2(t_1-) + \frac{1}{2}(x_1(2) - 1)^2 + \frac{1}{2}(x_2(2) + 4)^2 \to \min .$$

Table 4 shows the results of the influence analysis of the MSOMA parameters for various types of control law approximations.

Table 4. Influence of MSOMA parameters for different types of approximation

Parameters					min I	min I	min I	min I
$NStep$	PRT	NP	$Migration$	$MinDist$	piecewise constant	piecewise linear	quadratic spline	cubic spline
20	0.4	30	50	10^{-4}	11.57030	10.8387	10.83212	10.8323884
70	0.6	70	50	10^{-6}	11.43335	10.7963	10.83194	10.8802788
30	0.6	40	50	10^{-6}	11.43360	10.7963	10.83212	10.880272
40	0.5	50	50	10^{-7}	11.42422	10.7963	10.83195	10.880272

Coefficients of expansion into controls at the best value (with piecewise linear approximation):

$$u(\cdot) = \{-2.09646176;\ -1.09902696;\ -0.57844942;\ 8;\ 5.55796566;\ 1.36358511\}.$$

The obtained control law and the phase trajectory are shown in figure 2. Switching moment (with piecewise linear approximation): $t_1 = 0.64045776$.

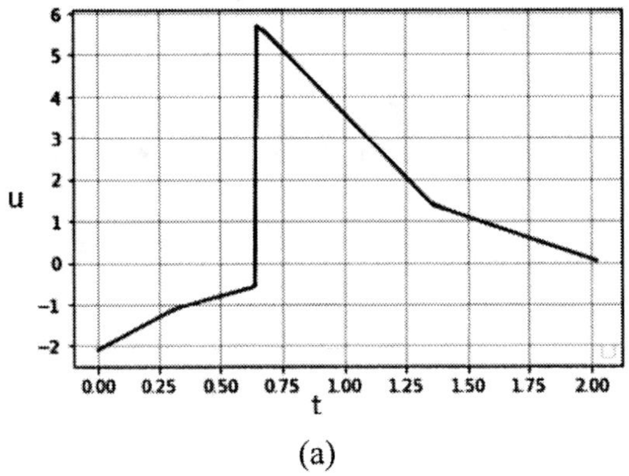

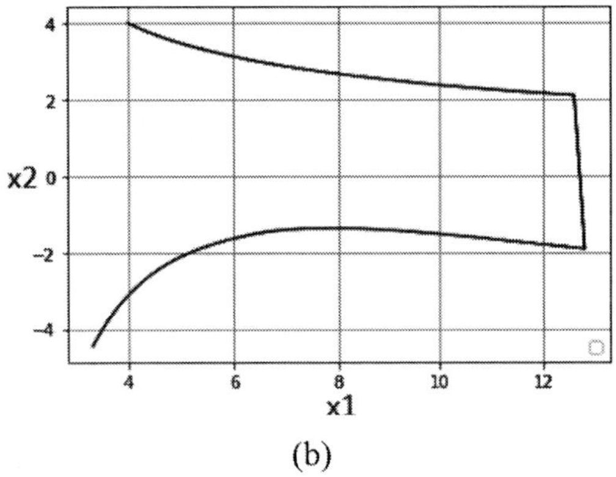

Figure 2. Control law using piecewise linear approximation (a) and phase trajectory (b)

The known solution: $t_1^{opt} = 0.6375$, $I_{opt} = 10.7165$. So, the best solution obtained by the presented algorithm is about 0.08 worse.

6 Conclusion

In this work, a strategy, a step-by-step algorithm and corresponding software for the approximate solution of the problem of optimal open-loop control of switching systems have been developed. The presented algorithm and program are tested on the examples of solving optimal control problems for switching systems with two and three subsystems. The influence of the parameters of the MSOMA algorithm on the quality of the obtained result is investigated. Comparison of the operation of the algorithm with a known solution, as well as with various methods of control laws approximation is given. Recommendations on the choice of algorithm parameters are given.

References

[1] A. Rao, *A survey of numerical methods for optimal control*, Preprint AAS 09-334 (2009)

[2] M. Karane, A. Panteleev, *Hybrid multi-agent optimization method of interpolation search*, AIP Conf. Proc., **2181(1)**, 020028 (2019)

[3] K. Rybakov, *Application of Walsh series to represent iterated Stratonovich stochastic integrals*, J. Phys.: Conf. Ser., **927**, 012080 (2020)

[4] A. Panteleev, Y. Stanovskaya, *The dragonfly algorithm for parametric optimization of open loop nonlinear dynamic systems control*, AIP Conf. Proc., **2181(1)**, 020027 (2019)

[5] K. Rybakov, A. Yushchenko, *Spectral method for solving linear Caputo fractional stochastic differential equations*, J. Phys.: Conf. Ser., **927**, 012077 (2020)

[6] A. Tyatyushkin, T. Zarodnyuk, A. Gornov, *Algorithms for nonlinear optimal control problems based on the first and second order necessary conditions*, J. of Math. Sci., **239(2)**, 185 (2019)

[7] C. Floudas, P. Pardalos, *Encyclopedia of Optimization* (Springer, New York, 2009)

[8] M. Gendreau, J. Potvin, *Handbook of Metaheuristics* (Springer, New York, 2019)

[9] A. Panteleev, V. Rakitianskii, *Development of a modified self-organizing migration optimization algorithm (MSOMA)*, Modelling and Data Analysis, **10(2)**, 62 (2020)

[10] D. Davendra, I. Zelinka, *Self-Organizing Migrating Algorithm: Methodology and Implementation*, in Studies in Computational Intelligence, **626** (Springer, New York, 2016)

[11] K. Rybakov, *Modeling and analysis of output processes of linear continuous stochastic systems based on orthogonal expansions of random functions*, J. Comput. Syst. Sci. Int., **59(3)**, 322 (2020)

[12] A.V. Panteleev, V.M. Rakitianskii, *Application of the modified self-organizing migration algorithm MSOMA in optimal open-loop control problems*, J. Phys.: Conf. Ser., **1925**, 012017 (2021)

[13] X. Xu, P.J. Antsaklis, *Optimal control of switched systems based on parameterization of the switching instants*, IEEE Trans. Aut. Contr., **49(1)**, 2 (2004)

Approximate methods of H-infinity control of nonlinear dynamic systems output

Andrei Panteleev[1] and *Aleksandra* Yakovleva[1,*]

[1]Moscow Aviation Institute (National Research University), 4, Volokolamskoe shosse, Moscow, 125993, Russia

Abstract. The problem of finding H_∞-regulators in the output control problem is considered. Two approaches for approximate synthesis of the closed loop systems are proposed. The paper considers the problem of finding the H-infinity control of a nonlinear continuous dynamical system output. The system is linear in control and perturbation, with a finite time of the system operation. Methods for finding H-infinity control are used to solve problems of synthesis of controllers, particularly for solving the problem of synthesis of controllers in conditions of incomplete state information. In such problems, it is difficult to find the structure of the regulator and its parameters. Sufficient H_∞-control conditions are formulated and proved. Two approximate methods for finding the H-infinity controller of closed loop nonlinear continuous systems were proposed. To illustrate the application of the methods, the problem of stabilization of the ZD559-Lynx helicopter, whose flight takes place during the imposed time period of the system operation, was considered.

1 Introduction

Problems and methods for finding H_∞-control laws are considered to be one of the most relevant problems of designing complex aerospace systems. The developing algorithms and software needed to find matrix gains of optimal controllers allow getting the desired quality of control over the object in conditions where the mathematical model and external perturbations contain indeterminacies. It's one of the main reasons why methods of finding H_∞-control have become so popular [1–6]. For many applied problems it is required to synthesize a controller in conditions of incomplete information about the state vector of the system. This article proposes applying methods like the ones described in [7, 8] for finding optimal regulator of a nonlinear system with a finite time of the system operation [7, 9]. As a result of the work, sufficient conditions of control of nonlinear dynamical system, which are required to design output feedback controller, will be formulated. To do so, we have to prove the sufficient conditions theorem, find controller in conditions of incomplete information and propose a solving method, which could be universal for various control problems and provide the desired quality of transient processes as well as the asymptotic stability of the system.

The ZD559-Lynx helicopter stabilization problem [10], which illustrates the effectiveness of the proposed approaches for solving applied control problems, is solved.

*e-mail: ayakovleva982@gmail.com

2 Methodology

2.1 Statement of the problem

The mathematical model of the plant is

$$\dot{x}(t) = a(t, x(t)) + B_1(t)w(t) + B_2(t)u(t), \quad x(0) = 0, \tag{1}$$

and model of a measuring system is

$$y(t) = C(t)x(t), \tag{2}$$

where $x \in R^n$ is a state vector, $u \in R^q$ is a control input, $w \in R^p$ is a disturbance, $y \in R^m$ is a measured output, $t \in T = [0, t_1]$ is a continuous time, t_1 is a given positive value. Continuous function $a(t, x)$ and matrices $B_1(t) \in R^{n\times p}$, $B_2(t) \in R^{n\times q}$, $C(t) \in R^{m\times n}$ are given. It is assumed that $w(.) \in L_2[0, \infty)$, $u(.) \in L_2[0, \infty)$; $m \leq n$, $\operatorname{rg} C = m \ \forall t \in T$; the system (1), (2) is fully controllable and observable.

Let's denote $\|z(t)\|^2 = f^0(t, y(t)) + u^T(t)Q(t)u(t)$ as performance output, where $f^0(t, y)$ is a given continuous function, $Q(t)$ is a positive definite symmetric matrix $q \times q$; $\|F(t_1)\|^2 = F(y(t_1))$ as performance output at the final moment of the control process.

It is required to ensure (if possible) the fulfillment of the inequality

$$\frac{\int_0^{t_1} \|z(t)\|^2 dt + \|F(t_1)\|^2}{\int_0^{t_1} \|w(t)\|^2_{P(t)} dt} = \frac{\int_0^{t_1} [f^0(t, y(t)) + u^T(t)Q(t)u(t)]\, dt + F(y(t_1))}{\int_0^{t_1} \|w(t)\|^2_{P(t)} dt} =$$
$$= \frac{\int_0^{t_1} [f^0(t, C(t)x(t)) + u^T(t)Q(t)u(t)]\, dt + F(C(t_1)x(t_1))}{\int_0^{t_1} w^T(t)P(t)w(t)\, dt} \leq \gamma^2, \tag{3}$$

where $P(t) \in R^{p\times p}$ is a positive definite symmetric matrix, $\gamma > 0$ is a given non-negative value, as well as the asymptotic stability of a closed loop system. It is preferred to find the minimum value of γ at which these properties are still valid, which can be achieved by minimizing the value of the fraction numerator while maximizing the denominator.

In other words, the cost functional must satisfy the condition

$$I(u, w) = \int_0^{t_1} \left[\|z(t)\|^2 - \gamma^2 \|w(t)\|^2_{P(t)}\right] dt + \|F(t_1)\|^2 =$$
$$= \int_0^{t_1} \left[f^0(t, C(t)x(t)) + u^T(t)Q(t)u(t) - \gamma^2 w^T(t)P(t)w(t)\right] dt + F(C(t_1)x(t_1)) \leq 0,$$

which will be fulfilled when the output vector tends to zero, minimizing control costs under the worst influence of disturbances possible.

2.2 Synthesis of full state vector H_∞-control

Let there be a function $V(t, x) \in C^{1,1}$ and the expression

$$R(t, x, u, w) = \frac{\partial V(t, x)}{\partial t} + \left(\frac{\partial V(t, x)}{\partial x}\right)^T [a(t, x) + B_1(t)w + B_2(t)u] +$$
$$+ f^0(t, C(t)x(t)) + u^T Q(t)u - \gamma^2 w^T P(t)w, \tag{4}$$
$$G(t_1, x) = V(t_1, x) - F(C(t_1)x),$$

where

$$\frac{\partial V(t,x)}{\partial x} = \left(\frac{\partial V(t,x)}{\partial x_1}, \dots, \frac{\partial V(t,x)}{\partial x_n}\right)^T.$$

Theorem. If a function $V(t,x) \in C^{1,1}$ exists, satisfying the condition $V(t,0) = 0$ and

$$\begin{gathered} R(t,x,u^*(x),w^*(x)) = \min_u \max_w R(t,x,u,w) = 0, \\ \forall x \in R^n, \quad \forall t \in T, \\ G(t_1,x) = 0 \quad \forall x \in R^n, \end{gathered} \tag{5}$$

where

$$u^*(t,x) = -\frac{1}{2}Q^{-1}(t)B_2^T(t)\frac{\partial V(t,x)}{\partial x}, \quad w^*(t,x) = \frac{1}{2\gamma^2}P^{-1}(t)B_1^T(t)\frac{\partial V(t,x)}{\partial x}, \tag{6}$$

and function $V(t,x) \in C^{1,1}$ satisfies the equation

$$\begin{gathered} \frac{\partial V(t,x)}{\partial t} + \left(\frac{\partial V(t,x)}{\partial x}\right)^T a(t,x) + \frac{1}{2}\left(\frac{\partial V(t,x)}{\partial x}\right)^T B_2(t)Q^{-1}(t)B^T(t)\frac{\partial V(t,x)}{\partial x} - \\ -\frac{1}{2}\left(\frac{\partial V(t,x)}{\partial x}\right)^T B_1(t)P^{-1}(t)B_1^T(t)\frac{\partial V(t,x)}{\partial x} + f^0(t,C(t)x) = 0, \\ V(t_1,x) = F(C(t_1)x), \end{gathered} \tag{7}$$

then the inequality (3) is true.

Proof. Let the assertion conditions be satisfied. Let's find $\min_u \max_w R(t,x,u,w)$ using the necessary conditions for an extremum:

$$\begin{gathered} \frac{\partial R(t,x,u,w)}{\partial u} = B_2^T(t)\frac{\partial V(t,x)}{\partial x} - 2Q(t)u = 0, \\ \text{therefore} \quad u^*(t,x) = -\frac{1}{2}Q^{-1}(t)B_2^T(t)\frac{\partial V(t,x)}{\partial x}, \\ \frac{\partial R(t,x,u,w)}{\partial w} = B_1^T(t)\frac{\partial V(t,x)}{\partial x} - 2\gamma^2 P(t)w = 0, \\ \text{therefore} \quad w^*(t,x) = \frac{1}{2\gamma^2}P^{-1}(t)B_1^T(t)\frac{\partial V(t,x)}{\partial x}. \end{gathered}$$

Here $u^*(t,x)$, $w^*(t,x)$ are structures for controlling the plant and disturbance (external influence).

Sufficient conditions for an unconstrained minimum with respect to u are fulfilled because $\frac{\partial^2 R(t,x,u,w)}{\partial u^T \partial u} = 2Q(t) > 0$, as well as sufficient conditions for maximum with respect to w are fulfilled because $\frac{\partial^2 R(t,x,u,w)}{\partial w^T \partial w} = -2\gamma^2 P(t) < 0$.

Then

$$\begin{aligned} R(t,x,u,w) = R(t,x,u^*(t,x),w^*(t,x)) - \gamma^2[w - w^*(t,x)]^T P(t)[w - w^*(t,x)] + \\ + [u - u^*(t,x)]^T Q(t)[u - u^*(t,x)]. \end{aligned}$$

Therefore

$$R(t,x,u^*(t,x),w(t,x)) \le R(t,x,u^*(t,x),w^*(t,x)) \le R(t,x,u(t,x),w^*(t,x)), \tag{8}$$

i.e. there is a saddle point.

Let the available function $V(t,x) \in C^{1,1}$ satisfy the condition $V(0,0) = 0$ and $R(t,x,u^*(t,x),w^*(t,x)) = 0$. Since the system is completely controllable and observable, for any $u(.) \in L_2[0,\infty)$, $w(.) \in L_2[0,\infty)$ we have $y(.) \in L_2[0,\infty)$.

Along the trajectories of the dynamical system it is true that

$$\frac{\partial V(t,x(t))}{\partial t} + \left(\frac{\partial V(t,x(t))}{\partial x}\right)^T [a(t,x(t)) + B_1(t)w(t) + B_2(t)u(t)] + \|z(t)\|^2 - \gamma^2\|w(t)\|^2_{P(t)} =$$

$$= \frac{dV(t,x(t))}{dt} + \|z(t)\|^2 - \gamma^2\|w(t)\|^2_{P(t)} = \underbrace{R(t,x(t),u^*(t,x(t)),w^*(t,x(t)))}_{0} -$$

$$-\gamma^2[w(t) - w^*(t,x(t))]^T P(t)[w(t) - w^*(t,x(t))] + [u(t) - u^*(t,x)]^T Q(t)[u - u^*(t,x(t))].$$

Let's consider the left hand side of inequality (8), namely $R(t,x,u^*(t,x),w(t,x)) \leq$ $\leq \underbrace{R(t,x,u^*(t,x),w^*(t,x))}_{0}$, i.e. when $u = u^*(t,x)$:

$$\frac{dV(t,x(t))}{dt} + \|z(t)\|^2 - \gamma^2\|w(t)\|^2_{P(t)} \leq 0.$$

We integrate the left and right hand parts in time from 0 to t_1:

$$V(t_1,x(t_1)) - V(0,x(0)) + \int_0^{t_1} \|z(t)\|^2 dt - \gamma^2 \int_0^{t_1} \|w(t)\|^2_{P(t)} \leq 0.$$

Since $G(t_1,x) = V(t_1,x) - F(C(t_1)x) = 0$, then $V(t_1,x(t_1)) = F(C(t_1)x(t_1))$. Since $x(0) = 0$, then $V(0,x(0)) = V(0,0) = 0$. It follows that

$$\int_0^{t_1} \|z(t)\|^2 dt + F(C(t_1)x(t_1)) \leq \gamma^2 \int_0^{t_1} \|w(t)\|^2_{P(t)} dt.$$

Therefore, condition (3) is fulfilled.

The proof is over.

Remark. If $P(t) = E$ and the energy of disturbances acting on the system is limited, i.e.

$$\int_0^{t_1} \|w(t)\|^2 dt \leq 1, \quad \text{then} \quad \int_0^{t_1} \|z(t)\|^2 dt + \underbrace{F(C(t_1)x(t_1))}_{F(t_1)} \leq \gamma^2.$$

2.3 Synthesis of Output H_∞-control

If $m = n$ and the matrix $C(t)$ is non-degenerate, then the state vector could be found directly from the output vector: $x = C^{-1}(t)y$. Then output control

$$u^*(t,y) = -\frac{1}{2}Q^{-1}(t)B_2^T(t)\frac{\partial V(t,C^{-1}(t)y)}{\partial x}.$$

If $m \leq n$, $\operatorname{rg} C(t) = m$ $\forall t \in T$, one can apply two approaches. The first one is related to finding a pseudo-solution $\tilde{x}$ of system $y = C(t)x$ and its application in control law, the second—to the synthesis of the state observer that develops the estimate $\hat{x}$ of the state vector, and using estimates in control.

First approach. Find system pseudo-solution $y = C(t)x$ using a pseudo-inverse matrix $C^{\sim 1}(t) = C^T(t)[C(t)C^T(t)]^{-1}$:

$$\tilde{x} = C^{\sim 1}(t)y,$$

i.e. column $\tilde{x}$ with the smallest modulus among all columns of x minimizing $|C(t)x-y|$, where the column modulus is $|x| = \sqrt{x_1^2 + \ldots + x_n^2}$. Then the output control has the form

$$u^*(t,y) = u^*(t,\tilde{x}) = -Q^{-1}(t)B_2^T(t)\frac{\partial V(t, C^T(t)[C(t)C^T(t)]^{-1}y)}{\partial x}. \tag{9}$$

Second approach. Synthesize an asymptotic full order observer

$$\frac{d\hat{x}}{dt} = a(t,\hat{x}(t)) + B_1(t)w(t) + B_2(t)u(t) + K(t)[y(t) - C(t)\hat{x}(t)], \quad \hat{x}(0) = x_0^*, \tag{10}$$

where $K(t) \in R^{n\times m}$ is the gain observer matrix, x_0^* is a column containing the a priori information about the initial state. Matrix $K(t)$ is selected with respect to condition, providing that the error $\varepsilon(t) = x(t) - \hat{x}(t)$ asymptotically tends to zero. Then the control law has the form

$$u^*(t,y_0^t) = u^*(t,\hat{x}) = -Q^{-1}(t)B_2^T(t)\frac{\partial V(t,\hat{x})}{\partial x}, \tag{11}$$

where $y_0^t = \{y(\tau),\ 0 \le \tau \le t\}$ is an accumulated information about the output measurement results.

2.4 Algorithm of Output H_∞-control

Step 1. Set parameter $\gamma > 0$. Find a solution to equation

$$\frac{\partial V(t,x)}{\partial t} + \left(\frac{\partial V(t,x)}{\partial x}\right)^T a(t,x) + \frac{1}{2}\left(\frac{\partial V(t,x)}{\partial x}\right)^T B_2(t)Q^{-1}(t)B^T(t)\frac{\partial V(t,x)}{\partial x} -$$
$$-\frac{1}{2}\left(\frac{\partial V(t,x)}{\partial x}\right)^T B_1(t)P^{-1}(t)B_1^T(t)\frac{\partial V(t,x)}{\partial x} + f^0(t,C(t)x) = 0,$$
$$V(t_1,x) = F(C(t_1)x)$$

at fixed γ, satisfying stability conditions of a closed loop system. Sequentially reducing γ, find the minimum value under which all conditions remain fulfilled.

Step 2. Find the control of a plant of the form (9) or (11) and the law of change of the disturbing effect $w^*(t,x) = \frac{1}{2\gamma^2}P^{-1}(t)B_1^T(t)\frac{\partial V(t,x)}{\partial x}$.

Step 3. Find output control law and closed-loop system trajectories.

Case 1. Control system with arbitrary disturbances satisfying the condition $\int_0^{t_1} \|w(t)\|^2 dt \le 1$:

$$\dot{x} = a(t,x(t)) + B_1(t)w(t) + B_2(t)u(t), \quad x(0) = 0, \quad y(t) = C(t)x(t),$$

first approach: $u(t,y) = u^*(t,\tilde{x}) = -Q^{-1}(t)B_2^T(t)\frac{\partial V(t, C^T(t)[C(t)C^T(t)]^{-1}y)}{\partial x}$;

second approach: $u(t,y_0^t) = u^*(t,\hat{x}) = -Q^{-1}(t)B_2^T(t)\frac{\partial V(t,\hat{x})}{\partial x}$,

$$\frac{d\hat{x}}{dt} = a(t,\hat{x}(t)) + B_1(t)w(t) + B_2(t)u(t) + K(t)[y(t) - C(t)\hat{x}(t)], \quad \hat{x}(0) = x_0^*.$$

Case 2. Control system with the worst perturbations $w(t,x) = \frac{1}{2\gamma^2}P^{-1}(t)B_1^T(t)\frac{\partial V(t,x)}{\partial x}$:

$$\dot{x} = a(t,x(t)) + B_1(t)w(t) + B_2(t)u(t), \quad x(0) = 0, \quad y(t) = C(t)x(t),$$

$$\text{first approach: } u(t,y) = u^*(t,\tilde{x}) = -Q^{-1}(t)B_2^T(t)\frac{\partial V(t, C^T(t)[C(t)C^T(t)]^{-1}y)}{\partial x};$$
$$\text{second approach: } u(t,y_0^t) = u^*(t,\hat{x}) = -Q^{-1}(t)B_2^T(t)\frac{\partial V(t,\hat{x})}{\partial x},$$

$$\frac{d\hat{x}}{dt} = a(t,\hat{x}(t)) + B_1(t)w(t) + B_2(t)u(t) + K(t)[y(t) - C(t)\hat{x}(t)], \quad \hat{x}(0) = x_0^*.$$

Verify the asymptotic stability of the closed loop system for each case.

3 Results and discussion

Based on the outlined algorithms, software was developed to find the parameters of synthesized controllers, to simulate processes in dynamic systems for a given period of time, to obtain information about the quality of transient processes and analyze the asymptotic stability of a closed loop system. We used the MATLAB computing environment.

Following is a model example, for which an analytical solution is obtained. Numerical experiment was carried out using the data of the model example.

3.1 Example

Let us consider the problem of stabilizing the ZD559-Lynx aircraft model [10, 11], where in (1)–(2) we set $n = 8$, $p = 1$, $q = m = 4$. It is important to note that $a(t,x(t)) = A(V(t))x(t)$ and $B_2(t) = B_2(V(t))$ in [10] are listed for the Lynx at flight speeds from hover to 140 knots in straight and level flight, i.e. trim angular velocities in fuselage axes system are equal to zero. Let us approximate these matrices. The equations of the linearized system are obtained for eight fixed values of the helicopter speed in [10] $V_i = (i-1)\Delta V, i = 1,\dots,8$, where $\Delta V = 20$ knots:

$$\dot{x}(t) = A_i x(t) + B_i u(t), \quad i = 1,\dots,8, \quad x(0) = x_0.$$

It is proposed to use an extended model

$$\dot{x}(t) = A(V)x(t) + B_2(V)u(t), \quad x(0) = x_0,$$

where $V(t)$ can be found by solving the equation $\dot{V}(t) = 140/T$, $V(0) = 0$, which is used for mode with linearly increasing speed or another equation $\dot{V}(t) = 280t/T^2$, $V(0) = 0$, which is used for mode with positive acceleration. As a result, $a(t,x(t)) = A(V)x$, $B_2(t) = B_2(V(t))$ can be approximated with law:

$$A(V) = \sum_{i=1}^{8} A(V_i)S_p^*(i,V) = \sum_{i=1}^{8} A(V_i)S_p^*\left(\frac{V}{\Delta V} - (i-1)\right),$$
$$B_2(V) = \sum_{i=1}^{8} B_2(V_i)S_p^*(i,V) = \sum_{i=1}^{8} B_2(V_i)S_p^*\left(\frac{V}{\Delta V} - (i-1)\right),$$

$$\text{where } S_p^*(t) = \begin{cases} 2^{p-1}(1+t)^p, & t \in [-1,-1/2],\\ 1 - 2^{p-1}|t|^p, & t \in [-1/2,1/2],\\ 2^{p-1}(1-t)^p, & t \in [1/2,1],\\ 0, & t \notin [-1,1] \end{cases} \text{—finite function with parameter } p.$$

Then the other matrix in plant model is described by the following one:

$$B_1 = \begin{pmatrix} 0 \\ 0 \\ 0 \\ 0 \\ 0.1 \\ 0.2 \\ 0 \\ 0 \end{pmatrix}.$$

The matrix in the equation of the measuring system has the form:

$$C = \begin{pmatrix} 0 & 0 & 0 & 57.2958 & 0 & 0 & 0 & 0 \\ 0 & 57.2958 & 0 & 0 & 0 & 0 & 0 & 0 \\ -16.26 & -0.9788 & 0.4852 & 0 & 0 & 0 & 0 & 0 \\ 0 & 0 & 0 & 0 & 1 & 0 & 0 & 0 \end{pmatrix}.$$

The state vector is as follows: x_1—velocity along aircraft x-axis, x_2—velocity along aircraft z-axis, x_3—angular velocity along aircraft y-axis, x_4—pitch angle, x_5—velocity along aircraft y-axis, x_6—angular velocity along aircraft x-axis, x_7—angle of attack, x_8—angular velocity along aircraft z-axis.

3.2 Simulations

Simulation Settings. For the correct computation of the output feedback control, it is necessary to select Q. During a search for suitable elements, we changed their design several times. After all we selected the design matrices as in table 1, which contains parameters of numerical simulations.

In addition, we choose function given in table 1 to describe disturbance, which affects the state vector.

Table 1. Numerical simulation parameters

Initial conditions x	Q	$T = [t_0; t_1]$	Disturbance
$\overline{0}$	$\begin{pmatrix} 0.169 & 0 & 0 & 0 \\ 0 & 0.169 & 0 & 0 \\ 0 & 0 & 0.78 & 0 \\ 0 & 0 & 0 & 0.1 \end{pmatrix}$	[0; 100]	$w(t) = \begin{cases} 0, & t < 0 \text{ or } t > t_w, \\ 1, & 0 \le t \le t_w,\ t_w = 1. \end{cases}$

Numerical Simulations. For one of the described cases, namely the case of control system with arbitrary disturbances, we performed solution of the model example with the parameter $\gamma = 0.05$, which satisfies (3) and still guarantees the fulfillment of the asymptotic stability property of a closed loop system.

Following are plots of the changing angle of attack for the first and second approaches (figure 1).

Analysis of plots indicates that problems of synthesis of suboptimal H^∞-controller can be solved by both approaches, which were proposed in the paper. Results for the both methods on the example presented on the figure 1 for the coordinate x_7 look similar due to the fact that

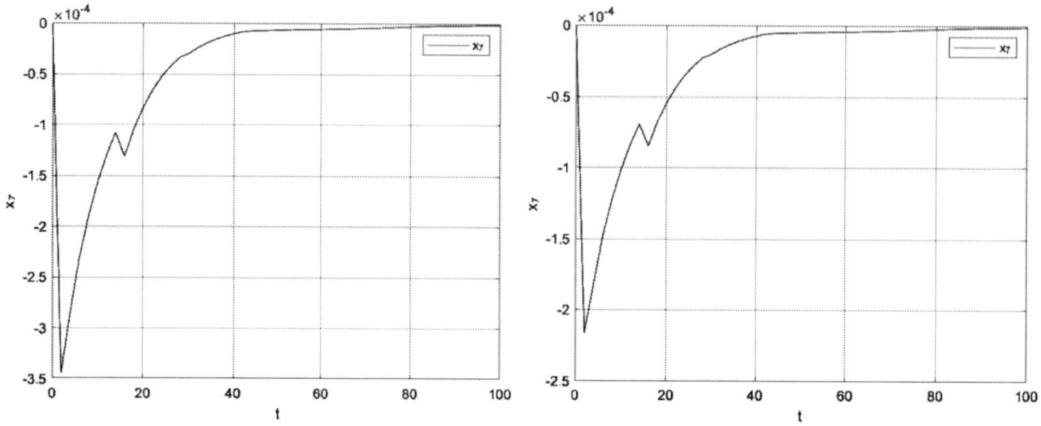

Figure 1. State response $x_7(t)$—angle of attack with $\gamma = 0.05$: (a)—for the first approach; (b)—for the second approach

each of the methods results in the optimal trajectory of the coordinate. Simulations give good results. A rather large influence is exerted by the elements of matrix Q in the quality criteria.

The coordinate x_7 and other coordinates of the state vector asymptotically converge to zero, which indicates the robustness of the system and the correct selection of parameters, for which the system retains its property of stability for any given disturbances.

Generally, both of the approaches produce comparable results observed with the processes change over time, which allows to make a conclusion about them both being a viable option in further research.

4 Summary

The purpose of this study was to design output feedback controller using the H^∞-control for nonlinear dynamical system. Based on the obtained results we can state that proposed algorithms of synthesis of controllers provide the desired quality of transient processes and asymptotic stability of closed loop systems. Using H^∞-control allows to effectively neutralize the negative influence of limited disturbances on the system.

H^∞-control method was successfully applied to solving the considered example. The described method successfully deals with the problem, demonstrating good results. The proposed methods could be useful in autopilot design problems for aircrafts and rotorcrafts.

Nowadays, methods such as LQG-synthesis or frequency response conditions are used to find H^∞-control. Often finding the control by these methods is reduced to solving linear-matrix inequalities, so the main difference between the methods proposed in the paper and other methods is that the proposed methods allow to avoid the solution of linear matrix inequalities and are distinguished by their simplicity for application.

References

[1] M. Green, D.J.N. Limebeer, *Linear Robast Control* (Dover Publications, 2012)

[2] J. Gadewadikar, F.L. Lewis, M. Abu-Khalaf, *Necessary and Sufficient Conditions for H-infinity Static Output–Feedback Control*, J. Guid. Control Dyn., **29(4)** (2006)

[3] S. Skogestad, I. Postlethwaite, *Multivariable Feedback Control: Analysis and Design* (John Wiley and sons, 2005)

[4] H.W. Koobloch, A. Isidori, D. Flockerzi, *Topics in control theory*, bd. 22 (Springer, Basel, 1993)

[5] J. Doyle, B. Francis, A. Tannenbaum, *Feedback Control Theory* (Publishing Co., 1990)

[6] D.V. Balandin, M.M. Kogan, *Minimax filtering: γ_0-optimal observers and generalized H_∞-optimal filters*, Autom. Remote Control, **74(4)**, 575 (2013)

[7] A.V. Panteleev, A.A. Yakovleva, *Sufficient conditions for H-infinity control on the finite time interval*, J. Phys.: Conf. Ser., **1925**, 012024 (2021)

[8] A.V. Panteleev, A.A. Yakovleva, *Approximate methods for H-infinity output feedback control*, IOP Conf. Ser., **927(1)**, 012026 (2020)

[9] M.M. Tchaikovsky, V.N. Timin, *Synthesis of anisotropic suboptimal control for linear time-varying systems on finite time horizon*, Autom. Remote Control, **78(7)**, 1203 (2017)

[10] G.D. Padfield, *Helicopter flight dynamics: the theory and application of flying qualities and simulation modeling*, 2nd ed. (Blackwell Publishing, 2007)

[11] H. Jinshuo, G. Hongbin, *Survey on flight control technology for large-scale helicopter*, Hindawi Int. J. Aerosp. Eng., **2017**, 5309403 (2017)

Simulation of collective self-management by the movement of robotic vehicles using the method of molecular dynamics

Vladislav Protasov[1] and *Zinaida* Potapova[1,*]

[1]Moscow Aviation Institute (National Research University), 4, Volokolamskoe shosse, Moscow, 125993, Russia

Abstract. The paper presents the results of computer modeling of traffic flows in a virtual city. The city has a local road, highway and road junction. The article proposes a system for autonomous driving of robotic vehicles in this virtual city. This system is based on the application of interaction rules taken from molecular dynamics modeling with kinetic energy dissipation. The proposed model uses the concept of a cellular automaton. The tests were carried out using computer simulations. The work has shown the efficiency of the proposed approach. The recommended rules of interaction allow you to organize traffic without traffic jams and without traffic lights in such a virtual city with the maximum possible traffic congestion.

1 Introduction

There are two approaches to the study of traffic flows. Each vehicle is viewed in interaction with each other—this is a micro-level approach. Transport streams are an object of study—this is a macro-level approach [1, 2]. Individual histories of the movement of vehicles are investigated depending on their characteristics and the structure of transport routes in the first case. Their integral characteristics are investigated in the second case. M.J. Lighthill and G.B. Whitham created in 1955 the first mathematical model designed to study traffic flows [3]. A hydrodynamic model was taken as a basis, which describes the dependence of the flow density on its intensity on a certain section of the road. The average speed of the traffic flow, its density and intensity were the main parameters of the model. The traffic flow is subject to phase transitions—sharp changes in speed and density, just like a liquid. This fact is shown within the framework of this model. The results made it possible to qualitatively describe the wave processes of the emergence and decay of plugs obtained in these works. However, specific recommendations are not given for managing traffic flows without congestion based on these models are not given in the discussed works.

2 Formulation of the problem

It is required to develop an autonomous system of self-government and find the rules of interaction for robotic vehicles. At the same time, traffic in the city should be organized without traffic jams, without traffic lights and without collisions, using these rules [4].

*e-mail: potapovaz@yandex.ru

The system should:

1. Provide the maximum possible traffic of the transport system.
2. The distance between two objects must not be less than a predetermined value.
3. Ensure that there are no traffic jams.

The task includes the formulation of recommendations for road users in a real city to improve safety and traffic optimization. The concept of *cellular* automaton was used as a modeling environment at the microlevel when solving the problem [5]. This approach has greatly simplified the computer simulation of traffic flows and made it possible to study a large number of implementations with different initial conditions [6–8].

3 Computer model

Research on the self-management system of robotic vehicles was carried out using computer simulation. The transport scheme of roads and junctions is set. Roads are arranged on two levels in the form of contiguous local ring roads, highways, and roundabouts. The quarters are located within the local ring road (figure 1). The traffic is one-way on all roads. There is a permanent parking place inside the quarters for each robotic vehicle. Vacant parking spaces are also available. All robotic vehicles have vision systems and can determine the coordinates of the nearest vehicles and road elements constantly. Traffic lights are not used. The constant movement of transport is organized so that the traffic flows do not cross. Each robotic vehicle has the ability to start moving at any time and to any vacant parking lot, stop there for the required time and return to the original parking lot.

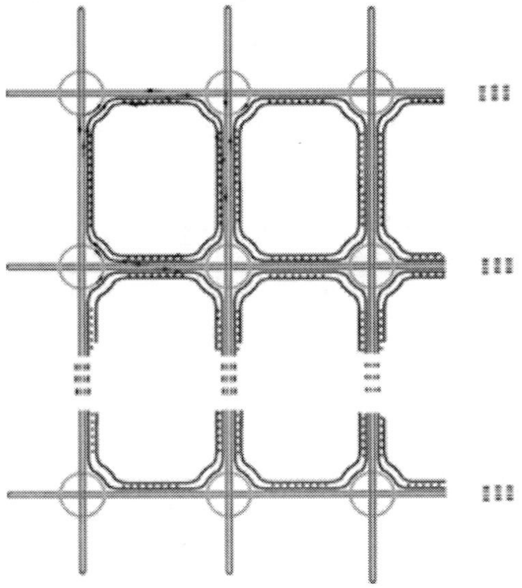

Figure 1. Transport scheme of the city

4 Transport Network Topology

The solution to the problem is divided into two parts. The topology of the network of roads and junctions is developed in the first part as follows:

- the arrangement of expressways is orthogonal and two-level;
- traffic streams do not cross.

Ring local roads exist next to each section of the highway. Travel speeds on these sections are lower than on highways. Roundabouts adjoin them. The speed of movement on them is even lower. This scheme of the organization of traffic flows is shown below in figure 1.

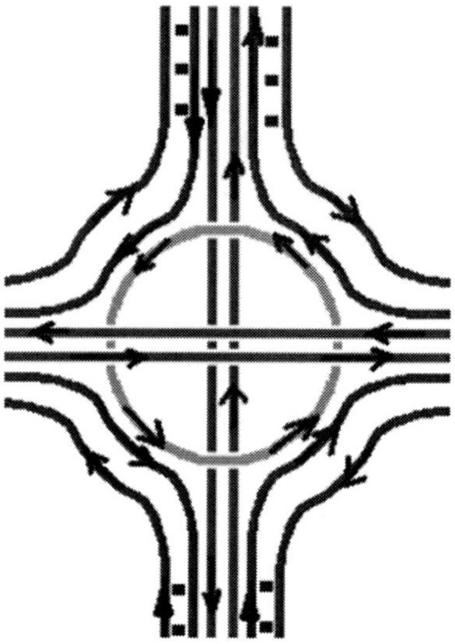

Figure 2. Two-level road junction scheme

Local roads—blue rings, roundabouts—green rings, long highways—red sections. The parking lots of the robotic vehicles are located between the blue roads. There are parking lots with even and odd numbers. Vehicles have an assigned parking space with odd numbers. Even numbered parking lots are shared property. Suppose a robotic vehicle chooses a destination. The optimal trajectory of movement is chosen from his own parking lot to any other, unoccupied parking lot with an even number. If this parking lot is occupied by another robot car, then the robotic vehicle selects the nearest unoccupied parking lot to it. The robotic vehicle can perform swaps in the following order: local road—roundabout—local road—highway—local road—roundabout—local road. It should be noted that the traffic order: roundabout—highway—roundabout is not provided. A diagram of a two-level movement with an indication of the direction of movement is shown in figure 2.

5 Interaction Rules

The second part of the assignment contains the organization of conflict-free movement. A system of rules for the interaction of moving robotic vehicles—intelligent agents—has been developed. Calculations of molecular dynamics are the basis. For each robotic vehicle, one-way traffic along the local road line and the roundabout is considered. In these calculations, the ring roads conditionally unfolds into a sequence of interconnected line segments. It is considered that only adjacent robotic vehicles affect the robotic vehicle. The quasi-cellular automaton implements the computer model of the city. A quasi-cellular automaton has two types of objects placed in cells.

The first type of objects is static—road elements and their types. The second type of objects—dynamic—imitates moving robotic vehicles. Each cell has a two-level link system. We will conventionally call local roads and ring intersections rings. Links of the first level provide an indication of the direction of movement between cells of one ring, and links of the second level provide transitions between rings.

The molecular dynamics method is the basis for the construction of a decentralized system of self-control of robotic vehicles. It is known that atoms are always located at distances not exceeding a certain value $r_{\min}$, during chaotic motion in various states of aggregation, as well as in cascades of atom-atom collisions [9]. Each robotic vehicle is conditionally represented by a macromolecule. The mass of this macromolecule is equal to the mass of the robotic vehicle, and the position of this macromolecule is related to the position of its center of mass. Force acts on a macromolecule, defined as the vector sum of all forces.

These forces act from the side of the nearest neighboring macromolecules and obstacles. The force is determined from the Lennard–Jones potential of the pairwise interaction known in molecular dynamics, acting on one macromolecule from the other [10]. The movement control of robotic vehicles does not allow the vehicles to collide with each other, built on the basis of such calculations. A special computational program was compiled and debugged to test the possibility of using the molecular dynamics method [9, 11, 12]. Computer simulation of the main modes of movement was carried out. Verification of the fulfillment of the law of conservation of energy in the system of macromolecules controlled the accuracy of the calculations.

The equations of motion have the form according to Newtonian mechanics:

$$\frac{d[\mathbf{r}_i]}{dt} = v_i, \quad m\frac{d[\mathbf{v}_i]}{dt} = f_i = \sum_{j=1, j\neq i}^{N} \mathbf{F}(\mathbf{r}_{i,j}).$$

Here t is the time, m is the mass of the macromolecule, f_i is the total force acting on the macromolecule with the number i from the side of other macromolecules of the system, $\mathbf{F}(\mathbf{r}_{i,j})$ is the force acting on the macromolecule with the number i from the side of the macromolecule with the number j.

The relationship between the force and energy of pair interaction is described by the equation

$$\mathbf{F}(\mathbf{r}_{i,j}) = -\nabla_i u(\mathbf{r}_{i,j}).$$

The equations are written as follows for the projections of the velocities along each coordinate axis:

$$V_{x(t+\Delta t)} = \frac{x(t+\Delta t) - x(t)}{\Delta t}, \quad V_{y(t+\Delta t)} = \frac{y(t+\Delta t) - y(t)}{\Delta t}.$$

Here Δt is the time step.

The forces acting in the system are described by Newton's II law:

$$\mathbf{F} = m\frac{d\mathbf{V}}{dt}.$$

The system of equations is obtained for the projections of the force acting on the macromolecule:

$$F_x = \frac{m}{\Delta t}(V_{x(t+\Delta t)} - V_{x(t)}), \quad F_y = \frac{m}{\Delta t}(V_{y(t+\Delta t)} - V_{y(t)}).$$

The equations are obtained after transformations for calculating the projections of the speeds of movement of robotic vehicles:

$$V_{x(t+\Delta t)} = V_{x(t)} + \frac{F_x \Delta t}{m}, \quad V_{y(t+\Delta t)} = V_{y(t)} + \frac{F_y \Delta t}{m}.$$

The Lennard–Jones potential was chosen to calculate the interactions in the work describing the Van der Waals interaction of neutral atoms. The Lennard--Jones potential has the form

$$U(r) = 4\varepsilon\left[\left(\frac{\sigma}{r}\right)^{12} - \left(\frac{\sigma}{r}\right)^{6}\right],$$

where σ is the value of the interatomic distance at which $U(\sigma) = 0$, ε is the depth of the potential well located at a distance $\sigma\sqrt[6]{2}$. In this expression, the term r^{-6} dominates at large distances and corresponds to dispersive dipole-dipole attraction. The term r^{-12} simulates a strong repulsion between a pair of atoms due to exchange interaction if they are very close to each other. The graph of this potential is shown in figure 3.

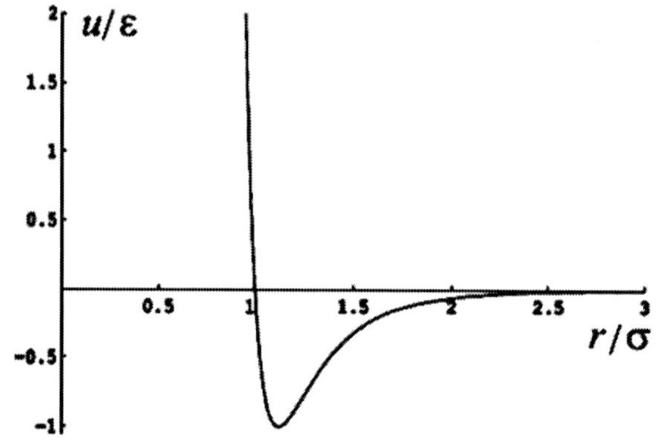

Figure 3. Lennard–Jones potential

The absolute value of the force is the first derivative of the potential energy of the pair interaction, taken with a minus sign:

$$F(r) = -\frac{d[U]}{dr}.$$

The so-called cutoff radius of the potential R_{ab} is introduced to speed up the calculations. In this case, the weak effect of distant macromolecules is not taken into account. Forces are calculated only within a sphere of this radius, acting on a given macromolecule. The issue of choosing the size of the time step Δt is important when carrying out calculations using schemes for the numerical integration of the equations of motion. As is known, a significant loss of calculation accuracy is observed when the value of Δt exceeds a certain threshold value, and the values of forces, coordinates, and velocities tend to infinity after some time.

Typically, researchers choose Δt by fit using molecular dynamics techniques. They increase the selected small time step value gradually. The procedure is carried out until a significant loss of accuracy occurs when repeatedly solving the same problem. Then the penultimate value Δt is chosen as the integration step. The change in the value of the integration step over time Δt can reach several orders of magnitude at the initial and final stages. The use of the traditional Δt selection scheme was an obstacle to computer simulation of the studied physical processes. An adaptive scheme for calculating Δt was proposed, which makes it possible to completely exclude the procedure of multiple empirical selection of this value. This allowed research [9, 11, 12]. The calculation scheme for Δt is constructed as follows. The minimum distance $r_{\min}$ is among all mutual distances between macromolecules. The potential of pair interaction is approximately replaced by a parabolic potential in the vicinity of this value. The second derivative of the potential energy of the pair interaction is equal to the constant of this parabolic potential

$$C = \left.\frac{d^2[U]}{dr^2}\right|_{r=r_{\min}}.$$

It is known that the period of harmonic oscillations of a macromolecule of mass m in such a field is

$$2\pi\sqrt{\frac{m}{C}}.$$

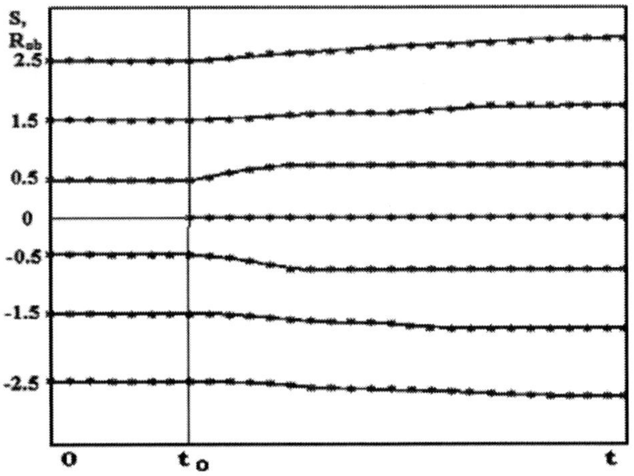

Figure 4. Transition of a tuple of robotic vehicles from a stressed state to a non-stressed state

We calculate the forces, coordinates and velocities of macromolecules only in a small vicinity of the potential acting on it. Therefore, we can take a rather small part of the vibrations of the macromolecule as the value of the time integration step. Experience shows that it is necessary to select this part in the range from 0.02 to 0.1 of the oscillation period. Then the divergence is not observed in the numerical scheme of calculations and there is no loss of the accuracy of calculations. The value of the time integration step is calculated in this work by the formula [9, 11]

$$\Delta t = 0.1\pi \sqrt{\frac{m}{\left.\frac{d^2[U]}{dr^2}\right|_{r=r_{\min}}}}.$$

Coefficient C is selected experimentally and introduced into the equations of motion. The coefficient is less than one and simulates the introduction of viscous forces into the system. These forces eliminate periodic oscillations of macro macromolecules near the positions of the potential minimum. In this case, all values of velocity projections are determined by integrating the motion of macroparticles. Projection values are multiplied by this factor. This leads to the dissipation of kinetic energy in the system. The values of the translational speed allowed for a given ring are subtracted from the velocities of all macromolecules. The interaction potential of a macromolecule is calculated as the sum of the interaction potentials of neighboring macromolecules. This potential can be zero. The road condition in this case is called unstressed. The interaction rules are formulated for the organization of accident-free movement of vehicles controlled by molecular dynamics calculations:

1. The robotic vehicle starts moving in accordance with the travel plan it has It can enter the road if a relaxed state is present at the moment. Suppose that a robotic vehicle has a need to rebuild to another road. He can enter it if the distance does not exceed the value $0.5R_{ab}$ to the nearest robotic vehicle.

2. The robotic vehicle can move around the ring at the speed allowed for that ring. The speed can only change according to the molecular dynamic's calculations when the point of the robotic vehicle on the ring is in a stressed state.

3. The robotic vehicle can change its speed from that allowed in a given place, if molecular dynamics calculations require it.

Suppose a robot vehicle enters another ring and puts the neighbors in a tense state. In this case, the process of calculating motion parameters begins with the use of molecular dynamics. This process clings until the entire cortege goes into a relaxed state. Let's say there are not enough ring sizes. Then the ring remains tense. Other vehicles cannot enter it. The unstressed state occurs when one or more robotic vehicles leave this ring. This process is shown in figure 4.

A tuple of six robotic vehicles travels along a certain ring during the time from 0 to t_0. The robotic vehicle enters from another ring between the third and fourth robotic vehicles at time t_0. The results of molecular dynamics calculations with kinetic energy dissipation show that the rearrangement processes proceed smoothly with oscillation damping.

6 Calculation results

Two groups of experiments were conducted. The city model consisted of $10 * 10$ local roads, and there were twenty thousand vehicles. The following values of the conditional velocities were chosen in the experiment: for the roundabouts—1, for the local roads—2, for the highway—4.

50 robotic vehicles and 50 vacancies were located along each side within the local road. The amount of traffic was measured in arbitrary units. $N_{\max}$ was the number of robotic vehicles that could simultaneously be at R_{ab} distance from each other. The total number of vehicles exceeded $N_{\max}$ by 3.2 times available in the city. Traffic congestion did not exist in all experiments with these parameters. The analysis of traffic flows was carried out in the case of the movement of robotic vehicles and in accordance with their individual travel histories in the situation of a virtual city model. Interaction rules were developed for car drivers based on this analysis. These rules were formulated as follows:

1. Drivers strive to always maintain a fixed distance between vehicles.
2. The distance must be greater than a fixed predetermined distance between his vehicle and vehicles already moving on the road. After that, the driver can start moving.
3. The driver can enter the local road only when the distance between his car and the cars already driving on the highway exceeds a fixed, predetermined value.
4. A car can go to another ring when changing lanes if the distance between cars is more than half of the fixed value.

The car tries to take a position in the middle between the front and rear vehicles until it moves away from them at a fixed distance.

Table 1. Results of computer experiments

	Cars			Robotic vehicles		
Traffic	0.14	0.28	0.44	0.18	0.34	0.58
Exceeding the time	1.22	1.28	1.47	1.08	1.15	1.35
Lengthening the Path	1.19	1.25	1.44	1.03	1.17	1.21

It should be noted that there is always a car located at the back and in front of each car in the case of a ring. The distances between them can be large. Traffic jams were not observed in the case of movement along the ring, subject to these rules. Therefore, rule 4 is universal. The results of computer experiments are shown in table 1 for robotic vehicles and cars.

The best movement parameters are observed in the case of robotic vehicles.

7 Conclusion

The obtained interaction rules allow organizing traffic without traffic jams, without traffic lights and without collisions of vehicles in the proposed city model. A large number of experiments have been carried out with different initial conditions. Traffic jams were not observed when the traffic intensity reached 58% of the theoretically possible in the case of robotic vehicles and 44% in the case of drivers driving cars while observing the rules of interaction. The molecular dynamics model has been proposed and tested in difficult traffic situations. The proposed model has shown its effectiveness. The methodical value of the work lies in this.

References

[1] A.O. Aristov, J. Phys.: Conf. Ser., **1392** 012064 (2019)

[2] A.O. Aristov, Mining Informational and Analytical Bulletin, **1**, 201 (2018), DOI: 10.25018/0236-1493-2018-1-0-201-214

[3] M.J. Lighthill, G.B. Whitham, Proc. Roy. Soc. Ser. A., **229**, 317 (1955), DOI: 10.1098/rspa.1955.0089.

[4] V.I. Protasov, Z.Ye. Potapova, A.V. Sharonov, Izvestiya Instituta inzhenernoy fiziki, **1(47)**, 78 (2008)

[5] A.O. Aristov, *Quasicellular Network Theory* (MISiS, Moscow, 2014)

[6] S. Psakhie, G.-P. Ostermeyer, A.I. Dmitriev, E.V. Shilko, Phys. Mesomech., **3(2)**, 5 (2000)

[7] W. Wang, W. Fu, H. Zhang, Y. Wanget, Res. J. Appl. Sci. Eng. Technol., **6**, 1598 (2013), DOI: 10.19026/rjaset.6.3876

[8] D. Inihov, Y. Kolesov, Y. Senichenkov, IFAC Proceedings Volumes, **45(2)**, 49 (2012), DOI: 10.3182/20120215-3-AT-3016.00008

[9] N. Moseev, V. Chudinov, B. Goshchitski, V. Protasov, Physica, Physica Status Solidi (a), **94(a)**, 147 (1986), DOI: 10.1002/PSSA.2210940116

[10] S. Stephan, J. Liu, K. Langenbach, W. Chapman, H. Hasse, The Journal of Physical Chemistry, **122(43)**, 24705 (2018), DOI: 10.1021/acs.jpcc.8b06332

[11] V. Chudinov, V. Protasov, Evolution of cascade region in Nb3Sn during thermal spice under fast neutron irradiation. Physica status solidi, **80(a)**, 349 (1983), DOI: 10.1080/00337578208211468

[12] M. Maximov, V. Protasov, Z. Potapova, A. Sharonov, I. Shatalov, Amazonia Investiga, **9(28)**, 404 (2020), DOI: 10.34069/AI/2020.28.04.45

MATEC Web of Conferences **362**, 01023 (2022) https://doi.org/10.1051/matecconf/202236201023
CMMASS 2021

On the question of the no symmetry of the stress tensor for open systems

Evelina Prozorova[1]

[1]Mathematics and Mechanics Faculty, St. Petersburg State University, 28, Universitetsky pr., Stary Peterhof, 198504, Russia

Abstract. The paper discusses the influence of stress no symmetry in continuum mechanics and in stochastic processes. The no symmetry of the stress tensor arises in connection with the influence of the moment and leads to the ambiguity of the solutions of the equations of classical mechanics and the transformation of two-dimensional structures into three-dimensional in the absence of compensating forces. The model for a continuous medium was constructed earlier from phenomenological considerations and from kinetic theory. The article proposes a modification of the classical Boltzmann equation with the inclusion of the diffusion process and the moment in addition to the convective operator. The importance of these terms in generalized kinetic equations and in stochastic processes (Brownian motion and Landau damping) is assumed. In the last equations the probabilistic terms are replaced by deterministic ones.

1 Introduction

It is known that a significant difference between theory and experiment occurs in those cases when the gradients of physical quantities (velocity, temperature, density, charge, etc.) are significant. The discrepancy between experiment and theory appears in problems of free oscillations, during wave propagation and forced high-frequency (ultrasonic) vibrations. Usually, kinetic and gas-dynamic mathematical modeling is carried out within the framework of a continuous medium model with the choice of a type as vector field description, i.e. mathematical points are considered and conservation laws are written out. An abstract point does not allow one to explore the structure, since there is no structure at the point. Extended information can be obtained on the basis of the generalized kinetic equation [1], the study of which is performed by the Fourier method. The equation is written out for a physically defined point. The analysis of the equation shows that this consideration does not take into account the effect of the moment in non-equilibrium processes and does not take into account the effect of flows through the boundary of the elementary volume, since the description of the process through the distribution function does not allow to take them into account through the boundary [2]. With high-frequency vibrations and sufficiently short wavelengths, the influence of the microstructure of the material inevitably affects.

The most common model that takes into account the microstructure of the material is the Cosserat moment model or similar models [3]. The proposed theories can appear only in the presence of a finite linear structural length parameter. In this case, the proper moment of the environment is actually taken into account. However, the law of conservation of moment is often not considered in this case. To construct conservation laws when using the Ostrogradsky–Gauss theorem and replacing the integral over the surface with an integral over the volume, we neglect the integral term outside, i.e. we neglect the circulation along the sides of the elementary volume (in the two-dimensional case, this is clearly seen) [4].

Circulation means the presence of rotation, which in turn means the presence of a moment of force (angular momentum). As a result, we have a symmetric stress tensor, a symmetric velocity tensor, etc. It does not follow from the Boltzmann equation that the pressure in the Euler and Navier–Stokes equations is equal to one third of the sum of the pressures on the corresponding coordinate axes. Using Pascal's law for equilibrium, the pressure is chosen equal to one third of the pressure on the coordinate pads. However, the theory remains the same when determining different pressures at each of the sites, i.e. p_x, p_y, p_z.

The use of one pressure is possible under equilibrium conditions (Pascal's law), but for nonequilibrium conditions the fact is not obvious. Attention is drawn to this in the textbook of L.G. Loityansky [5]. The role of the moment is clearly traced in meteorology through the formation of vortex formations of various types. In the theory of elasticity, when considering the relationship between the components of the strain tensor and the stress tensor, is used the experimental fact of the change in the components p_x, p_y, p_z (stresses) normal to the elementary area (stresses) that are proportional to the sum of the stresses of other components, and they all differ. It should be noted that in the nonequilibrium case there will always be a moment around the axis outside the plane. If it is not balanced, the problem will become three-dimensional.

The existing theory is connected with the fact that when deriving conservation laws in the theory of elasticity, the contribution of the distributed moment to the conditions of equilibrium of forces is excluded. As a result, the law of balance of forces and moments of forces are considered separately. The inaccuracy in determining the velocities in the stress tensor does not greatly affect the results at low velocities. The peculiarity of classical equations is the absence of "nesting" of simpler equations into more complex ones. In stochastic processes of open systems, the movement of fast molecules is accompanied by a change in the position of the center of inertia, which is accompanied by the appearance of a moment. For the processes under study, a binary interaction is selected that depends only on the distance between particles. The appearance of a moment leads to a change in the direction of velocities and the formation of local structures. It can be assumed that the moment will make a significant contribution to the equation of state for liquids and gases at high pressures (virial equation), since there are fast and slow molecules even for equilibrium distribution functions. The effect of the velocity distribution is most clearly manifested for locally equilibrium conditions in the Langevin and Fokker–Planck equations and in the Landau damping in plasma, in the self-organization of nanostructures. In the general case, during the dynamic formation of the structure, the position of the center of inertia changes, which entails a change in the angular momentum. In fact, when considering the collision integral in the Boltzmann equation, the distribution function changes not only in space, but also in velocities. These changes are taken into account in the listed equations.

A modification of the theory is proposed that eliminates the noticed inaccuracies. In this work, a mathematical analysis of the equations of continuum mechanics, including the system of equations of the theory of elasticity with the original asymmetric stress tensor, is carried out. It turned out that in the plane case for four unknowns in the classical formulation we have three equations: two equations from the stress equilibrium condition and one equation - the moment equilibrium condition.

Unknown voltages: σ_{xx}, σ_{yy}, τ_{xy}, τ_{yx}. Thus, we need to close the problem with an additional condition. In the classical version, this condition is the symmetry condition of the stress tensor. However, you can think of other closure options. The absence of symmetry, as noted, is also indicated by the use of the Ostrogradsky–Gauss theorem for a medium with rotation, although a stationary medium is usually considered. It is possible to obtain a solution to a problem with an no symmetric tensor using an iterative procedure. To do this, you must first solve the problem with a symmetric tensor, leave one solution that was obtained, then use the second equation and solve it as a differential one. This will be one of the possible solutions. In three-dimensional problems, three equations of stress equilibrium and three equations for moments are solved. The algorithm for solving a three-dimensional problem is the same as for a two-dimensional one. Sometimes, when solving problems of the theory of elasticity, the first invariants are used, but this can be done if the tensor is symmetric. When considering vortex motion, the tensor is not symmetric (Love's method) [5].

The equilibrium condition is met if there are no internal and external forces, however, any surface forces can be converted by changing variables into internal forces and deformations will occur. It can be shown that the stress tensor will be symmetric only under equilibrium conditions, but it is impossible to transfer the results to nonequilibrium processes. In equilibrium in continuum mechanics, equilibrium means the uniformity of the distribution of all macroparameters. However, any additional external influence leads to the emergence of certain disturbances and creates a distributed moment, which creates additional force.

The processing of numerical calculations in early works was carried out without averaging the values along the sides of the elementary volume. In the modern theory of elasticity, problems are solved by the finite element method, after which the results are averaged. The finite volume method is often used to numerically solve the Euler and Navier–Stokes equations. However, its formulation is based on differential equations with a symmetric stress tensor. In the mechanics of liquid and gas, it is not customary to average the calculation results over the calculated cells. Consequently, since the mesh has finite dimensions, in old works, the asymmetry of the tensor is partially taken into account in numerical calculations. With a large number of steps, averaging gives significant errors. The main argument in favor of choosing a symmetric tensor is the use of the force equilibrium equation without the influence of the moment. However, the gradient of the moment is power. Consequently, the requirement of simultaneous fulfillment of the laws of the balance of forces and moments of forces changes the equations. In modern experiments with rods, measurements are taken at the center of the rod and do not provide information about the action of the moment, since they are actually one-dimensional measurements. For a model with a symmetric tensor, it is necessary to fulfill the law of conservation of the moment (equality of the mixed derivatives of the projections of the corresponding moments) or the conjugation condition. However, this is rarely done.

It can be assumed that the moment will make a significant contribution to the equation of state for liquids and gases at high pressures, since there are fast and slow molecules even for equilibrium distribution functions. The movement of fast molecules is accompanied by a change in the position of the center of inertia, which is accompanied by the appearance of a moment. The appearance of a moment leads to a change in the direction of velocities and the formation of local structures. This is shown in [6–8]. It seems unnecessary to separate the divergent and rotational components for speed. At low speeds and the previous level of computer technology, this had to be done, especially since the speeds for with and without vortex in no compressible fluid coincide. The physical sense is also well tracked. Currently, there is no need for such a division. The main goal of the work is to show the role of moment in all processes of collective interaction of particles in open systems, while in classical mechanics closed systems are considered and the role of moment is often not taken into account. In the proposed version of the theory, the reversibility of the equations remains the same, but the role of boundary conditions increases. They also receive little attention in modern theory. For example, for the Liouville equation.

2 Examples of the influence of the moment in the theory of elasticity

The problems are taken from the book [9]. Consider the problem when a concentrated force is given on a half-plane. Let us find the stress distribution with and without taking into account the influence of the moment (figure 1(a)). Consider two equilibrium equations forces and the equation of equilibrium of moments

$$\frac{\partial\sigma_x}{\partial x}+\frac{\partial\sigma_{zx}}{\partial z}=0,\quad \frac{\partial\sigma_z}{\partial z}+\frac{\partial\sigma_{xz}}{\partial x}=0,$$
$$x\left(\frac{\partial\sigma_{xz}}{\partial x}+\frac{\partial\sigma_{zz}}{\partial z}\right)-z\left(\frac{\partial\sigma_{xx}}{\partial x}+\frac{\partial\sigma_{zx}}{\partial z}\right)+\sigma_{zx}-\sigma_{xz}=0.$$

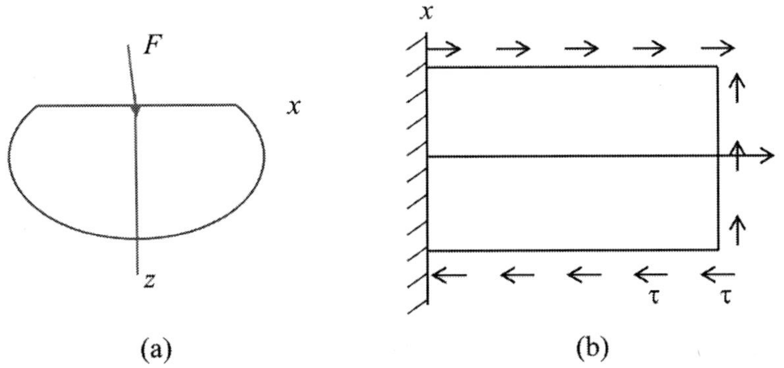

Figure 1. Formulation of the problem one (a), and the problem two (b)

Suggested that $\sigma_{zx}=-A\frac{x}{x^2+z^2}$, option for case $\sigma_{xz}=\sigma_{zx}$ and

$$\sigma_z=-A\frac{z^3}{(x^2+z^2)^2}$$

was considered in [4]. First, we solve the problem when the symmetry condition of the stress tensor is satisfied $\sigma_{xz}=\sigma_{zx}$

$$\frac{\partial\sigma_{zx}}{\partial z}=-A\frac{2xz}{(x^2+z^2)^2},$$
$$\sigma_x=-2Az\int_0^x\frac{x\,dx}{(x^2+z^2)^2}-Az\int_0^y\frac{dy}{(y+z^2)^2}=Az\frac{1}{(z^2+y)}+f(z),$$
$$\sigma_x=Az\frac{1}{(z^2+x^2)}+f(z),$$
$$\frac{\partial\sigma_{xz}}{\partial x}=-A\left(\frac{1}{x^2+z^2}-\frac{2x^2}{(x^2+z^2)^2}\right)=-A\left(-\frac{z^2-x^2}{(x^2+z^2)^2}\right),$$
$$\sigma_z=-A\int_0^z\left(\frac{1}{x^2+z^2}-\frac{2x^2}{(x^2+z^2)^2}\right)dz=$$
$$=-A\frac{1}{x}\operatorname{arctg}z-A\left(\frac{z}{x(1+z^2)}+\frac{1}{x}\operatorname{arctg}z\right)=-A\frac{z}{x(1+z^2)}+f(x).$$

Let σ_x and σ_{zx} remain the same. Then from the equation for the moments it is possible to determine the difference a new value σ_{xz}

$$\frac{\partial\sigma_{xz}}{\partial x} - \frac{\sigma_{xz}}{x} + \frac{\partial\sigma_{zz}}{\partial z} = 0,$$

$$x\left(\frac{\partial\sigma_{xz}}{\partial x} + \frac{\partial\sigma_{zz}}{\partial z}\right) + \sigma_{zx} - \sigma_{xz} = 0.$$

Then a new value σ_z ($\sigma_z = \sigma_{zz}$) determined. An example is given in order to demonstrate the many possible solutions in the two-dimensional case. Consequently, the solution of the two-dimensional problem under the condition of symmetry of the stress tensor is only one of the options for solving the problem. Algorithm for finding stresses with an no symmetric tensor easily programmed using previously obtained results. Consider the combined action of shear stresses τ (figure 1(b)).

The solution of e equations with a symmetric tensor is the function

$$\tau_{xz} = \tau_{zx} = \varphi(x), \quad \sigma_z = \varphi' z + R(x).$$

From the equation for the moments and iterating, we obtain

$$\tau_{xz} = \varphi(x) + f(x).$$

The functions $\varphi(x)$, $f(x)$ are determined from the boundary conditions.

Algorithm for finding stresses with an asymmetric tensor easily programmed using previously obtained results. Earlier, the no symmetry of the stress tensor was proved in [8], proceeding from the generalization of the Ostrogradsky–Gauss theorem for continuum mechanics. The proposed study complements the previously obtained results. Thus, the symmetry condition for the stress tensor is one of the possible conditions for the closure of the system of equations of the theory of elasticity. Consequently, for structural particles, in addition to the system of internal moments, there are moments associated with the action of stress gradients, and they need be taken into account.

3 The influence of the angular momentum in stochastic processes

Consider the classical problems of nonequilibrium statistical mechanics [10–14]. One of the problems, the problem of Brownian motion, is associated with the motion of a heavy colloidal particle placed in a liquid consisting of light particles. The solution to the problem is associated with the introduction of a probabilistic process. Let's try to replace it with a deterministic one. The process of motion of a Brownian particle is referred to as stochastic. Let us first discuss the changes in the Boltzmann equation taking into account the influence of the moment. For the generalized kinetic equation in the Landau representation, it is also required to introduce new terms.

The classical kinetic theory (modified and classical) also does not follow Pascal's law. Consider the classical Boltzmann equation (the law of momentum does not hold)

$$f(t+dt, r+\xi_j dt, \xi_j + F_j dt)\, dr\, d\xi_j = f(r_i, \xi_j, t)\, dr\, d\xi_j + \left(\frac{\partial f}{\partial t}\right)_{coll} dt,$$

$$u(t,x) = \frac{1}{n}\int \xi f(t,x,\xi)\, d\xi, \quad P_{ij} = m\int c_j c_i f(t,x,\xi)\, d\xi,$$

$$q_i = m\int c^2 c_i f(t,x,\xi)\, d\xi, \quad c = \xi - u.$$

f—is the distribution function, r—is the radius vector; x—point coordinate; ξ—is the velocity of a point, is the molecular weight, and, according to the definition of the distribution function f_N, the probability of finding the system at points (x, ξ) in the intervals $dx_i\, d\xi_i$ is

$$f_N(t, x_1, x_2, \ldots, x_N, \xi_1, \xi_2, \ldots, \xi_N)\, dx_1 \ldots dx_N\, d\xi_1 \ldots d\xi_N.$$

Taking into account rotation and diffusion, the equation has the form

$$f\left(t + dt, r + \xi_i\, dt + r \times \omega, \xi_i + F_i\, dt + \frac{\partial M}{\partial r} dt\right) dr\, d\xi_i + G_2(t + dt, r + \xi_i\, dt + r \times \omega\, dt),$$

$$\xi_i + F_i\, dt + \frac{\partial M}{\partial r} dt = f(t, r, \xi_i)\, dr\, d\xi_i + G_1(t, r, \xi_i) + \left(\frac{\partial f}{\partial t}\right)_{coll} dt.$$

M is the moment associated with the collective action of all particles on each other as a result of the displacement of the center of inertia, which is the result of the movement of particles with different speeds. G_1 and G_2—flows through the boundaries of the considered elementary volume. These values $G_i = m\xi_i \frac{\partial f}{\partial r}$. They are responsible for the additional terms obtained by S.V. Vallander from phenomenological considerations [15]. When calculating macroparameters through the distribution function and projecting values on the coordinate axis, the symmetry of some quantities not may be violated. This can happen when calculating the pressure and the pressure tensor:

$$P_{ij} = m \int c_i c_j f(t, x, \xi)\, d\xi.$$

The symmetry of the stress tensor is postulated on the basis of this form. In aeromechanics, the projections of the calculated values used, and not the indices of the velocities included in the formula. Therefore, there is no way to speak unambiguously about symmetry.

The theory of Brownian motion is one of the main branches of the statistical theory of open systems. In the theory of Brownian motion elementary objects are small particles, while in kinetic theory, the main objects are molecules. Both models are macromodels, but the level of description of the structure of the environment is different. The kinetic equation corresponds to a more detailed description. We believe that the environment is in equilibrium. We will consider two approaches to solve problems: the equation for a single particle and for an ensemble of particles (the Fokker–Planck equation) To take into account the atomic structure of a liquid, Langevin introduced an additional force into the equations of motion

$$F_L = mv(t), \quad F = -m\gamma v, \quad \gamma = m\eta, \quad \eta = \rho v.$$

In the classical case, equilibrium is possible between Brownian particles and the medium; the particles can be distributed evenly. However, such an assumption can be considered unlikely due to the distribution of particles over velocities and the formation of new moments for individual particles due to the motion of the center of inertia. The fact is that in this case the action of the moment creates a force that distributes the particles not only in terms of velocities, but also in coordinates. The proposed modified Fokker–Planck equation has the form:

$$\frac{\partial f}{\partial t} + v\frac{\partial f}{\partial r} - \frac{1}{m}\frac{\partial U}{\partial r}\frac{\partial f}{\partial v} + \frac{1}{m}\frac{\partial M}{\partial r}\frac{\partial f}{\partial v} = D\frac{\partial^2 y}{\partial v^2} + \frac{\partial}{\partial v}(\gamma v f).$$

Thus, in the kinetic theory for a gas, for the Landau damping and for the motion of Brownian particles, the nonuniform distribution of particles in velocities and coordinates is supported by the angular momentum and creates fluctuations in physical quantities that must be taken into account. The action of the moment is local in space and time. Therefore, we can say that the process will still remain "Markov" as in the probabilistic approach.

Let us turn to Landau damping in plasma. The peculiarity of plasma is not the binary interaction of two particles, but the collective one. In calculations, ions can be considered as heavy particles and, consequently considered as Brownian particles.

Therefore, let's move on to the simplest problem. Let us consider oscillations in a plasma without collisions, that is, let us proceed to the study of waves propagating in a plasma, the frequency of which is high in comparison with the frequency of pair collisions of electrons and ions. In this case, there are several options to consider. Landau collisional damping for large Knudsen numbers; for small Knudsen numbers in unbounded plasma; for small Knudsen numbers in a confined plasma. They differ from each other. When studying oscillations, we will consider small deviations from equilibrium Consider an unbounded plasma for small Knudsen numbers $L \ll \lambda$. Diffusion works here. Let us consider the dispersion and damping of longitudinal oscillations of an electron plasma under the influence of the thermal motion of plasma particles [1].

Let us investigate a variant of a limited plasma, a free-molecular flow with a region of wavelengths (values of wave numbers) for which the contribution corresponding to Landau damping is the main one.

$$l \to L^0 \quad (l \gg L), \qquad r_D \ll \gamma \ll \sqrt{r_D L} \quad (r_D \ll L \ll l).$$

We must use the Vlasov kinetic equations with a self-consistent field. Since we are interested in high-frequency oscillations, for which $\omega\tau \gg 1$, where τ is the average time between pair collisions of particles, we can ignore the integrals of particle collisions in the kinetic equations. Longitudinal oscillations of an electron plasma in the classical case are described by the following two equations (collisionless case, Vlasov equation)

$$\frac{\partial \delta f}{\partial t} + v\frac{\partial \delta f}{\partial r} + e\delta E\frac{\partial f_0}{\partial p} = 0,$$

$$div\,\delta E = 4\pi \int dp\,\delta f,$$

$$\varepsilon_l(\omega, k)e\int dc\,\delta f(p, k, \omega) = i\int dre^{-ikr}e\int dp\frac{\delta f(p, r, t_0)}{\omega - kv}.$$

Proposed option

$$\frac{\partial \delta f}{\partial t} + v\frac{\partial \delta f}{\partial r} + e\delta E\frac{\partial f_0}{\partial p} + \frac{\partial \delta M}{\partial r}\frac{\partial f_0}{\partial p} + \frac{\partial}{\partial r}D\frac{\partial \delta f}{\partial r} = 0,$$

$$div\,\delta E = 4\pi \int dp\,\delta f,$$

$$\varepsilon_l(\omega, k)e\int dc\delta f(p, k, \omega) = i\int dre^{-ikr}e\int dp\frac{\delta f(p, r, t_0)}{\omega - kv} + i\int dre^{-ikr}\int dp\frac{\delta M(p, r, t)}{\omega - kv}\frac{\partial f_0}{\partial p}.$$

Qualitatively, we can say that for this case, diffusion plays a small role and, since part of the energy is converted into rotational motions (the action of the moment), the reversible operator will act as a dissipative one. Note that at the initial moment, the distributed moment of force also exists and concentrates a certain amount of energy. For monochromatic waves of large amplitude, the action (principles) can lead to the formation of a vertical velocity component, forming complex plane flows. Despite the collisionless nature of the movement binary collisions exist [16]. They create additional dissipation. It should be noted that the generalized equation for a unified description of kinetic and gas-dynamic processes is suitable for "weak" interactions. As before, the contribution of the moments in the motion of molecules is not taken into account. Most likely, the difference between the most probable and average values is due precisely to the lack of taking into account the rotational movements for which the moment is responsible.

4 Conclusion

Open systems are among the most real systems encountered in practice. Building models of such systems is an important task. The paper analyzes the effect of stress no symmetry in models of continuum mechanics and in stochastic processes of open systems. The no symmetry of the stress tensor associated with the work of the moment leads to ambiguity in the solutions of the equations of classical mechanics and the transformation of two-dimensional structures into three-dimensional in the absence of compensating forces.

The substantiation of the no symmetry of the strain tensor on the basis of continuum mechanics and kinetic theory is considered. A modification of the classical Boltzmann equation with the inclusion of a diffusion process and a moment is proposed. The importance of these terms in generalized kinetic equations and in stochastic processes (Brownian motion and Landau damping) is assumed. In the last equations, the probabilistic terms are replaced by deterministic ones.

References

[1] Y.L. Klimontovich, *Statistical theory of open systems*, vol. 1, 2 (Lenand, Moscow, 2019)
[2] E.V. Prozorova, JP Journal of Heat and Mass Transfer, **17**, 1 (2019)
[3] E. Cosserat, F. Cossert, *Theories des corps deformables* (Hermann, Paris, 1909)
[4] E. Prozorova, *Ostrogradsky–Gauss Theorem for Problems of Gas and Fluid*, J. Phys.: Conf. Ser., **1334**, 012009 (2019)
[5] L.G. Loytsyansky, *Mechanics of liquid and gas* (Nauka, Moscow, 1970)
[6] E. Prozorova, WSEAS Transaction on Applied and Theoretical Mechanics, **15** (2020)
[7] E. Prozorova, WSEAS Transaction on Fluid Mechanics, **15** (2020)
[8] E. Prozorova, WSEAS transactions on Applied and Theoretical Mechanics, **16** (2021)
[9] E.A. Bulanov, *Moment stresses in the mechanics of solid, bulk and liquid bodies* (University book, Moscow, 2012)
[10] R. Balescu, *Equilibrium and nonequilibrium statistic mechanics*, (A Wiley-Intersciences Publication John Willey and Sons, New-Yourk–London, 1975)
[11] P. Resibois, M. De Lener, *Classical kinetic theory of fluids* (John Wiley and Sons, New-York–London, 1977)
[12] N.G. Van Kampen, *Stochastic processes in physics and chemistry* (North-Holland, Amsterdam, 1991)
[13] *Physics of Simple Liquids*, ed. by H.N.V. Temperley, I.S. Rowlinson, G.S. Rushbrooke (North Holland Publishing Company, Amsterdam, 1968)
[14] A.A. Vlasov, *Nonlocal statistical mechanics* (Nauka, Moscow, 1978)
[15] S.V. Vallander, M.P. Elovskih, Reports of the ANAS, **79**, 37 (1951)
[16] F. Goodman, G. Vachmann, *Dynamics of gas scattering by a surface* (Mir, Moscow, 1980)

Representation of the thermo-stress state of a plate based on the 3D elasticity theory

Victor Revenko[1,*] and *Vladimir* Bakulin[2]

[1]Pidstryhach Institute for Applied Problems of Mechanics and Mathematics NASU, 3-b, Naukova st., Lviv, 79060, Ukraine

[2]Institute of Applied Mechanics of Russian Academy of Sciences, 7, Leningradsky prospekt, Moscow, 125040, Russia

Abstract. The general formulation of the 3D temperature static problem of the theory of elasticity is considered. A new representation of the general solution of static equations of thermoelasticity is proposed. The three-dimensional stress state of a thick or thin plate are divided into three parts: bending of the plate, symmetric compression of the plate on it's flat end surfaces and symmetric temperature problem. Third thermoelastic stress state of the plate is studied in detail. It have been reduced to a two-dimensional state after integration the stresses components along the normal to the middle surface. The displacements and stresses in the plate are expressed in terms of two two-dimensional harmonic functions and a particular solution, which is determined by a given temperature on the flat surfaces of the plate. Only the assumption has been used that the normal stresses perpendicular to the middle surface are insignificant. The introduced harmonic functions are determined from the boundary conditions on the lateral surface of the plate. The formula for the experimental determination of the sum of normal stresses has been found through the measured deflections of the end surfaces of the plate.

1 Introduction

Elastic materials are widely used in aerospace and other engineering, therefore, solving the equations of the theory of thermoelasticity and studying their stress-strain state is an important scientific problem [1–3]. Thin and thick plates, to which forces loads, given temperature fields and internal heat sources are applied, are also widely used in power engineering, technological and engineering structures [3, 4]. In [2, 5], a review of the literature on the representation of the equations of the plane theory of thermoelasticity is made using the hypotheses of plane stresses for thin plates .

The purpose of research is to use a three-dimensional thermoelastic stress state to construct a two-dimensional theory of thin and thick plates, which are under the influence of power and temperature loads, without using the hypotheses of plane stresses.

*e-mail: victor.rev.12@gmail.com

2 Formulation of the problem and construction of basic equations

Let us consider the general formulation of a three-dimensional temperature static problem of the theory of elasticity for thin or thick plates of constant thickness h, the middle surface of which occupies a region S with a contour L and is located in the plane Oxy of the Cartesian coordinate system: $x_1 = x,\ x_2 = y,\ x_3 = z$. The lateral surface of the plate is denoted by Z. Normal loads $q_j(x, y)$ are applied to the flat end surfaces of the plate ($z = h_j$, $h_1 = h/2$, $h_2 = -h/2$) and temperatures $t_j(x, y)$, $j = \overline{1,2}$ are set. But there are no tangential loads. We will assume that the temperature field $T(x, y, z)$ in the plate is known.

We use the symmetry of the equations of the theory of elasticity and divide the three-dimensional problem into two parts [6]. For the first task describing the symmetrical bend of the plate, and normal loads of the flat surfaces of the plate are asymmetrical:

$$\sigma_z(x, y, h_1) = g^+(x, y), \quad \sigma_z(x, y, h_2) = -g^+(x, y),$$

and for the second, which describes symmetric compression, are respectively equal

$$\sigma_z(x, y, h_1) = p^+(x, y), \quad \sigma_z(x, y, h_2) = p^+(x, y), \tag{1}$$

where $g^+ = \frac{1}{2}(q_1 - q_2)$, $p^+ = \frac{1}{2}(q_1 + q_2)$, signs "+", "−" describe the values of the corresponding functions on the upper and lower surfaces of the plate. The temperature field is known for each task. For the first problem on the lateral surface Z, the normal stresses will be asymmetric with respect to the middle surface, and for the second, they will be symmetric.

Let us divide the second problem into two problems: 2) symmetric force compression of the plate on flat end surfaces [7]; 3) a symmetric plane temperature problem, when temperatures are known $t^- = t$, $t^+ = t$, $t = \frac{1}{2}(t_1 + t_2)$. There are no forces loads.

Let us consider in detail the third problem. It is described by normal stresses symmetric with respect to the middle surface, and in conditions (1) must be set $p^+ = 0$. On the lateral surface of the plate Z, we have the following three-dimensional boundary conditions:

$$[\sigma_y \sin^2\alpha + \sigma_x \cos^2\alpha + \tau_{xy}\sin 2\alpha]\,|_Z = \tau_{1n}|_Z, \quad \left.\left[\frac{\sin 2\alpha}{2}(\sigma_y - \sigma_x) + \tau_{xy}\cos 2\alpha\right]\right|_Z = \tau_{2n}|_Z,$$

$$[\tau_{xz}\cos\alpha + \tau_{yz}\sin\alpha]\,|_Z = \tau_{3n}|_Z, \tag{2}$$

where $\tau_{jn}(x, y, -z) = \tau_{jn}(x, y, z)$, $j = \overline{1,2}$, $\tau_{3n}(x, y, -z) = -\tau_{3n}(x, y, z)$ are normal and tangential loads, α is the angle between the normal to the contour L and the axis Ox. From conditions (1), (2) it follows that normal stresses will be symmetrical: $\sigma_j(x, y, -z) = \sigma_j(x, y, z)$, $j = \overline{1,3}$.

The general representation of thermoelastic stresses is expressed for the three considered problems in terms of deformations [1, 3]

$$\sigma_k = 2G\left[\varepsilon_k + \frac{\nu}{1-2\nu}e - \frac{1+\nu}{1-2\nu}\alpha T(x, y, z)\right], \quad \tau_{kj} = G\gamma_{kj}, \quad k \neq j, \tag{3}$$

where $e = \frac{1-2\nu}{E}\Theta + 3\alpha T$ is volume deformation, $\Theta = \sigma_1 + \sigma_2 + \sigma_3$.

We substitute relation (3) into the equilibrium equations and write down the equations of stationary thermoelasticity in displacements [1, 3]

$$(1-2\nu)\nabla^2 u_k + \frac{\partial e}{\partial x_k} = 2(1+\nu)\alpha\frac{\partial T}{\partial x_k}, \quad k=\overline{1,3}, \tag{4}$$

$$\nabla^2 T(x,y,z) = 0, \quad \nabla^2 = \frac{\partial^2}{\partial x_1^2} + \frac{\partial^2}{\partial x_2^2} + \frac{\partial^2}{\partial x_3^2}, \tag{5}$$

where ν is the Poisson's ratio, α is the coefficient of thermal expansion.

We use the general solution of the Lamé equations [8], add to it the well-known thermoelastic potential [1, 3] and write down a new representation of the general solution of the thermoelasticity equations (4), which describes the thermoelastic stress state of the plate

$$u_x = \frac{\partial P}{\partial x} + \frac{\partial Q}{\partial y}, \quad u_y = \frac{\partial P}{\partial y} - \frac{\partial Q}{\partial x}, \quad u_z = \frac{\partial P}{\partial z} - 4(1-\nu)\Phi, \tag{6}$$

where $P = z(\Phi + \beta\Omega) + \Psi$; Φ, Ψ, Q are three-dimensional harmonic displacement functions; $\Omega = \int_0^z T\,dz$; $\beta = \frac{1+\nu}{2(1-\nu)}\alpha$; Ω, T are known functions that satisfy equation (5); $T(x,y,-z) = T(x,y,z)$. The biharmonic function P, Q satisfie the equations

$$\Delta P + \frac{\partial^2}{\partial z^2}P = 2\frac{\partial}{\partial z}(\Phi + \beta\Omega), \tag{7}$$

where $\Delta = \frac{\partial^2}{\partial x^2} + \frac{\partial^2}{\partial y^2}$ is two-dimensional Laplace operator.

We use the displacement representation (6) and find the deformations. Using formula (3), we determine the general expression for the normal stresses

$$\sigma_j = 2G\left[\frac{\partial^2 P}{\partial x_j^2} - (-1)^j\frac{\partial^2 Q}{\partial x_1\,\partial x_2} - 2\nu\frac{\partial\Phi}{\partial x_3} - 2\beta T\right], \quad j=\overline{1,2},$$

$$\sigma_3 = 2G\left[\frac{\partial^2 P}{\partial x_3^2} - 2(2-\nu)\frac{\partial\Phi}{\partial x_3} - 2\beta T\right], \tag{8}$$

and shear stresses

$$\tau_{12} = G\left[2\frac{\partial^2 P}{\partial x_1\,\partial x_2} + \frac{\partial^2 Q}{\partial x_2^2} - \frac{\partial^2 Q}{\partial x_1^2}\right],$$

$$\tau_{j3} = G\left[\frac{\partial}{\partial x_j}\left[2\frac{\partial P}{\partial x_3} - 4(1-\nu)\Phi\right] - (-1)^j\frac{\partial^2 Q}{\partial x_{3-j}\,\partial x_3}\right], \quad j=\overline{1,2} \tag{9}$$

where $G = E/2(1+\nu)$, E are shear and Young's moduli. Let us express the volume deformation e and the sum of normal stresses Θ

$$e = -2(1-2\nu)\frac{\partial}{\partial z}\Phi + 2\beta T, \quad \Theta = -2E\left(\frac{\partial}{\partial z}\Phi + \frac{\alpha T}{1-\nu}\right). \tag{10}$$

According to the problem statement and conditions (1), there will be zero stresses on the plate surfaces. Let us take into account the symmetry conditions, relations (8), (10) and the fact that the normal stresses $\sigma_3 \ll \sigma_1$, $\sigma_3 \ll \sigma_2$, after its integration, we obtain

$$\frac{\partial P^+}{\partial z} = 2(2-\nu)\Phi^+ + 2\beta\Omega^+, \tag{11}$$

where $\Omega^+ = \int_0^{h_1} T\,dz$.

We use representations (9) and write down the conditions for the absence of tangential loads on the plate surfaces

$$\frac{\partial}{\partial x_j}\left[\frac{\partial P^+}{\partial x_3} - 2(1-\nu)\Phi^+\right] - \frac{(-1)^j}{2}\frac{\partial^2 Q^+}{\partial x_{3-j}\partial x_3} = 0, \quad j = \overline{1,2}. \tag{12}$$

Let us take into account relation (11) and simplify equations (12)

$$4\frac{\partial}{\partial x_j}(\Phi^+ + \beta\Omega^+) = (-1)^j\frac{\partial^2 Q^+}{\partial x_{3-j}\partial x_3}, \quad j = \overline{1,2}. \tag{13}$$

From equations (13), after mathematical transformation, we obtain that the values of the functions on the surface of the plate will be harmonious functions:

$$\Delta(\Phi^+ + \beta\Omega^+) = 0, \quad \Delta\frac{\partial Q^+}{\partial z} = 0. \tag{14}$$

After integration of equation (7) along the normal to the middle surface, taking into account relations (11), (14), we write down the key equations of the theory of plates on the basic functions $\tilde{P}$, $\tilde{Q}$, Φ^+

$$\Delta\tilde{P} = -4(1-\nu)\Phi^+, \quad \Delta\tilde{Q} = -2\frac{\partial}{\partial z}Q^+, \tag{15}$$

where $\tilde{P} = \int_{-h_1}^{h_1} P\,dz$, $\tilde{Q} = \int_{-h_1}^{h_1} Q\,dz$. Equations (13)–(15) and the known function Ω^+ are sufficient to represent these functions.

We had used equations (6)–(10) and reduced the three-dimensional problem for a symmetrically loaded thick plate to a two-dimensional problem, without had used the hypotheses of the plane stress state. We used only the statement physically substantiated for the plate that the normal stresses perpendicular to the middle surface are insignificant in comparison with the longitudinal stresses.

We use relations (8), (9), (14), (15) and construct a two-dimensional mathematical theory of a symmetrically loaded heat-sensitive thick plate. To do this, we substitute three-dimensional stresses (8), (9) into the well-known expressions of normal and tangential efforts [7] and after integrating them over the variable we obtain:

$$T_1 = 2G\left[\frac{\partial^2\tilde{P}}{\partial x^2} + \frac{\partial^2\tilde{Q}}{\partial x\,\partial y} - 4\nu\Phi^+ - 4\beta\Omega^+\right], \quad T_2 = 2G\left[\frac{\partial^2\tilde{P}}{\partial y^2} - \frac{\partial^2\tilde{Q}}{\partial x\,\partial y} - 4\nu\Phi^+ - 4\beta\Omega^+\right],$$

$$S_{12} = S_{21} = 2G\left[\frac{\partial^2\tilde{P}}{\partial x\,\partial y} + \frac{1}{2}\left(\frac{\partial^2\tilde{Q}}{\partial y^2} - \frac{\partial^2\tilde{Q}}{\partial x^2}\right)\right]. \tag{16}$$

Relationships (15) (16) coincide with the key equations of the plane stress state found in [7], when $\Omega^+ = 0$.

Note that if we substitute relations (16) into the equilibrium equation of the plate in the efforts [9], then we obtain the equation in partial derivatives of the fourth order

$$\Delta\Delta\tilde{P} = 4\beta(1-\nu)\,\Delta\Omega^+. \tag{17}$$

Equation (17) also follows from the obtained relations (14), (15).

3 Representation of thermoelastic stresses in terms of harmonic functions

Let us take into account the first harmonic equation of relations (14) and determine the representation of the function Φ^+

$$\Phi^+ = -h\frac{\partial^2 \varphi}{\partial y^2} - \beta\Omega^+, \tag{18}$$

where φ is unknown harmonic function. We use expression (18), relation (13) and obtain the following simple dependence:

$$\frac{\partial Q^+}{\partial z} = 4h\frac{\partial^2 \varphi}{\partial x\,\partial y}. \tag{19}$$

Let us take into account representations (18), (19) and write down the general solution of equations (15)

$$\tilde{P} = 2(1-\nu)h\left[y\frac{\partial \varphi}{\partial y} + \beta\omega_1\right] + hg_1(x,y), \quad \tilde{Q} = -4yh\frac{\partial \varphi}{\partial x} + hg_2(x,y), \tag{20}$$

where ω_1 is a particular solution of the equation

$$h\,\Delta\omega_1 = 2\Omega^+, \tag{21}$$

g_j, $j = \overline{1,2}$ are harmonic functions that can be represented

$$g_1 = (1+\nu)h\left[\frac{\partial \phi}{\partial x} - \frac{\partial \psi}{\partial x}\right], \quad g_2 = (1+\nu)h\left[\frac{\partial \phi}{\partial y} + \frac{\partial \psi}{\partial y}\right], \tag{22}$$

ϕ, ψ are harmonic functions.

We substitute functions (20), (22) into relations (16) and see that the function ϕ is not included in the representation of efforts (15), so that it can be ignored.

Thus, we can write functions (20) in such a simple form:

$$\tilde{P} = 2(1-\nu)h\left[y\frac{\partial \varphi}{\partial y} + \beta\omega_1\right] - (1+\nu)h\frac{\partial \psi}{\partial x},$$

$$\tilde{Q} = -4hy\frac{\partial \varphi}{\partial x} + (1+\nu)h\frac{\partial \psi}{\partial y}, \quad \Phi^+ = -h\frac{\partial^2 \varphi}{\partial y^2} - \beta\frac{h}{2}\Delta\omega_1, \tag{23}$$

where functions ω_1, Ω^+ describe the effect of temperature on the stress state of the plate, and harmonic functions φ, ψ correspond to the plane stress state.

Let us express the efforts in terms of functions that are determined by the known temperature field. Let us substitute expressions (23) into relations (16) and obtain simple formulae

$$T_1 = -Eh\alpha\frac{\partial^2 \omega_1}{\partial y^2}, \quad T_2 = -Eh\alpha\frac{\partial^2 \omega_1}{\partial x^2}, \quad S_{12} = S_{21} = Eh\alpha\frac{\partial^2 \omega_1}{\partial x\,\partial y}. \tag{24}$$

Let's introduce the biharmonic function $U = 2E\left(y\frac{\partial \varphi}{\partial y} + \frac{\partial \psi}{\partial x} - \frac{\alpha}{2}\omega_1\right)$. From relations (16), (23), (24) we obtain a general representation of thermoelastic stresses, which coincides in form with the known expressions for the stresses of the plane problem of the elasticity theory [7]

$$\sigma_x = \frac{\partial^2 U}{\partial y^2}, \quad \sigma_y = \frac{\partial^2 U}{\partial x^2}, \quad \tau_{xy} = -\frac{\partial^2 U}{\partial x\,\partial y}. \tag{25}$$

We have written the sum of normal stresses (25) and have got

$$\sigma_x + \sigma_y = 4\mathrm{E}\frac{\partial^2 \varphi}{\partial y^2} - \frac{2\alpha E}{h}\Omega^+. \tag{26}$$

We determine the harmonic functions φ, ψ from the boundary conditions (2) specified on the lateral surface of the plate, which, after integrating them, will be reduced to the conditions on the contour L of the region S

$$[\sigma_y \sin^2\alpha + \sigma_x \cos^2\alpha + \tau_{xy}\sin 2\alpha]\,|_L = \sigma_n,$$
$$\left[\frac{\sin 2\alpha}{2}(\sigma_y - \sigma_x) + \tau_{xy}\cos 2\alpha\right]\Bigg|_L = \tau_n, \tag{27}$$

where $\sigma_n = \frac{1}{h}\int_{-h_1}^{h_1} \sigma_n(x, y, z)\,dz\Big|_L$, $\tau_n = \frac{1}{h}\int_{-h_1}^{h_1} \tau_n(x, y, z)\,dz\Big|_L$, α is the angle between the contour normal L and the axis Ox. The third condition (2) is identically satisfied.

We will assume that the partial solution of the equation (21) is known function ω_1. The obtained expressions for stresses (25) and boundary conditions (27) make it possible to solve various boundary value problems for thermoelastic plates. Methods for satisfying boundary conditions (27) were developed in [10–12].

We find the displacements u_x, u_y in the plate after averaging formulas (6)

$$u_x = \frac{1}{h}\left[\frac{\partial \tilde{P}}{\partial x} + \frac{\partial \tilde{Q}}{\partial y}\right] = 2(1+\nu)\left\{\frac{\partial^2 \psi}{\partial y^2} - y\frac{\partial^2 \varphi}{\partial x \partial y}\right\} - 4\frac{\partial \varphi}{\partial x} + 2(1-\nu)\beta\frac{\partial \omega_1}{\partial x},$$
$$u_y = \frac{1}{h}\left[\frac{\partial \tilde{P}}{\partial y} - \frac{\partial \tilde{Q}}{\partial x}\right] = 2(1-\nu)\frac{\partial \varphi}{\partial y} - 2(1+\nu)\left[y\frac{\partial^2 \varphi}{\partial y^2} + \frac{\partial^2 \psi}{\partial x \partial y}\right] + 2(1-\nu)\beta\frac{\partial \omega_1}{\partial y}.$$

From the last formula (6) we will find a deflection of the upper flat surface of the plate

$$u_z^+ = -2h\nu\frac{\partial^2 \varphi}{\partial y^2} + (1+\nu)\alpha\Omega^+. \tag{28}$$

We use formula (26) and write the deflection (28) in another form

$$u_z^+ = -\frac{h\nu}{2\mathrm{E}}(\sigma_x + \sigma_y) + (1+2\nu)\alpha\Omega^+. \tag{29}$$

If we do not take into account the temperature, then formula (29) will have a simple form

$$u_z^+ = -\frac{h\nu}{2E}(\sigma_x + \sigma_y). \tag{30}$$

Formulas (29), (30) can be used to experimental determination of the sum of normal stresses in the plate. They allow experimentally determine the stress concentration on the free contour ($\sigma_n = 0$) of the hole in the plate. We take into account that the equality is performed on the load-free contour of the hole

$$\sigma_\tau = \sigma_x + \sigma_y, \tag{31}$$

where σ_τ is the unknown stress along the contour of the hole, directed perpendicular to the normal to the contour. The stresses is determined based on equations (29), (31)

$$\sigma_\tau = -\frac{2E}{h\nu}u_z^+ + \frac{2(1+2\nu)}{h\nu}E\alpha\Omega^+. \tag{32}$$

We experimentally have measured the deflection u_z^+ and have defined the temperature $\Omega^+ = \int_0^{h_1} T\,dz$. By formula (32) we find the concentration of stresses σ_τ along the free contour of the hole.

4 Conclusions

On the basis of the three-dimensional theory of elasticity, a two-dimensional theory of thermoelastic, symmetrically loaded thick plates is constructed without the use of hypotheses about flat stresses distributions. The found representations of stresses and displacements are exactly equal to the corresponding averaged values of the three-dimensional theory of thermoelasticity. The obtained formulas imply the representation of the stresses of the plane problem of the theory of elasticity. A linear relationship between the normal deflection of the flat surface of the plate and by the value of the sum of normal stresses has been established. This formula is used to experimentally determine the concentration of stresses along the free contour of the hole in the plate.

References

[1] A.D. Kovalenko, *Basics of thermoelasticity* (Naukova dumka, Kiev, 1970)

[2] V.V. Meleshko, *Selected topics in the history of the two-dimensional biharmonic problem*, Appl. Mech. Rev., **56(1)**, 33 (2003), DOI: 10.1115/1.1521166

[3] R.B. Hetnarski, M.R. Eslami, *Thermal Stresses—Advanced Theory and Applications* (Springer, Switzerland AG, 2019)

[4] M.H. Sadd, *Elasticity. Theory, applications, and numeric* (Academic, Burlington, 2009)

[5] Y. Tokovyy, *Plane Thermoelasticity of Inhomogeneous Solids*, in *Encyclopedia of Continuum Mechanics*, ed. by H. Altenbach, A. Öchsner (Springer, Berlin–Heidelberg, 2019), DOI: 10.1007/978-3-662-53605-6_361-1

[6] V.P. Revenko, Mater. Sci., **51(6)**, 785 (2015), DOI: 10.1007/s11003-016-9903-7

[7] V.P. Revenko, A.V. Revenko, *Separation of the 3D stress state of a loaded plate into two-dimensional tasks: bending and symmetric compression of the plate*, Scientific Journal of TNTU, **103(3)**, 53 (2021)

[8] V.P. Revenko, Int. Appl. Mech., **45**, 730 (2009), DOI: 10.1007/s10778-009-0225-4

[9] L.H. Donell, *Beams, plates and shells* (McGraw-Hill, New York, 1976)

[10] V.P. Revenko, V.N. Bakulin, J. Phys.: Conf. Ser., **1392**, 012021 (2019), DOI: 10.1088/1742-6596/1392/1/012021

[11] V.P. Revenko, V.N. Bakulin, *Solving equations of 3D elasticity for orthotropic bodies*, IOP Conf. Ser.: Mater. Sci. Eng., **927**, 012052 (2020), DOI: 10.1088/1757-899X/927/1/012052

[12] V.N. Bakulin, V.P. Revenko, Russian Mathematics, **60**, 1 (2016), DOI: 10.3103/S1066369X16060013

MATEC Web of Conferences **362**, 01025 (2022) https://doi.org/10.1051/matecconf/202236201025
CMMASS 2021

Effect of vacuuming on gas-dynamic parameters

Larisa Ruleva[1,*] and *Sergey* Solodovnikov[1]

[1]Ishlinsky Institute for Problems in Mechanics of Russian Academy of Sciences, 101(1), Vernadsky prospekt, Moscow, 119526, Russia

Abstract. The peculiarity of end-to-end temperature registration on models at the beginning of the experiment, then during high-speed pumping of the test unit into a vacuum and when a high-speed flow is applied to the model is experimentally revealed. The dependence of temperature registration on the vacuum depth and the independence of heat flow registration is shown.

1 Introduction

Computer modeling of aerothermodynamic parameters of high-speed aircraft [1–4] is associated with physical modeling on models. The necessary conditions for the implementation of validation of calculations were created at the laboratory installation "Hypersonic aerodynamic shock tube" (HAST) [5].

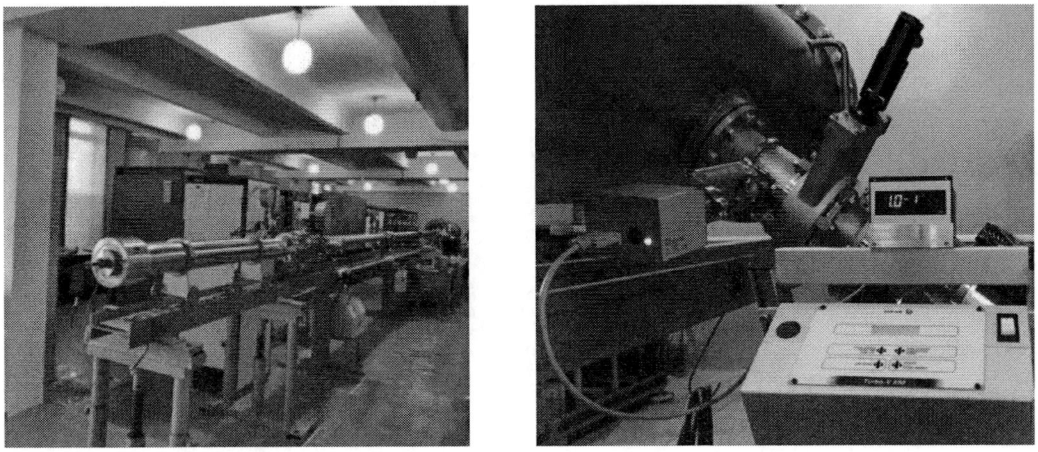

(a) (b)

Figure 1. The laboratory hypersonic aerodynamic shock tube (HAST) and a display view "0.1 mbar" in high-speed pumping station

The test tasks for the HAST behind the nozzle section are given in [6]. The flow outflow from hypersonic nozzles at Mach numbers M = 5–9 on the model is initiated by a shock wave within a wide range of experimental parameters. Setting the pressure in the high (HPC) and

*e-mail: ruleva@ipmnet.ru

low (LPC) pressure chambers allows you to create a flow braking pressure at the nozzle inlet up in a wide range [7]. The arguments given below relate to the rapid vacuuming of a HAST test block by high-speed pumping stations to a high vacuum short-term. It is revealed the temperature on the models changes in the process from rapid vacuuming. The HAST and high-speed pumping stations are shown in figure 1.

The readings of the heat flow sensors located on the models change during vacuuming. In fast experiments involving vacuuming and starting the nozzle, it is necessary to take into account the temperature change during vacuuming.

2 Research techniques

In [8], the measurement of heat fluxes in high-speed flows was carried out using sensors on anisotropic thermoelements and thin-film resistance sensors and was calculated from the change in the temperature of the working surface in time.

There are known studies [9] in which, to determine the initial temperature, calorimetric heat flow sensors were first heated to 294–302 K. Then the heated model with sensors is inserted into the flow using a lateral movement device. This method of registering the initial temperature cannot be applied in HAST.

The gas-dynamic parameters measured in the flow by sensors located on the models behind the nozzle depend on the initial conditions in the vacuum unit. When the aerodynamic block is evacuated, for example, to 1 Pa, the temperature drops significantly. The test unit is pumped to a vacuum of 10^2–10^{-2} Pa by high-performance pumping stations. Then the flow flows out to the model with Mach numbers M = 5–9. Heat flow sensors of the thermocouple and calorimetric type are installed on the model "blunted edge" flush with its surface. Photos of the model with heat flow sensors on the blunted edge are shown in figure 2.

Figure 2. Photo models

The calorimetric heat flow sensor CT1 and the coaxial thermocouple certified MCT sensor are installed on the model in the same plane and at the same distance from the nozzle cut. The calorimetric sensor CT2 is at a distance of 20 mm from the sensor CT1.

A feature of the measurement of the heat flow in the flow in HAST is the limited time of the quasi-stationary initial flow outflow from the nozzle. These times are 0.5–7 mc for different HAST regimes.

Then the specific heat flow is determined by expression:

$$q = \rho \cdot c \cdot h \frac{\Delta T}{\Delta \tau},$$

where ρ, c, h—density, heat capacity and its thickness, respectively; ΔT—is the change in sensor temperature during the exposure time interval $\Delta\tau$. In this case, there $\Delta\tau$ should be less time for quasi-stationary: $\Delta\tau \leq t_{gf}$, the parameter $\rho \cdot c \cdot h$ is a constant-valued quantity that characterizes properties of the heat flow sensor.

The temperature was recorded as follows. At first, before vacuuming, the zero signal was driven in accordance with the room temperature by applying a transfer coefficient. So, for the MST sensor, it is equal to 60 μV/K. When the sensor signal passes through an amplifier with a coefficient of K_{amp} = 100 and a low-pass filter, the transfer coefficient in the experiments became K_{mst} = 6 mV/K.

The sensors registered: first, the room temperature, the temperature after 2–3 minutes of switching on the pre-vacuum pump, then turning on the turbomolecular pump of the pumping station, then during the operation of the pumping station, after it is turned off and when the flow affects the model.

3 Results and discussion

It was experimentally revealed that vacuuming an aerodynamic block with a volume of 800 liters (4 meters long, 0.5 m in diameter) to a high vacuum takes 20–40 minutes. Procedures for switching off vacuum pumps, monitoring the pressures in the HPC and LPC, pressing the "Start" buttons of the computer and the high-speed valve [10], were taken 5–7 minutes.

The short time of through-vacuuming and starting the nozzle led to the following features of the temperature sensors readings by installed in the model in front of the nozzle. In the first 3–5 minutes of vacuuming, the readings of the certified MCT sensor and the sensors of laboratory manufacture CT1 and CT2-in all modes increase, and then decrease until the end of vacuuming from 10^5 Pa to 1 Pa (figure 3).

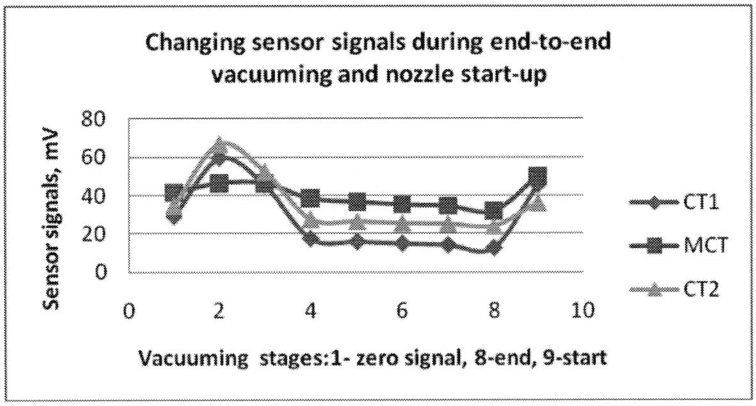

Figure 3. Graphs of the MCT, CT1 and CT2 sensors during end-to-end vacuuming of the HAST and starting the nozzle

The first sensor readings in mV correspond to room temperature. At the beginning of rapid vacuuming from 10^5 Pa, when the free path of gas particles is small, an air flow sweeps over the surface of the model. In this flow, the gas molecules collide, the temperature rises. In the future, when the turbomolecular pump is turned on after the action of the forevacuum pump, the pumping goes faster, the free path of the molecules increases, the density, temperature and pressure drop. At the same time, the temperature is lower at a higher vacuum. Then the vacuum pumps are stopped and after 3–5 minutes the discharge from the nozzle is carried out. The temperature measured by the sensors is rising.

For the fast processes described above, the heat exchange of the gas cooled by vacuuming inside the aerodynamic block with the external environment through the receiver wall does not have time to happen in one experiment. When the HAST test block is standing under vacuum for a long time, the temperature on the sensors is equalized relative to room temperature. The temperature before starting the nozzle is not equal to room temperature; it must be recalculated to the temperature corresponding to the degree of vacuum. Temperature measurements during vacuuming showed that first it is necessary to correlate the room temperature and the sensor readings before the experiment. Then recalculate it at the end of vacuuming and compare it with the temperature when it flows out of the nozzle.

The data on the thermal response of a certified MCT sensor in different starts and under different initial conditions are shown in figure 4.

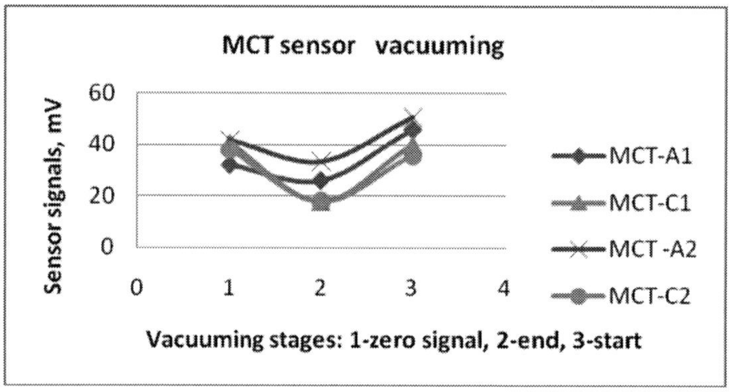

Figure 4. The MCT vacuuming graph for various initial conditions

The temperatures in the test mode $P_{test} = 1$ Pa are close (figure 4, point 1, MCT-C). When the test mode is $P_{test} = 10^2$ Pa, the temperature is higher (figure 4, point 1, MCT-A).

The heat flows measured by a certified MCT sensor at $P_{HPC} = 36 \cdot 10^5$ Pa, test regimes in the LPC and test block are shown in figure 5.

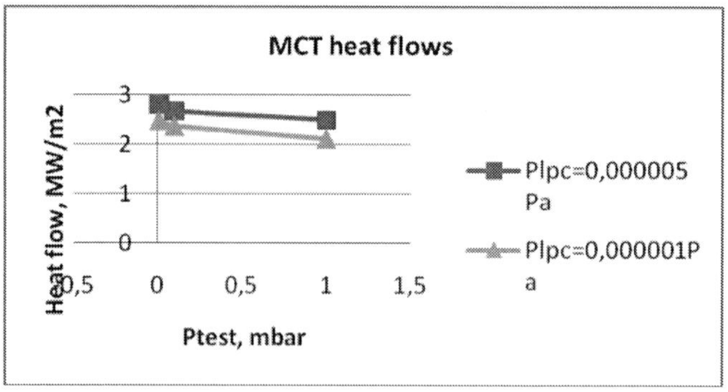

Figure 5. Heat flow to the MCT sensors

The advantage of a more accurate determination of the heating temperature of the model in a high-speed gas flow, taking into account the vacuuming in figure 4 will show. Heat flow is measured traditionally by temperature increment between stage 2 and stage 3 and a known transmission coefficient. The determination of the temperature of the surface heated by the flow in a high vacuum environment has changed.

Previously, there was no graph part from stage 1 to stage 2. Two sensor signals were measured before (stage 2) and after (stage 3) starting the nozzle. At the same time, the signal at point 2 was adjusted to the room temperature, for example, 293 K. Then, at $k = 6$ mV/K, when starting the nozzle, the temperature for MCT-A2 ($P_{test} = 10^2$ Pa) and MCT-C2 ($P_{test} = 1$ Pa) increased by 3 K and 3.3 K, respectively, and was determined as 296 K and 296.3 K, respectively. But linking stage 2 to room temperature to determine the nominal temperature is incorrect in fast processes and deep vacuuming of the model location environment.

Now the signal in stage 1 (before vacuuming) is adjusted according to room temperature, then the temperature is calculated before (stage 2) and after (stage 3) starting the nozzle. For example, at $k = 6$ mV/K after vacuuming for MCT-A2 ($P_{test} = 10^2$ Pa) and MCT-C2 ($P_{test} = 1$ Pa), the temperature decreased in stage 2 by 2 K and 2.83 K, respectively. Temperature increase when starting the nozzle between stage 2 and stage 3 increased by 3 K and 3.3 K, respectively. However, in stage 3, the measured nominal temperatures were 294 K and 293.47 K, respectively.

If vacuuming and starting the nozzle is not fast, it takes a lot of time, during which heat exchange with the external environment occurs, the difference in the sensor readings in stage 2 and stage 1 figure 4 is leveled. This is also typical for low vacuum or without vacuuming of the model placement environment.

For fast experiments with high vacuum, the effect of the difference in sensor readings in stage 2 and stage 1 figure 4 is to 3 K.

The error of measuring the MCT temperature in different phases from the sensor nozzle was up to 1.2 percent.

The physical causes of temperature changes during vacuuming are as follows. At the beginning of rapid vacuuming from 1 atm, when the free path of the gas particles is small, an air stream sweeps over the surface of the model. In this flow, the gas molecules collide, the temperature rises. In the future, when the turbomolecular pump turns on after the action of the forevacuum pump, pumping goes faster, the free path of molecules increases, density, temperature and pressure drop. At the same time, the temperature is lower at a higher vacuum. Then the vacuum pumps stop and the nozzle is start after 3–5 minutes. The temperature measured by the sensors is rising.

For the fast processes described above, the heat exchange of the gas cooled by vacuuming inside the aerodynamic block with the external environment through the receiver wall does not have time to happen in one experiment.

4 Conclusion

During the experiments, a temperature change was detected by quickly checking the temperature on the models at the beginning, at the end of vacuuming before exposure to a high-flow model, and then directly into the flow. At the same time, the temperature on the model after vacuuming was not equal to room temperature, which corresponds to zero signals.

The presented study was supported by Government program of basic research of Russian academy of science (contract # AAAAA 20-120011690135-5).

References

[1] S.T. Surzhikov, *Computer aerophysics of descent spacecraft Two-Dimensional Models* (FIZMATLIT, Moscow, 2018)

[2] S.T. Surzhikov, *Calculation analysis of the experimental data of HIFiRE-I using the computer code NERAT-2D*, J. Phys.: Conf. Ser., **1009**, 012001 (2018)

[3] V.G. Candler, M.S. Holder, L.D. Boyd, W.L. Wang, *CFD validation for hypersonic flight: hypersonic double-cone flow simulations*, in 40th AIAA Aerospace Sciences Meeting & Exhibit, RTO-TR-AVT-007-V3, Reno NV, USA (2006)

[4] D.S. Yatsukhno, *Computational study of the waverider aerothermodynamics by the UST3D computer code*, J. Phys.: Conf. Ser., **1009**, 012002 (2018)

[5] L.B. Ruleva, S.I. Solodovnikov, *Heat flow measurements on high-speed aircraft models*, IOP Conf. Ser.: Mater. Sci. Eng., **927**, 012083 (2020)

[6] S.T. Surzhikov, *Calculated initial data for solving test problems in the measuring section of the hypersonic shock wind tunnel (HSWT) of the RadGDLaboratory of IPMech RAS*, Physical-Chemical Kinetics in Gas Dynamics, **22(1)** (2021)

[7] S.T. Surzhikov, *Calculated initial data for solving test problems in the working area of the hypersonic shock wind tunnel HAST of theRadGDLaboratory of the IPMech RAS*, Physical-Chemical Kinetics in Gas Dynamics, **22(1)** (2021)

[8] V.I. Zapryagaev et al., *Heat transfer in supersonic separated flow of the compression corner*, J. Phys.: Conf. Ser., **1382**, 012049 (2019)

[9] P. Popov, V. Sakharov, T. Lapushkina, S. Poniaev, N. Monakhov, *Heat Flux Measurements by Sensors Based on Anisotropic Thermoelements in a Gasdynamic Experiment on Shock Tubes*, Physical-Chemical Kinetics in Gas Dynamics, **22(3)** (2021)

[10] S.N. Isakov, S.V. Yurkin, *Method of bringing to readiness an inflatable airbag of safety device for vehicle valve device*, Patent US 7.232.152 B2 (2007)

Empirical model of the viscosity of an epoxy binder at the initial stage of the polymerization process

Sergey Rusakov[1,*], *Valeriy* Gilev[1], and *Alexey* Rakhmanov[1]

[1]Perm State University, 15, Bukireva st., Perm, 614990, Russia

Abstract. This article presents the results of an experimental study and numerical simulation of the kinetics of curing of an epoxy binder at the initial stage of the polymerization process. Object of research: two-component composition of epoxy resin L and hardener EPH 161, certified as a binder for composite materials for aerospace purposes. The ordinary differential equation is used as the basic model for describing the polymerization process. The values of the material parameters required to perform the calculation operations were obtained by the rheological method using the analysis of changes in the viscosity of the binder depending on the time and temperature of polymerization. Isothermal and non-isothermal polymerization cases are considered. The modes in which the increase in the temperature of the process leads to a decrease in the viscosity of the epoxide composition, despite the polymerization process, were highlighted. The estimates of the model parameters obtained by processing the data of full-scale experiments are also given.

1 Introduction

Epoxy adhesive compositions are widely used due to their high strength properties, good adhesion to various materials, resistance to external factors and low shrinkage. In practical use of such compositions and composite materials on their basis, the main parameters are temperature dependences of viscosity and degree of conversion determining "viability" of binder and duration of product formation.

In the general formulation, the kinetics of curing of reactive oligomers is one of the central problems, the solution of which allows, firstly, to consider the chemistry of the process; secondly, to introduce quantitative parameters for comparison between the composition of different materials, and, thirdly, to give reasonable forecasts of the productivity and efficiency of real technological processes [1].

Two significantly different approaches are possible to evaluate the kinetics of curing: microkinetic and macrokinetic. In the first case, direct experimental methods are used to determine the change in the concentration of certain reactive groups. So traditionally, the degree of curing is estimated by the content of the gel fraction, which is determined by the extraction of cured binders with organic solvents [2]. However, this indicator is not sensitive to the presence of reaction centers in polymer chains. Chromatographic methods of analysis based on the different solubility of substances in the mobile and stationary phases are close to this approach. IR analysis [3], NMR methods [4], and Raman spectroscopy [5] are used to determine the amount of reactive groups in a polymer matrix.

*e-mail: rusakov@psu.ru

More effective from the practical point of view is macrokinetic approach based on the observation of changes in time of one or another integral parameter reflecting the totality of chemical transformations occurring during oligomerization. For this purpose, various physicochemical methods are used, including rheological analysis.

The curing of epoxy oligomers by the rheological method has been studied in a number of works [6–11]. Thus, [6, 7] defines the activation energy and discusses the applicability of different models to describe viscosity versus curing time. Kinetic characteristics of the curing process of ED-20 oligomeric resin were established in [8]. As a result, an equation was obtained that establishes the change in viscosity as a function of the average molar mass, density, and mass fraction of the cured product. Mathematical models of the kinetics of curing of thermosetting resins and the increase in viscosity during curing are considered in [9–11]. In [9], the main conclusions are based on the analysis of units of measurement. In [10, 11], the considered model takes into account the change in the concentration of fragments of the molecules of the reaction mixture during curing. However, these models make predictions mainly on the transition of the binder to the gel fraction. In a real technological process, the most interesting is the first (initial) stage of changing the viscosity of the binder, when the curable material can be molded into the required shapes. A significant factor in the course of the actual curing process is a change (usually an increase) in temperature. Even if some constant curing temperature is selected, the actual chemical process turns out to be non-isothermal.

The purpose of this work is to consider a method for estimating changes in viscosity at the initial (technological) stage of polymerization, both in the isothermal case and under conditions of programmable temperature change.

2 Object of study. details of the experiment

In the experiments was used "cold-curing" epoxy composition, which consists of epoxy resin L and hardener EPH 161, certified for use in structural composite materials for aviation purposes. The weight ratio of epoxy resin to hardener was 4:1, which is the manufacturer's recommended ratio. After preparation, the mixture was thoroughly mixed for 1–2 minutes with an electromechanical mixer and for an additional 5–10 minutes in a Digital Ultrasonic Cleaner CD 4820 ultrasonic bath at 40 kHz at room temperature. The last of these operations also contributes to the degassing of the mixture. Measurements of the mass of solutions and their components necessary to calculate the concentration of the mixture were performed using analytical scales LV-210A of the 2nd accuracy class, the absolute error of measurement of which was ±0.4 mg.

Rheological measurements were performed on a Physica MCR 501 rotary rheometer. The "cone-plate" system with cone diameter $d = 25$ mm and cone angle $\alpha = 1°$ was used in all measurements. Such geometry ensures uniformity of the shear velocity gradient in the measurement gap. A special temperature device H-PTD 200 based on the Peltier effect was used to maintain and change the temperature mode. During the experiment, approximately 0.07 ml of the solution was placed on the working surface of the rheometer plate. The thickness of the solution layer along its outer radius was 0.047 mm, which ensured the rapid establishment of the operating temperature in the sample even under heat release conditions due to the polymerization reaction.

The measurements were performed under the conditions of shear deformation according to a harmonic law with a frequency of 1 Hz in the mode of control of the value of deformation, which allows us to estimate not only the viscous but also the viscoelastic characteristics of the samples.

The change in binder viscosity at the initial stage of the polymerization process is shown in figure 1(a).

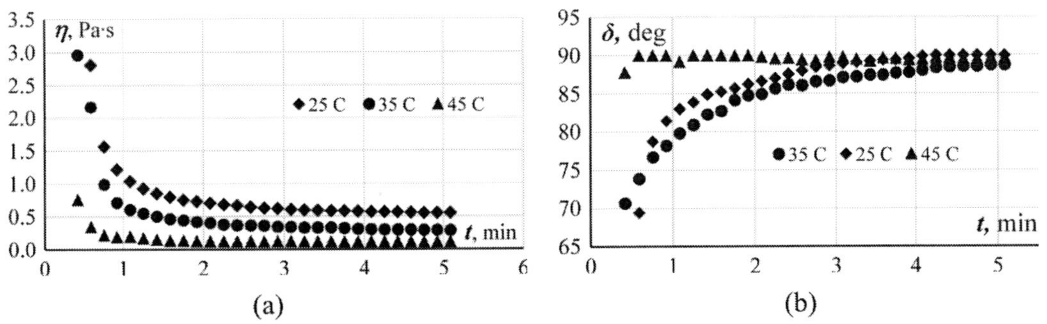

Figure 1. The initial stage of the polymerization process. (a) Viscosity change at the beginning of the polymerization process. From top to bottom: 25, 35, 45°. (b) Change of the phase angle at the initial stage of the polymerization process. From top to bottom: 25, 35, 45°C

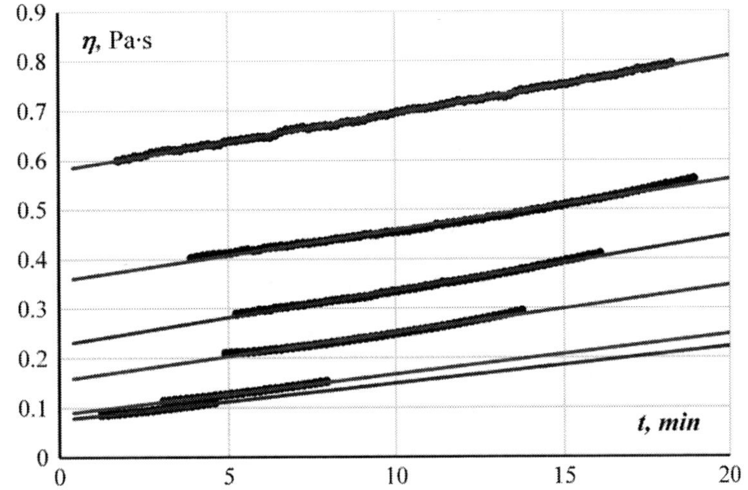

Figure 2. Change in viscosity at the initial stage of the polymerization process. Top–down: $T = 25$, 30, 35, 40, 45, 50°C

As you can see from figure 1(a), immediately after the start of measurements for all samples, the effective viscosity decreases with increasing shear rate, but eventually reaches a constant value practically independent of the shear rate, which can be conditionally interpreted as the second Newtonian viscosity.

The oscillation mode allows us to evaluate not only the viscous but also the viscoelastic characteristics of the samples. In this mode, the deformation γ is given by the harmonic law $\gamma = \gamma_0 \sin(\omega t)$, where γ_0 is its amplitude value, ω is the cyclic frequency. In the region of small deformations, the stress τ also varies according to the harmonic law $\tau = \tau_0 \sin(\omega t + \delta)$, where δ is the phase shift angle between the shear stress and deformation. If the phase shift angle differs by $\pi/2$, such body is called viscous. If the phase shift angle is in the range $\pi/2 > \delta > 0$, such a body is called viscoelastic.

Figure 1(b) shows the change in the phase shift angle δ between the shear stress and the deformation γ depending on the measurement time. Immediately after the start of measurements, the binder exhibits viscoelastic properties: $\delta < 90°$, but as the shear rate increases, the phase shift angle tends to 90°, which corresponds to a purely viscous behavior. This corresponds to the viscosity curves in terms of reaching the second Newtonian viscosity.

The second Newtonian viscosity is commonly used to correctly compare the behavior of non-Newtonian fluids. When analyzing the viscosity curves measured at different temperatures, it was found that reaching the second Newtonian viscosity provides pre-mixing for no more than 5 minutes. Accordingly, the following test mode was chosen: first, the flow curve was measured in the range of shear rates from 0.3 to 30, with which the output to a constant viscosity value was additionally controlled. The measurements were then continued at a shear rate of 30 s up to the required moment of time.

The results of viscosity measurements at the initial stage of the isothermal regime are shown in figure 2.

It can be seen that at the initial stage of polymerization in the considered temperature range, in all cases, there is a section where the viscosity behaves almost linearly. Note that the duration of this quasi-linear section decreases with increasing temperature.

3 Isothermal polymerization analysis

Earlier work by the authors [12] measured the temperature dependences of the viscosity for the resin (ep index) and hardener (am index) separately:

$$\eta_{ep}(T) = 4.512 \cdot 10^{-5} \exp\left(\frac{937.6}{72.42 + T}\right), \quad \eta_{am}(T) = 1.277 \cdot 10^{-5} \exp\left(\frac{1120.3}{87.42 + T}\right). \tag{1}$$

Here and in the future, T is expressed in Celsius degrees. We construct the dependence of the viscosity of the epoxy composition at the zero moment of time, taking into account the ratio of stoichiometric equilibrium 4:1 in multiplicative form [12]:

$$\eta(0, T) = \eta_0(T) = \eta_{ep}^{0.8}(T) \cdot \eta_{am}^{0.2}(T). \tag{2}$$

We approximate the quasi-linear sections of viscosity changes (figure 2) by a linear dependence of the form

$$\eta(t, T) \approx \eta_0(T)(1 + \alpha_1(T)t). \tag{3}$$

The calculated values of $\eta_0(T)$ (marked by dots) in comparison with the empirical functional dependence (2) are shown in figure 3(a). It can be seen that there is good agreement between them.

The authors [13] proposed and experimentally tested the simplest kinetic model, which describes the process of polymerization of similar binder. This model is a Cauchy problem for ordinary differential equations, which has an analytical solution, which in the case of stoichiometric equilibrium looks like:

$$\frac{d\psi(t, T)}{dt} = K(T)(1 - \psi(t, T))^2, \quad \psi(0, T) = 0, \quad \psi(t, T) = \frac{K(T)t}{1 + K(T)t}, \tag{4}$$

where $\psi(t, T)$ is the fraction of the reacted resin molecules (or hardener). In the same place, based on the processing of the experimental data, the function $K(T)$ was approximated by the Arrhenius function:

$$K(T) = 51590 \cdot \exp\left(-\frac{5020}{T + 273.15}\right). \tag{5}$$

Let's write down the model of change of viscosity of epoxy composition from temperature and fraction of reacted molecules, characterizing the degree of curing in the form:

$$\eta(t, T) = \eta(\psi(t, T), T) = \eta_0(T)\,\eta_1(\psi(t, T), T), \quad \eta_1(\psi(0, T), T) = 1. \tag{6}$$

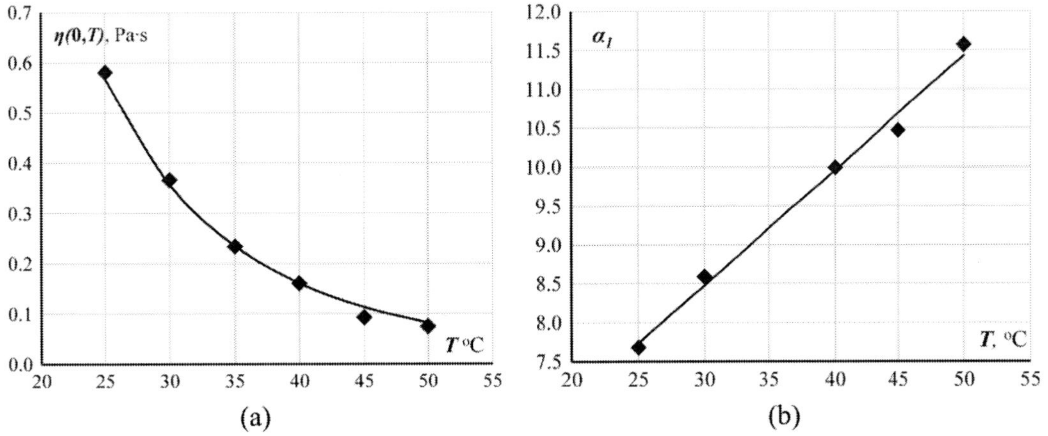

Figure 3. Change in viscosity and parameter α_1 as a function of temperature. (a) Initial viscosity values depending on temperature. (b) Change of parameter α_1 from temperature: squares are calculated values obtained on the basis of experimental data according to formula (3), a solid curve is an approximation of the calculated data

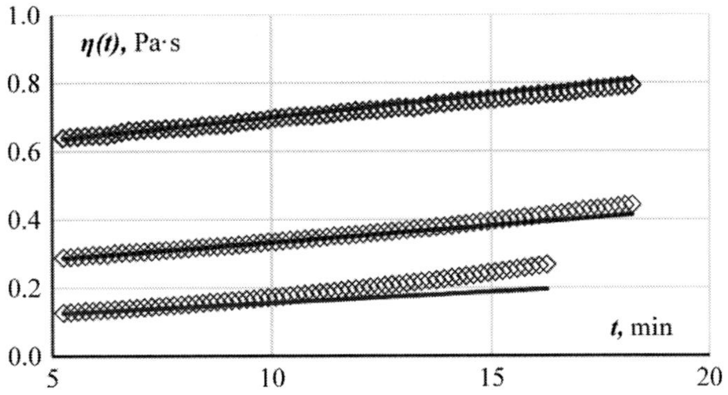

Figure 4. Comparison of experimental data with model (7). From top to bottom: T = 25, 35, 45°C. Rhombuses—experiment, solid lines—calculation

Decomposing function (6) into Taylor series in the vicinity of point $t = 0$, we obtain

$$\eta(t,T) = \eta(0,T) + \frac{d\eta(0,T)}{dt}t + \ldots = \eta_0(T)\left(1 + \frac{\partial\eta_1(\psi(0,T))}{\partial\psi}\frac{d\psi(0,T)}{dt}t\right) + \ldots$$

According to model (4): $\dfrac{d\psi(0,T)}{dt} = K(T)$. Comparing this result with the empirical dependence (3) we have: $\dfrac{\partial\eta_1(\psi(0,T))}{\partial\psi} = \dfrac{\alpha_1(T)}{K(T)}$. The results of this comparison are shown in figure 3(b).

The result of approximation of the experimental data: $\dfrac{\partial\eta_1(0,T)}{\partial\psi} = 4.04 + 0.15T$. Analytical solution of the problem (4) in linear approximation is $\psi(t,T) \approx K(T)t$. Thus, dependence (6) at the initial stage of curing can be written in the form:

$$\eta(t,T) \approx \eta_0(T)\left(1 + \frac{\partial\eta_1(0,T)}{\partial\psi}\psi(t,T)\right). \quad (7)$$

Figure 4 shows the results of comparing the experimental data with approximations according to model (7). Note that, the area of the linear section decreases with increasing temperature, according to the model.

4 Analysis of nonisothermal polymerization

The temperature change was modeled by the following law to analyze the behavior of the binder viscosity at the initial stage of nonisothermal polymerization:

$$T(t) = \begin{cases} T_0, & t \in [0, t_0), \\ \alpha(t - t_0), & t \in [t_0, t_1), \\ T_1 = \alpha(t_1 - t_0), & t \geq t_1. \end{cases} \tag{8}$$

In all cases $t_0 = 5.5$ min, $T_0 = 25, 30, 35$°C, $T_1 = 50$°C were used and the rate of temperature change was varied: $\alpha = 0.1, 0.2, 0.3, 0.4, 0.5$ deg/min. The non-isothermal polymerization revealed the influence of two mechanisms on viscosity change: a) viscosity increase as a result of chemical reaction, b) viscosity decrease as a consequence of temperature increase.

Figure 5(b) shows the change in viscosity during the first 45 minutes of the polymerization process at $T_0 = 25$°C, $\alpha = 0.1$ and $\alpha = 0.5$ deg/min.

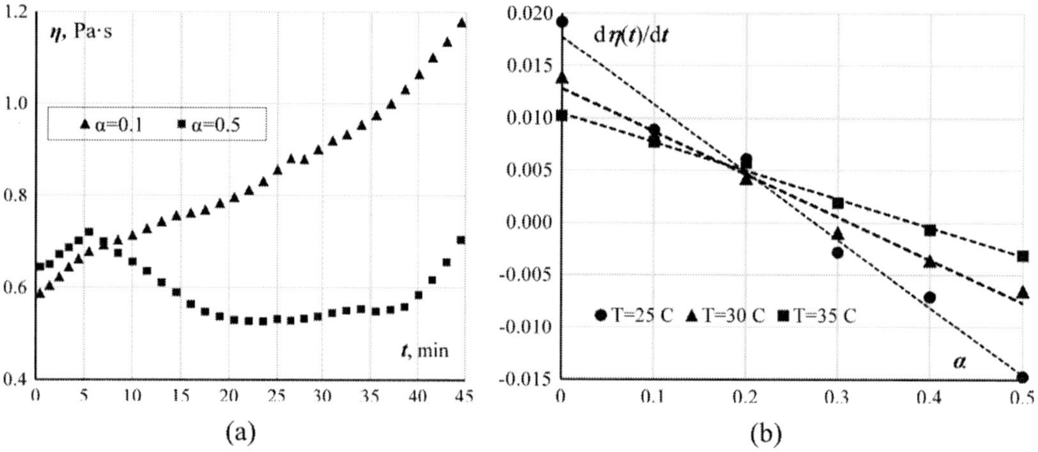

Figure 5. Viscosity change during the first 45 minutes of the polymerization process. (a) Change in viscosity depending on the heating rate. $T = 25$°C. (b) Dependence of the rate of change of viscosity on the parameter α and temperature T_0

At $\alpha = 0.1$, there is a monotonic increase in viscosity over the entire time interval (the first mechanism dominates). At $\alpha = 0.5$ in the isothermal region ($t < 5.5$, $T(t) = T_0$) the viscosity increases and then decreases till about $t = 20$ min (the second mechanism dominates).

Processing of experimental data shows that, here, as well as in the isothermal case, during the first 10–15 minutes, the viscosity of the non-isothermal area from the time can be approximated by decomposition of the form:

$$\eta(t, T(t); \alpha) = \eta(t_0, T_0) + (t - t_0)\frac{d\eta(t_0, T_0; \alpha)}{dt} + \ldots$$

The change in the value $\dfrac{d\eta(t_0, T_0; \alpha)}{dt}$ depending on the rate of temperature increase (parameter α) for $T_0 = 25$, 30, and 35°C is shown in figure 5(b). Such curves can be approximated by the expression $\dfrac{d\eta(t_0, T_0; \alpha)}{dt} = b_0 + b_1\alpha$.

Note that at this stage of the polymerization process, at $\alpha < 0.25$ the viscosity increases and at $\alpha > 0.4$ it decreases. We will evaluate the parameters b_0 and b_1 using a modified model (6): $\eta(t, T(t)) = \eta(\psi(t, T(t))) = \eta_0(T(t))\eta_1(\psi(t, T(t)))$.

In this case:

$$\frac{d\eta(t, T(t))}{dt} = \frac{d\eta_0(T(t))}{dT}\frac{dT(t)}{dt}\eta_1(\psi(t, T(t))) + \\ + \eta_0(T(t))\frac{d\eta_1(\psi(t, T(t)))}{d\psi}\left(\frac{\partial\psi(t, T(t))}{\partial t} + \frac{\partial\psi(t, T(t))}{\partial T}\frac{dT(t)}{dt}\right).$$

Taking into account that $\dfrac{dT(t_0)}{dt} = \dfrac{dT(t_0 + 0)}{dt} = \alpha$ the model expression for the parameters b_0 and b_1 takes the form:

$$\tilde{b}_0 = \eta_0(T_0)\frac{d\eta_1(\psi(t_0, T_0))}{d\psi}\frac{\partial\psi(t_0, T_0)}{\partial t}, \tag{9}$$

$$\tilde{b}_1 = \frac{d\eta_0(T_0)}{dT}\eta_1(\psi(t_0, T_0)) + \eta_0(T_0)\frac{d\eta_1(\psi(t_0, T_0))}{d\psi}\frac{\partial\psi(t_0, T_0)}{\partial T}. \tag{10}$$

Note that in the non-isothermal case the analytical solution of problem (4) has the form:

$$\psi(t, T(t)) = \frac{\hat{K}(T)t}{1 + \hat{K}(T)t}, \quad \hat{K}(T(t)) = \frac{1}{t}\int_0^t K(\tau)\,d\tau. \tag{11}$$

To determine the parameters $\tilde{b}_0$ and $\tilde{b}_1$ it is enough to differentiate equations (2), (5) by T and for the functions $\eta_0(T)$, $\psi(t, T)$ and $\hat{K}(T(t))$.

The results of processing experimental data (see figure 5(b)) and calculations by formulas (9)–(10) are presented in table 1, while it was assumed that $\dfrac{\partial\eta_1(t_0, T_0)}{\partial\psi} \approx \dfrac{\partial\eta_1(0, T_0)}{\partial\psi}$.

Table 1. Comparison of experimental and calculated data

T_0, °C	b_0 − exp	$\tilde{b}_0$	b_1 − exp	$\tilde{b}_1$
25	0.0178	0.0170	−0.0647	−0.0640
30	0.0128	0.0139	−0.0409	−0.0382
35	0.0105	0.0118	−0.0273	−0.0241

Comparison of the calculated and experimental data shows their sufficient agreement (maximum deviation is not more than 10%).

Thus, the viscosity behavior at the initial stage of polymerization (the first 15–20 min of the process) for the epoxy composition under consideration is largely determined by the viscosity of the resin and hardener separately and by the degree of hardening using model (4).

5 Conclusions

This artical has detailed the experimental study and numerical simulation of changes in the viscosity of an epoxy binder at the initial stage of the polymerization process. It is shown that a linear approximation is possible at the initial period of time, which allows quantifying the mutual influence of two mechanisms of viscosity change:

a) viscosity increase as a result of chemical reaction;

b) viscosity decrease as a result of temperature increase.

The ranges of the rate of temperature change, in which one or another mechanism of viscosity behavior is the main one, were also determined. It is shown that the proposed mathematical model that takes into account the degree of curing of the binder allows us to describe the real process in good approximation.

The study was carried out with the financial support of the Government of the Perm region in the framework of the research project No. C-26/793 and the RFBR grant No. 17-41-590649.

References

[1] V.I. Irzhak, S.M. Mezhikovskii, Uspekhi Khimii, **77**, 78 (2008)
[2] S. Tonogai, Y. Sakaguchi, S. Seto, Appl. Polym. Sci., **22**, 3225 (1978)
[3] L. Moraes, R. Rocha, L. Menegazzo, E. Araújo, K. Yukimitu, J. Moraes, Appl. Oral. Sci., **16**, 145 (2008)
[4] J. Barton, G. Buist, I. Hamerton, B. Howlin, J. Jones, S. Liu, Polym. Bull., **33**, 215 (1994)
[5] C. Pianelli, J. Devaux, S. Bebelman, G. Leloup, Biomed. Mater. Res. (Appl. Biomat.), **48**, 675 (1999)
[6] D.V. Onuchin, R.I. Sopotov, Yu.A. Kerber et al., Plast. Massy, **8**, 6 (2012)
[7] V.S. Osipchik, I.Yu. Gorbunova, N.V. Kostromina, Yu.V. Olikhova, D.M. Bui, Khim. Khim. Tekhnol. E. **57**, 19 (2014)
[8] A.V. Kosarev, V.N. Studentsov, Plast. Massy, **1–2**, 3 (2019)
[9] V.N. Studentsov, Plast. Massy, **1–2**, 24 (2019)
[10] A.L. Svistkov, A.Yu. Eliseeva, A.V. Kondyurin, Vest. Perm. Univ. Fizika, **1**, 9 (2019)
[11] A. Kondyurin, L.A. Komar, A.L. Svistkov, Composites. Part B: Engineering, **43**, 616 (2012)
[12] V.G. Gilev, A.V. Kondyurin, S.V. Rusakov, *Investigation of epoxy matrix viscosity in the initial stage of its formation*, IOP Conf. Ser.: Mater. Sci. Eng., **208**, 012014 (2017), DOI: 10.1088/1757-899X/208/1/012014
[13] V.G. Gilev, S.V. Rusakov, V.S. Chudinov, A.Yu. Rakhmanov, A.V. Kondyurin, Mech. Compos. Mater., **57**, 361–372 (2021)

Spectral representation of stochastic integration operators

Konstantin Rybakov[1,*]

[1]Moscow Aviation Institute (National Research University), 4, Volokolamskoe shosse, Moscow, 125993, Russia

Abstract. The spectral representation for stochastic integration operators with respect to the Wiener process is proposed in the form of a composition of spectral characteristics used in the spectral form of mathematical description for control systems. This spectral representation can be defined relative to the various orthonormal bases. For given deterministic square-integrable kernels, the spectral characteristic of a stochastic integration operator is determined as an infinite random matrix. The main applications of such a representation suppose solving linear stochastic differential equations and modeling multiple or iterated Stratonovich stochastic integrals. Specific formulas are provided that allow to represent the spectral characteristic for the stochastic integration operator, the kernel of which is the Heaviside function, relative to Walsh functions and trigonometric functions.

1 Introduction

The article considers a stochastic integration operator with respect to the Wiener process [1], it is associated with a spectral characteristic. For the chosen orthonormal basis of the square-integrable function space, it is an infinite random matrix depended on this orthonormal basis. It is proposed to represent the spectral characteristic of the stochastic integration operator by the spectral characteristic of the integration operator and the spectral characteristic of the multiplier, which are used in the spectral form of mathematical description in the control theory [2].

The main idea of the spectral form of mathematical description is to represent functions and random processes in the form of ordered sets of their expansion coefficients for the chosen orthonormal basis, i.e., in the form of spectral characteristics [2]. The relations for calculating such spectral characteristics and transforming them are implemented as packages of elementary and specialized algorithms and programs of the spectral method [2–5]. For the spectral representation of stochastic integration operators, it is proposed to use orthonormal bases that are usually applied in the spectral form of mathematical description: Legendre polynomials, Walsh and Haar functions, trigonometric functions, etc.

The main application of the spectral representation of stochastic integration operators is the modeling of solutions to linear stochastic differential equations and systems of such equations [6]. In particular, it can find the application for the representation and modeling of multiply or iterated Stratonovich stochastic integrals [7, 8]. It can also be used for modeling solutions to nonlinear stochastic differential equations and systems of such equations using high-order numerical methods based on the Taylor–Stratonovich expansion [8–10].

*e-mail: rkoffice@mail.ru

2 Definitions

Let be $(\Omega, \mathcal{S}, \mathrm{P})$ a complete probability space. The random process is a function $f(t,\omega)\colon [0,T]\times\Omega \to \mathrm{R}$ that is $\mathcal{S}$-measurable for each $t \in [0,T]$, and we denote it by $f(t)$ for brevity. We restrict ourselves to random processes that satisfy the condition

$$\mathrm{E}\int_0^T f^2(t)\,dt < \infty, \tag{1}$$

where E is the mathematical expectation. Realizations of the random process $f(t)$ are elements of the space $L_2([0,T])$.

We call the spectral characteristic of the random process $f(t)$ defined relative to the orthonormal basis $\{q_i(t)\}_{i=0}^{\infty}$ of the space $L_2([0,T])$ the ordered set of expansion coefficients

$$F_i = \int_0^T q_i(t) f(t)\,dt, \quad i = 0,1,2,\ldots, \tag{2}$$

and this spectral characteristic is represented as the infinite random column matrix F. In fact, the column matrix F defines the random process $f(t)$:

$$f(t) = \sum_{i=0}^{\infty} F_i q_i(t), \quad t \in [0,T], \tag{3}$$

or more strictly: the column matrix F defines the equivalence class that consists of random processes $f(t)$, for which

$$\lim_{n\to\infty} \mathrm{E}\int_0^T \left(f(t) - \sum_{i=0}^{n} F_i q_i(t)\right)^2 dt = 0.$$

For the orthonormal basis $\{q_i(t)\}_{i=0}^{\infty}$, we require the additional condition: the quadratic integrability of products $q_l(t)q_j(t)$, $l, j = 0,1,2,\ldots$

Let be $\mathcal{A}$ the linear operator (Hilbert–Schmidt operator):

$$\mathcal{A}f(t) = \int_0^T k(t,\tau) f(\tau)\,d\tau, \quad f(\cdot) \in L_2([0,T]), \tag{4}$$

where $k(t,\tau)$ is the kernel satisfying the condition

$$\int_0^T\int_0^T k^2(t,\tau)\,d\tau\,dt < \infty.$$

Its spectral characteristic defined relative to the orthonormal basis $\{q_i(t)\}_{i=0}^{\infty}$ is the infinite matrix A with entries [2, 6]

$$A_{ij} = \int_0^T q_i(t)\mathcal{A}q_j(t)\,dt = \int_0^T\int_0^T q_i(t)k(t,\tau)q_j(\tau)\,d\tau\,dt, \quad i,j = 0,1,2,\ldots$$

Next, we define the random linear operator $\mathcal{A}_W$:

$$\mathcal{A}_W f(t) = \int_0^T k(t,\tau) f(\tau)\,dw(\tau), \quad f(\cdot) \in L_2([0,T]), \tag{5}$$

where $w(t)$ is a standard Wiener random process defined on the probability space $(\Omega, \mathcal{S}, \mathrm{P})$, and the infinite random matrix $A^{\mathcal{V}}$ with entries

$$A_{ij}^{\mathcal{V}} = \int_0^T q_i(t)\mathcal{A}_W q_j(t)\,dt = \int_0^T\int_0^T q_i(t)k(t,\tau)q_j(\tau)\,dw(\tau)\,dt, \quad i,j = 0,1,2,\ldots, \tag{6}$$

is called the spectral characteristic of the random linear operator $\mathcal{A}_W$.

3 Spectral Representation of Stochastic Integration Operators

Let us introduce an additional notation. Let V be the infinite three-dimensional matrix with entries

$$V_{ijk} = \int_0^T q_i(t)q_j(t)q_k(t)\,dt, \quad i,j,k = 0,1,2,\ldots$$

In the spectral form of mathematical description, this matrix is called the spectral characteristic of the multiplier [2, 6].

Theorem 1. Let $A^{\mathcal{V}}$ be the spectral characteristic of the random linear operator $\mathcal{A}_W$ with the kernel $k(t,\tau)$, A be the spectral characteristic of the Hilbert–Schmidt operator with the same kernel, V be the spectral characteristic of the multiplier, $\mathcal{V}$ be an infinite random column matrix, whose elements are independent random variables with the standard normal distribution. Then spectral characteristics $A^{\mathcal{V}}$ and A satisfy the relation

$$A^{\mathcal{V}} = A(V\mathcal{V}),$$

or in the expanded form

$$A^{\mathcal{V}}_{ij} = \sum_{l=0}^{\infty}\sum_{k=0}^{\infty} A_{il}V_{ljk}\mathcal{V}_k.$$

Proof. Let us write the formula (6) for entries of the spectral characteristic $A^{\mathcal{V}}$ of the random linear operator $\mathcal{A}_W$, representing the kernel $k(t,\tau)$ as a series relative to the orthonormal basis $\{q_i(t)\}_{i=0}^{\infty}$:

$$k(t,\tau) = \sum_{k,l=0}^{\infty} A_{kl}q_k(t)q_l(\tau), \quad t,\tau \in [0,T],$$

i.e.,

$$A^{\mathcal{V}}_{ij} = \int_0^T q_i(t)\int_0^T \sum_{k,l=0}^{\infty} A_{kl}q_k(t)q_l(\tau)q_j(\tau)\,dw(\tau)\,dt =$$

$$= \lim_{n,m\to\infty}\sum_{k=0}^{n}\sum_{l=0}^{m} A_{kl}\int_0^T q_i(t)q_k(t)\,dt\int_0^T q_l(\tau)q_j(\tau)\,dw(\tau) =$$

$$= \lim_{n,m\to\infty}\sum_{k=0}^{n}\sum_{l=0}^{m} A_{kl}\delta_{ik}\int_0^T q_l(\tau)q_j(\tau)\,dw(\tau) = \lim_{m\to\infty}\sum_{l=0}^{m} A_{il}\int_0^T q_l(\tau)q_j(\tau)\,dw(\tau),$$
$$i,j = 0,1,2,\ldots,$$

where δ_{ik} is the Kronecker delta.

Further, we rewrite the product $q_l(t)q_j(t)$ as a series relative to the orthonormal basis $\{q_i(t)\}_{i=0}^{\infty}$:

$$q_l(t)q_j(t) = \sum_{k=0}^{\infty} V_{ljk}q_k(t),$$

then

$$A^{\mathcal{V}}_{ij} = \lim_{m\to\infty}\sum_{l=0}^{m} A_{il}\int_0^T \sum_{k=0}^{\infty} V_{ljk}q_k(\tau)\,dw(\tau) = \lim_{n,m\to\infty}\sum_{l=0}^{m} A_{il}\sum_{k=0}^{n} V_{ljk}\int_0^T q_k(\tau)\,dw(\tau) =$$

$$= \lim_{n,m\to\infty}\sum_{l=0}^{m} A_{il}\sum_{k=0}^{n} V_{ljk}\mathcal{V}_k = \sum_{l=0}^{\infty}\sum_{k=0}^{\infty} A_{il}V_{ljk}\mathcal{V}_k, \quad i,j = 0,1,2,\ldots,$$

where $\mathcal{V}_k$ are independent random variables with the standard normal distribution. ◂

Note that the definition of the random linear operator $\mathcal{A}_W$ can be extended to the space of random processes $f(t)$ satisfying condition (1). However, it should be highlighted that the stochastic integral in this case is understood in the Stratonovich sense [8–10]. Then the following theorem can be formulated.

Theorem 2. Let X and F be spectral characteristics of random processes $x(t)$ and $f(t)$, respectively, satisfied the relation $x(t) = \mathcal{A}_W f(t)$ be the random linear operator with the spectral characteristic $A^{\mathcal{V}}$. Then spectral characteristics X, F and $A^{\mathcal{V}}$ satisfy the relation

$$X = A^{\mathcal{V}} F,$$

or in the expanded form

$$X_i = \sum_{j=0}^{\infty} A_{ij}^{\mathcal{V}} F_j.$$

Proof. Let us write the formula (2) for entries of the spectral characteristic X:

$$X_i = \int_0^T q_i(t)\, x(t)\, dt = \int_0^T q_i(t) \mathcal{A}_W f(t)\, dt, \quad i = 0, 1, 2, \ldots,$$

and represent the random process $f(t)$ in the form (3), and also use the definition (6) for entries of the spectral characteristic $A^{\mathcal{V}}$:

$$X_i = \int_0^T q_i(t) \mathcal{A}_W \sum_{j=0}^{\infty} F_j q_j(t)\, dt = \lim_{n\to\infty} \sum_{j=0}^{n} F_j \int_0^T q_i(t) \mathcal{A}_W q_j(t)\, dt =$$

$$= \lim_{n\to\infty} \sum_{j=0}^{n} A_{ij}^{\mathcal{V}} F_j = \sum_{j=0}^{\infty} A_{ij}^{\mathcal{V}} F_j, \quad i = 0, 1, 2, \ldots \quad ◄$$

Consider the particular case $k(t,\tau) = 1(t-\tau)$, where $1(t-\tau)$ is the Heaviside function. Here, $\mathcal{A}_W$ is the stochastic integration operator, the spectral characteristic of which we denote by $P^{-1,\mathcal{V}}$. Therefore, if

$$x(t) = \int_0^t f(\tau)\, dw(\tau), \quad t \in [0, T],$$

where the stochastic integral is understood in the Stratonovich sense, and X and F are spectral characteristics of random processes $x(t)$ and $f(t)$, respectively, then

$$X = P^{-1,\mathcal{V}} F = P^{-1}(V\mathcal{V})F,$$

where P^{-1} is the spectral characteristic of the integration operator [2, 6], i.e., the spectral characteristic of the linear operator (4) for $k(t,\tau) = 1(t-\tau)$.

The main application of the above theorems is the represention of solutions to linear stochastic differential equations and systems of such equations [6], including the representation of multiply or iterated Stratonovich stochastic integrals [7, 8]. The advantage of this representation is based on the fact that the calculation for entries of the spectral characteristic P^{-1}, as well as entries of the spectral characteristic V relative to the various orthonormal bases, such as Legendre polynomials, Walsh and Haar functions, trigonometric functions, is often used to solve analysis and synthesis problems for different control systems by the spectral form of mathematical description [2, 11], i.e., algorithms for the calculation are known.

For example, for the Walsh functions

$$q(i,t) = \frac{1}{\sqrt{T}} \begin{cases} 1 & \text{for} \quad i = 0, \\ \prod_{\{k:\, \gamma_k = 1\}} r(k,t) & \text{for} \quad i = 1, 2, \ldots, \end{cases}$$

where $\gamma_k \in \{0,1\}$ are coefficients in the binary representation of the number i, and $r(k,t)$ are the Rademacher functions defined by $r(k,t) = 2^{\lfloor 2^k \pi t/T \rfloor}$ ($\lfloor . \rfloor$ is the floor function, $k = 1,2,\ldots$), we have

$$P^{-1}_{00} = \frac{T}{2}, \quad P^{-1}_{k,2^m+k} = -P^{-1}_{2^m+k,k} = \frac{T}{2^{m+2}},$$

$$P^{-1}_{ml} = 0 \quad \text{otherwise}, \quad m,l = 0,1,2,\ldots, \quad k = 0,1,\ldots,2^m-1;$$

$$V_{ijk} = \frac{1}{\sqrt{T}} \begin{cases} 1 & \text{for} \quad i \oplus j \oplus k = 0, \\ 0 & \text{for} \quad i \oplus j \oplus k \neq 0, \end{cases} \quad i,j,k = 0,1,2,\ldots,$$

where $\oplus$ means the bitwise addition of numbers modulo 2 (XOR bitwise operation). Moreover, it is not difficult to obtain an explicit representation for the product $E^{\mathcal{V}} = V\mathcal{V}$ taking into account the group property of the Walsh functions [12]:

$$E^{\mathcal{V}}_{ij} = \frac{1}{\sqrt{T}} \mathcal{V}_{i \oplus j}, \quad i,j = 0,1,2,\ldots,$$

i.e., $E^{\mathcal{V}}$ is the Walsh–Toeplitz infinite matrix [13].

Another example is the use of trigonometric functions

$$q(i,t) = \frac{1}{\sqrt{T}} \begin{cases} 1 & \text{for} \quad i = 0, \\ \sqrt{2}\cos\dfrac{i\pi t}{T} & \text{for} \quad i = 2,4,6,\ldots, \\ \sqrt{2}\sin\dfrac{(i+1)\pi t}{T} & \text{for} \quad i = 1,3,5,\ldots, \end{cases}$$

as an orthonormal basis (Fourier basis):

$$P^{-1}_{00} = \frac{T}{2}, \quad P^{-1}_{0m} = -P^{-1}_{m0} = \frac{\sqrt{2}T}{(m+1)\pi}, \quad P^{-1}_{l-1,l} = -P^{-1}_{l,l-1} = \frac{T}{l\pi},$$

$$k = 1,2,3,\ldots, \quad m = 2k-1, \quad l = 2k;$$

$$V_{ijk} = V_{jik} = \frac{1}{\sqrt{T}} \begin{cases} \delta_{jk} & \text{for } i = 0, \\ \dfrac{\sqrt{1+\delta_{ij}}\,\delta_{j-i,k} + \delta_{i+j,k}}{\sqrt{2}} & \text{for } i = 2n, \\ \dfrac{\delta_{i+j,k} - \delta_{j-i-2,k}}{\sqrt{2}} & \text{for } i = 2n-1 \text{ and } j = 2l, \\ \dfrac{\sqrt{1+\delta_{ij}}\,\delta_{j-i,k} - \delta_{i+j+2,k}}{\sqrt{2}} & \text{for } i = 2n-1 \text{ and } j = 2l-1, \end{cases} \quad \begin{matrix} i,k = 0,1,2,\ldots, \\ j = i,i+1,i+2,\ldots, \\ n,l = 1,2,3,\ldots, \end{matrix}$$

and for the product $E^{\mathcal{V}} = V\mathcal{V}$ we can write the following expression if $i \leq j$:

$$E^{\mathcal{V}}_{ij} = E^{\mathcal{V}}_{ji} = \frac{1}{\sqrt{T}} \begin{cases} \mathcal{V}_j & \text{for } i = 0, \\ \dfrac{\sqrt{1+\delta_{ij}}\,\mathcal{V}_{j-i} + \mathcal{V}_{i+j}}{\sqrt{2}} & \text{for } i = 2n, \\ \dfrac{\mathcal{V}_{i+j} - \mathcal{V}_{j-i-2}}{\sqrt{2}} & \text{for } i = 2n-1 \text{ and } j = 2l, \\ \dfrac{\sqrt{1+\delta_{ij}}\,\mathcal{V}_{j-i} - \mathcal{V}_{i+j+2}}{\sqrt{2}} & \text{for } i = 2n-1 \text{ and } j = 2l-1, \end{cases} \quad \begin{matrix} i = 0,1,2,\ldots, \\ j = i,i+1,i+2,\ldots, \\ n,l = 1,2,3,\ldots \end{matrix}$$

Above formulas are sufficient to represent multiply or iterated Stratonovich stochastic integrals of arbitrary multiplicity with unit weight functions (for integrals of multiplicity 2, relations to calculate entries of the spectral characteristic P^{-1} are sufficient). For other weight functions, the spectral characteristic V is additionally required.

4 Conclusion

The article gives the spectral representation for stochastic integration operators with respect to the Wiener process in the form of a composition of spectral characteristics used in the spectral form of mathematical description for control systems. Results of the article can be applied for the exact or approximate representation of multiply or iterated Stratonovich stochastic integrals and for approximate modeling in the implementation of methods for the numerical solution to stochastic differential equations with high orders of strong convergence [7, 8, 14], where the transition from infinite matrices to finite ones using "truncation" [2, 6] is necessary. Taking into account how the Stratonovich stochastic integral is related to the Itô stochastic integral [8–10], as well as in the general case with the θ-integral [10], $0 \leq \theta \leq 1$, we can obtain the corresponding spectral representations. Obtained results can be generalized to represent the stochastic integration operator with fractional order by defining the kernel $k(t,\tau)$, for example, taking into account the results from [15]. Obtained results can also be applied for modeling optimal control systems with random disturbances [16–18].

References

[1] A.V. Skorohod, *Random Linear Operators* (Springer, Dordrecht, 1984)
[2] W.W. Solodownikow, W.W. Semjonow, M. Peschel, D. Nedo, *Berechnung von Regelsystemen auf Digitalrechnern. Anwendung von Spektral- und Interpolationsmethoden* (Verlag Technik, Berlin, 1979)
[3] A.V. Panteleev, K.A. Rybakov, Autom. Remote Control, **72(2)**, 393 (2011)
[4] J.P. Boyd, *Chebyshev and Fourier Spectral Methods* (Dover Publ., New York, 2000)
[5] B.-Y. Guo, *Spectral Methods and Their Applications* (World Scientific, Singapore, 1998)
[6] K.A. Rybakov, IOP Conf. Ser.: Mater. Sci. Eng., **1864**, 012025 (2021)
[7] D.F. Kuznetsov, Ufa Math. J., **11(4)**, 49 (2019)
[8] D.F. Kuznetsov, Differ. Uravn. Protsessy Upr., **4**, A.1 (2020)
[9] P.E. Kloeden, E. Platen, *Numerical Solution of Stochastic Differential Equations* (Springer, Berlin, 1992)
[10] S.S. Artemiev, T.A. Averina, *Numerical Analysis of Systems of Ordinary and Stochastic Differential Equations* (VSP, Utrecht, 1997)
[11] K.A. Rybakov, Differ. Uravn. Protsessy Upr., **3**, 98 (2020)
[12] B. Golubov, A. Efimov, V. Skvortsov, *Walsh Series and Transforms: Theory and Applications* (Springer, Dordrecht, 1991)
[13] K. Morrison, New York J. Math., **1**, 75 (1995)
[14] M.D. Kuznetsov, D.F. Kuznetsov, Differ. Uravn. Protsessy Upr., **1**, 93 (2021)
[15] K. Rybakov, A. Yushchenko, IOP Conf. Ser.: Mater. Sci. Eng., **927**, 012080 (2020)
[16] E.V. Karachanskaya, J. Math. Sci., **248**, 67 (2020)
[17] I. Kudryavtseva, A. Efremov, A. Panteleev, IOP Conf. Ser.: Mater. Sci. Eng., **927**, 012059 (2020)
[18] A.V. Panteleev, A.V. Lobanov, Autom. Remote Control, **81(11)**, 2026 (2020)

Computer technologies for forming professional competences in Applied mathematics

Konstantin Rybakov[1], *Vladimir* Vinogradov[1], and *Tatyana* Volkova[1,*]

[1]Moscow Aviation Institute (National Research University), 4, Volokolamskoe shosse, Moscow, 125993, Russia

Abstract. The article is devoted to the use of computer technologies for forming the professional competences. We describe the study of basic and special courses for bachelor's or master's degrees in the field "Applied mathematics" at the Moscow Aviation Institute (National Research University). All kinds of practices during the study are considered. The continuous forming technology of professional competences at each stage of study with using the computer technologies is shown.

1 Introduction

To form the Federal State Educational Standards of Higher Education based on the competence approach it is determined the list of competences that young specialists should have [1]. This allows creating a list of academic courses and tools that provide these competences and creating methodologies and criteria for determining objectively the accordance level of study with respect to requirements of specified competences.

The competence is the ability to apply knowledge and skills to a successful activity in the chosen field. Results of the education for this approach are no longer formulated in terms of knowledge, skills and abilities, but in the form of competences [1–3]. The competence model of the graduate in the Federal State Educational Standard of Higher Education is the most important requirement. Within the "National Research University" framework the Moscow Aviation Institute has developed the Self-Established Educational Standard with the requirement structure similar to Federal State Educational Standard. It allows performing requirement in terms of ensuring the education quality [4]. The list of courses and their study order may be changed, but the competences from the Federal State Educational Standard must be required [1, 2].

The main goal of this article is to describe the education system at the Moscow Aviation Institute within the "National Research University" framework.

The rest of this article is structured as follows. The next section provides detailed description of the learning process for bachelors and masters at each stage of study with using the computer technologies. The section lists all basic and special courses, as well as all types of practices in the learning process. The final section presents the conclusions for this article.

*e-mail: tbvolkova@mail.ru

2 Forming professional competences

Graduates with bachelor's or master's degrees should have professional competences corresponding to professional activities such as researches, technological processes, the organization and management.

To provide them, students, starting from the first year, in the study of basic courses such as:

- Complex Analysis;
- Control Theory;
- Discrete Mathematics;
- Linear Algebra and Analytical Geometry;
- Mathematical Analysis;
- Numerical Methods;
- Optimization Methods;
- Ordinary Differential Equations;
- Partial Differential Equations

apply computer technologies for laboratory works and courseworks with studying the new algorithms, techniques, methods and with using the textbooks in traditional and online forms [5].

The Moodle learning platform [6] is used to teach students in all basic courses. This platform has become most relevant in the past 2 years due to the Covid-19 pandemic, since training is partially or completely carried out remotely. In the Moscow Aviation Institute, the learning management system based on Moodle contains necessary educational materials, videoconference services, etc. To assess the knowledge level, both intermediate and final testing are used.

Let us describe the technology for preparing tests, which provides the automated generation of variants and filling the bank of questions for testing for basic courses. A special program is developed by Delphi or C/C++ programming language (the choice is not limited to the listed ones) that generates tasks with numerical parameters that cover all themes of the studied discipline. For this purpose the LaTeX format with syntax defined in the moodle package is used. Numerical parameters are set either by a deterministic algorithm or randomly. Different types of questions are used: multiple choice, numerical, matching, etc. The answer options in multiple choice questions are formed in such a way as to minimize the possibility of guessing the correct answer. Using the LaTeX translator the question file is converted into an XML file, which is imported into the learning platform Moodle, and a PDF file (documentation for tests). This technology allows one to create tests for basic courses with a large number of questions, minimizing repetitive options for different students, with a set of correct and incorrect answers, including complex math formulas.

To ensure the technological processes activity the graduates should have the ability to use standard applications for solving the practical problems. Graduates should be able to debug and to test software, to have knowledge of modern programming languages, operating systems, office applications, methods of the data management, principles of operating systems. Such knowledge at the Mathematical Cybernetics Department of the Moscow Aviation Institute is provided by following special courses [7]:

- Computer Linguistics and Information Technologies;
- Development of Information Systems;
- Hypertext Technologies in Information Systems;

- Intelligent Systems;
- Languages and Technologies of Information Exchange;
- Professional Software Environments.

The content of these courses covers a wide range of issues related to the possession for currently available means of creating hypertext systems, principles of intelligent systems and modern mathematical methods used for the development of information systems. The content is also covers issues related to the ability for solving the problems of the knowledge representation and knowledge processing as well as the computer linguistics.

The organization and management activity assumes that graduates are capable and ready to solve practical problems, to take the responsibility, to determine economic expediency of technical and organizational decisions.

The mandatory participation of students in the researches develops abilities to work independently. Students participate in researches from the second year of study and sometimes from the first year of study. They should make presentations about their research work at student scientific conferences and they can participate in student scientific competitions.

The research activity involves the ability to identify the essence of science problems arising in the professional work, the willingness to use mathematical tools for solving the practical problems. It also involves the ability to apply the appropriate mathematical model and to check its adequacy, to analyze results of modeling and to make a decision based on these results.

Such competences correspond to following special courses [7]:

- Analysis and Synthesis of Nonlinear Stochastic Systems in Aviation Technology and Economics;
- Application of Global Optimization Methods in Control Theory;
- Computer Methods in Control Theory Problems;
- Information Technologies in Decision-Making Theory;
- Mathematical Methods for Synthesis of New Control Systems Classes;
- Methods of Solving Problems for Information Processing;
- Modern Theory of Optimal Control;
- Operations Research;
- Optimization of Hybrid Control Systems;
- Prediction of States in Dynamical Systems.

The content of these courses associates with the use of mathematical methods and mathematical modeling on computers to perform laboratory works and courseworks.

In the course "Information Technologies in Decision-Making Theory" it is proposed to use the software for the aggregation of preferences. This software has been developed at the Mathematical Cybernetics Department of the Moscow Aviation Institute and it can also be applied in the course "Discrete Mathematics" [8, 9], since the described algorithms are based on procedures of the graph theory.

When forming the methodologies and criteria that allow objectively determining the compliance level for students to requirements of specified competences it is necessary to rely on all kinds of practices: educational, computational, research and pre-diploma as well as courseworks, which allow evaluating how students can use the gained knowledge for solving the practical problems.

Students after the first year of study have the educational practice. Its purpose is to teach students the use of special mathematical packages for solving the practical problems. At the educational practice the students learn to use computer algebra systems such as Mathcad and Matlab. Further, these knowledge and skills will be used in the courses "Analysis and Synthesis of Nonlinear Stochastic Systems in Aviation Technology and Economics" and "Computer Methods in Control Theory Problems" that involve the use of the spectral form of mathematical description for the complex control systems [10].

The main goal of the computational practice is to summarize knowledge gained in two years of study. Students should apply mathematical tools for solving one of the practical problem and should develop computer programs on the specified programming language for calculations.

The research practice takes place after the third year of study. It allows not only to write a computer program for calculations that arise in a lot of practical problems, but also to formulate a mathematical statement of the problem as well as to analyze the data obtained from calculations.

Each practice brings the student closer to the final stage of study. The last but one stage is the pre-diploma practice, for which students usually takes place at the company or enterprise. They get acquainted with the research subject, algorithms for solving practical problems and learn to work on a team. The final stage of study is the thesis and its defense. Students get a task at the company or enterprise, where they will start a career. They can also receive a task related to the practical problem at the Mathematical Cybernetics Department of the Moscow Aviation Institute.

Graduates of bachelor's degree should publicly defend the thesis with demonstrating the ability to formulate a mathematical statement of the problem, to describe the mathematical model, to explain methods for solving the problem, to make the computer program for calculations and to analyze obtained results. They should see the progress of their work and practical applications in the workplace thereby demonstrating the professional competences obtained as a result of the special courses study. The bachelor's degree is only the first stage of a specialist formation.

Graduates of master's program to ensure research activities should have the ability to develop and to explore mathematical models of objects, systems, and processes, to develop software, to provide scientific experiments, to evaluate the research results. These competences are supported by the study of special courses [7]:

- Analysis and Synthesis of Nonlinear Stochastic Systems in Aviation Technology and Economics;
- Application of Global Optimization Methods in Control Theory;
- Applied Methods of Optimization and Information Processing;
- Computer Linguistics and Information Technologies;
- Information Security;
- Information Technologies of Decision-Making Theory;
- Intelligent Systems;
- Languages and Technologies of Information Exchange;
- Methods of Solving Problems of Information Processing;
- Modern Technologies for Parallel and Distributed Programming;
- Modern Theory of Optimal Control;
- Optimization of Hybrid Control Systems.

With these courses the students get knowledge about new directions of computer technologies: features of modern SQL and NoSQL database management systems, languages and technologies of the information exchange based on XML, intelligent analysis of big data based on data mining and machine learning algorithms, computer linguistics, decision-making theory, principles of the symmetric and asymmetric encryption and their using to protect information. Students use electronic signatures for their practical works and coursework.

The study of special courses includes laboratory works and courseworks. Laboratory works is carried out in display classes and require not only theoretical knowledge, but also the ability to work with special mathematical software (e.g. computer algebra systems), to make practical decisions and to analyze obtained results.

Let us show the practical applicability of the for forming the professional competences by the laboratory work "Search for Association Rules" in the course "Intelligent Systems". When performing this work, students receive the analyzed dataset and perform a preliminary transformation to apply the Apriory method. Then they apply this method by two stages:

(1) search for frequently encountered sets of elements that meet the minimum level of minimum support *MinSupport* offered by the tutor;

(2) generation of the found frequently found sets of association rules that satisfy the minimum level of the minimum confidence *MinConfidence* also proposed by the tutor at the initial stage of solving the problem.

Next, students analyze these association rules and change the above values of the indicators to obtain interesting rules. Then students repeat the stage (2) of the algorithm, changing *MinConfidence* to the *Lift* and *Leverage* indicators. Performing such work allows students to form the ability to make decisions, draw conclusions about the applicability of methods and analyze results.

To ensure the technological processes activity the students should be able to analyze and to synthesize complex control systems, for which they should be able to make the scientific and technical documentation, scientific and technical reports, and reviews. To do this, there are exists different forms of studying such as courseworks and both practices research and professional. In courseworks and practices the students, getting a new task, should independently look for ways to solve it, showing its relevance and novelty. Students should develop a computer program and should analyze obtained results. Then, they also should make a report on the problem solution in the form of abstracts at the scientific conference or research articles (see, for example, [11–14]).

Knowledge and skills obtained during all kinds of practices are applied by students in courses "Prediction of States in Dynamical Systems" and "Methods of Solving Problems of Information Processing" with respect to solving the optimal filtering and prediction problems in complex dynamical systems [15–18], and especially in courses "Application of Global Optimization Methods in Control Theory", "Optimization of Hybrid Control Systems", "Mathematical Methods for Synthesis of New Control Systems Classes", and "Modern Theory of Optimal Control", in which various modern models and methods of the optimal control are considered [19–25]. These knowledge and skills are also very important for courses "Computer Linguistics and Information Technologies", "Information Security", "Intelligent Systems", "Languages and Technologies of Information Exchange", and "Modern Technologies for Parallel and Distributed Programming".

One of the significant task in the training is the mastery of modern instrumental software by students. This software may require programming skills in modern languages and may not require special skills. The software used in the training may be free to distribution or proprietary if the training process does not violate the restrictions on its use. In many cases, proprietary software allows the noncommercial use for the training. Therefore, for laboratory works the students actively use Microsoft SQL Server Management Studio [26] to study capabilities of modern database management systems working with XML or full text data. Students use XML processors to perform laboratory works on validating XML documents, generating XSD schema, creating XPath queries, XSLT and Schematron scripts [27]. The study of machine learning algorithms is provided by using Python [28] with following libraries: Scikit-learn, Numpy, Pandas, Matplotlib, etc. Students widely apply machine learning platforms such as Weka [29] and Knime [30], which do not require programming skills.

The master's thesis ensure the competence of the organizational and management activity, including the willingness to take the responsibility for decisions, to make non-standard decisions and to solve problem situations. On the master's thesis defense the students make a report on the work results, answer questions from members of the State Examination Commission (Thesis Committee) demonstrating that they can take the responsibility for their decisions and can solve problem situations.

3 Conclusion

The article provides detailed description of the learning process for bachelors and masters in the field "Applied mathematics" at the Moscow Aviation Institute within the "National Research University" framework. The main focus is on using the information technologies in the learning process. The study of basic and special courses for bachelor's or master's degrees is described. All kinds of practices during the study are also considered. The continuous forming technology of professional competences at each stage of study with using the computer technologies is shown.

References

[1] Federal State Educational Standard of Higher Education. Direction "Applied Mathematics" in Federal State Educational Standards of Higher Education, URL `http://www.fgosvo.ru`

[2] *New Educational Technologies in Engineering*, ed. by A.N. Gerashchenko, M.Yu. Kuprikov, A.Yu. Sidorov (MAI Publishing House, Moscow, 2012)

[3] T.B. Volkova, T.I. Korotkova, *Application of information technologies when training in the master's degree program*, IOP J. Phys.: Conf. Ser., **891**, 012372 (2017)

[4] Moscow Aviation Institute (National Research University), URL `http://www.mai.ru`

[5] A. Alekseychuk, A. Panteleev, T. Korotkova, *Usage of the software system based on web conferences for distance online learning*, in Proc. 11th Int. Conf. Application of Information and Communication Technologies, IEEE, (2017)

[6] Moodle, URL `http://www.moodle.org`

[7] Mathematical Cybernetics Department of the Moscow Aviation Institute, URL `http://www.dep805.ru`

[8] V.N. Nefyodov, V.A. Osipova, S.O. Smerchinskaya, N.P. Yashina, Russian Mathematics, **62(5)**, 61 (2018)

[9] S.O. Smerchinskaya, N.P. Yashina, Int. J. Model. Simul. Sci. Comput., **9(1)**, 1850006 (2018)

[10] G. Baghdasaryan, M. Mikilyan, A. Panteleev, K. Rybakov, AIP Conf. Proc., **2181**, 020030 (2019)

[11] T. Korotkova, Ya. Kuzmina, AIP Conf. Proc., **2181**, 020004 (2019)

[12] T.I. Korotkova, S.A. Kolesnik, B.A. Garibyan, S.Y. Luneva, Y.V. Kuzmina, IOP J. Phys.: Conf. Ser., **1565**, 012035 (2020)

[13] A.V. Panteleev, T.A. Letova, E.A. Pomazueva, Autom. Remote Control, **79(1)**, 153 (2018)

[14] E. Pegachkova, A. Boboshina, AIP Conf. Proc., **2181**, 020021 (2019)
[15] T.A. Averina, K.A. Rybakov, Rus. J. Numer. Anal. Math. Modelling, **35(2)**, 55 (2020)
[16] E.A. Rudenko, J. Comput. Sys. Sci. Int., **57(1)**, 43 (2018)
[17] E.A. Rudenko, J. Comput. Sys. Sci. Int. **57(4)**, 505 (2018)
[18] K. Chugai, I. Kosachev, K. Rybakov, AIP Conf. Proc., **2181**, 020001 (2019)
[19] A.S. Bortakovskii, J. Comput. Sys. Sci. Int., **57(4)**, 543 (2018)
[20] A.S. Bortakovsky, I.V. Uryupin, J. Comput. Sys. Sci. Int., **60(5)**, 701 (2021)
[21] M.M. Khrustalev, *Terminal invariance of stochastic diffusion systems*, Autom. Remote Control, **79(8)**, 1434 (2018)
[22] L. Kondratieva, AIP Conf. Proc., **2181**, 020002 (2019)
[23] I. Kudryavtseva, A. Efremov, A. Panteleev, AIP Conf. Proc., **2181**, 020008 (2019)
[24] A.V. Panteleev, V.N. Panovskiy, Adv. Sci. Technol. Eng. Syst. J., **4(5)**, 23 (2019)
[25] A.V. Panteleev, A.V. Lobanov, Algorithms, **14(11)**, 332 (2021)
[26] Microsoft SQL Server Management Studio, URL http://www.microsoft.com
[27] W3SCHOOLS, URL http://www.w3schools.com
[28] Python, URL http://www.python.org
[29] Weka (Waikato Environment for Knowledge Analysis), URL http://www.cs.waikato.ac.nz/ml/weka
[30] KNIME (Konstanz information miner), URL http://www.knime.com

Image-based simulation of dispersed composites and porous rocks

Sergey Sheshenin[1,*], *Nina* Artamonova[1], *Peter* Klement'ev[1], *Egor* Orlov[1], and *Julia* Frolova[1]

[1]Lomonosov Moscow State University, 1, Leninskiye gory, Moscow, 119991, Russia

Abstract. The article presents the results of numerical simulation of the effective elastic properties and stress concentration of the dispersed composite and porous rocks as well as the elastoplastic diagrams of tensile stress of the B_4C/2024Al composite using asymptotic averaging method. 2D and 3D simulations of uniaxial tension was carried out using the image-based models of the real structure obtained by X-ray tomography. We also used model structures of the B_4C/2024Al composite, consisting of an aluminum matrix with boron carbide inclusions in the form of ellipsoids with a random orientation. It turned out that in the elastic region 3D real structure and 3D ellipsoidal structure lead to the same effective moduli. On contrary, the stress concentrations are much higher in the real structure model than in the model of ellipsoidal inclusions. Also, two types of the models behave completely differently in the plastic region. Comparison of 2D and 3D calculations with experimental data shows that 2D models can be also used for determining the effective elastic moduli of B_4C/2024Al composite, but the use of 3D models is preferable for porous geomaterials and dispersed composites.

1 Introduction

Asymptotic averaging is a mathematically rigorous approach for determining the effective properties of heterogeneous and porous composites. Initially, the method was developed for determining the elastic moduli of periodic structures [1–3]. However, this method can be used for non-periodic structures [4, 5] and for determining the other properties, for example, Biot's tensor parameter of water-saturated porous rocks [6–8]. In this case, it is important to define a representative volume element (RVE) and formulate boundary conditions at the boundary of the RVE. The first definitions of the RVE and effective elastic moduli as well as the method for their calculation were given by R. Hill [9, 10]. In our work, we used the generalized definition of a representative volume element formulated in [5]. RVE is the volume for which the dependence of the effective property under consideration (for example, Young's modulus or Biot's coefficient) on the typical microstructure parameter (for example, inclusion volume ratio or porosity) lies on the same curve with acceptable accuracy.

In the first part of this paper, the solution of the local problems is performed for the B_4C/2024Al composite in both elastic and plastic regions. The calculations were carried out for an image-based composite structure obtained using X-ray tomography. Further in this paper we call such a structure as the real one. For comparison, we also used the generated

*e-mail: sergey.sheshenin@math.msu.ru

structure with ellipsoidal inclusions, further called as the model structure. Both material microstructures were used for subsequent finite element (FE) mesh generation. The effective stress-strain diagrams obtained using both structures were compared with the experimental results. The calculations took into account the progressive destruction of the matrix in the plastic region.

The second part of the article is devoted to the determination of the effective elastic moduli and Biot parameter of porous rocks. The development of X-ray tomography method and software for digitizing tomography images made it possible to obtain high-quality 3D models of the real structure of porous materials, which can be used for numerical experiments with the help of finite element codes. The results of 2D and 3D calculations were also compared with experimental data.

2 Study of stress-strain state in B_4C/2024Al composite in computational uniaxial tension experiments

Aluminum alloys are widely used as a matrix for metal matrix composites because of its good machinability, wear resistance and mechanical properties. Boron carbide (B_4C) is promising reinforcing inclusion in composites with an aluminum matrix due to its attractive properties such as ultra-high hardness, low density and good thermal and chemical stability [11]. Boron carbide inclusions improve the mechanical properties of the composite, such as elastic moduli, tensile strength, compressive strength, hardness and yield strength [12]. As a result, aluminum matrix composites reinforced with boron carbide (B_4C) particles are popular in automotive and aerospace industry, and especially in nuclear industry [11–13]. The results of the experimental determination of the mechanical properties of metal matrix composites with reinforcing inclusions are widely presented in the literature. In addition to experimental studies, it is advisable to develop numerical modeling of such composites based on X-ray images.

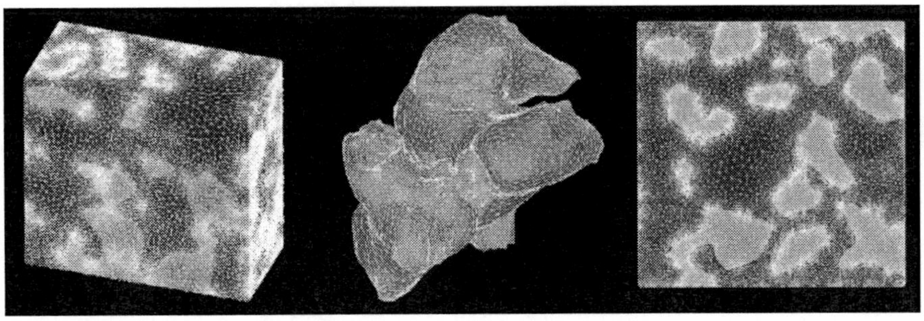

(a) (b) (c)

Figure 1. 3D model (a), B_4C inclusions (b) and 2D model (c) of the B_4C/2024Al composite

In our work, we studied the elastic/plastic properties of the dispersed B_4C/2024Al composite depending on the volume ratio of boron carbide based on X-ray images of the internal structure (figure 1). Figure 1(a) shows a 3D two-component model of the B_4C/2024Al composite constructed on the basis of X-ray images obtained at the Shanghai Institute of Applied Physics and processed in the VG MAX 3.3 program [14] at Sovtest-Service LLC (Kursk). The images were supplied by Prof. Qiang Zhang from Harbin Institute of Technology, China in framework of collaborative project. In figure 1(b) B_4C inclusions are depicted, and figure 1(c) shows a 2D model also built from X-ray images. As one can be seen, the inclusions have an irregular angular shape. In order to study the influence of the inclusion shape on the effective properties of the B_4C/2024Al composite, we also used models in which boron carbide inclusions were modeled by ellipsoids with a random orientation.

2.1 Determination of elastic moduli and stress concentration in B_4C/2024Al composite

The effective elastic moduli of dispersed B_4C/2024Al composites were determined by the well-known asymptotic averaging method [3] based on 2D and 3D models of the real structure (figure 1) and 3D models with ellipsoids. The calculated values of Young's modulus were compared with experimental ones. According to the method mentioned above, it is necessary to solve a local problem in the RVE with a special boundary condition in the form of a linear function of coordinates on the external border Σ_{RVE} [4, 5]:

$$\left(C_{ijkl}(\boldsymbol{\xi})u_{k,l}(\boldsymbol{\xi})\right)_{,j} = 0, \quad \boldsymbol{\xi} \in \mathrm{V}_{\mathrm{RVE}}; \quad u_k = \varepsilon\varepsilon^0_{kq}\xi_q, \quad \sigma_\tau = C_{ijkl}(\boldsymbol{\xi})u_{k,l}(\boldsymbol{\xi})n_j\tau_i = 0, \quad \boldsymbol{\xi} \in \Sigma_{\mathrm{RVE}}.$$

Here, $\boldsymbol{\xi}$ are fast coordinates in the RVE, $\xi_q = x_q/\varepsilon$, $\varepsilon = l/L$, ε is a small geometric parameter, l is the characteristic linear size of the RVE, L is the characteristic linear size of the studied composite volume, n_j and τ_i are the components of normal and tangent vectors to Σ_{RVE}, respectively. The following elastic properties were set in the calculations for the boron carbide inclusions: $E = 450$ GPa, $\nu = 0.16$ and for the aluminum matrix: $E = 74.7$ GPa, $\nu = 0.32$.

Numerical experiments on uniaxial tension were carried out in a finite element (FE) program. Elastic moduli are calculated by the formulas:

$$E^{eff}_\alpha = \frac{\langle\sigma_{\alpha\alpha}\rangle}{\langle\varepsilon_{\alpha\alpha}\rangle} = \frac{\langle\sigma_{\alpha\alpha}\rangle}{\varepsilon^0_{\alpha\alpha}}; \quad \langle\sigma_{\alpha\alpha}\rangle = \frac{1}{\mathrm{V}_{\mathrm{RVE}}}\int_{\mathrm{V}_{\mathrm{RVE}}} \sigma_{\alpha\alpha}\, dV.$$

Figure 2 shows the results of calculating Young's modulus and comparison with the experimental data.

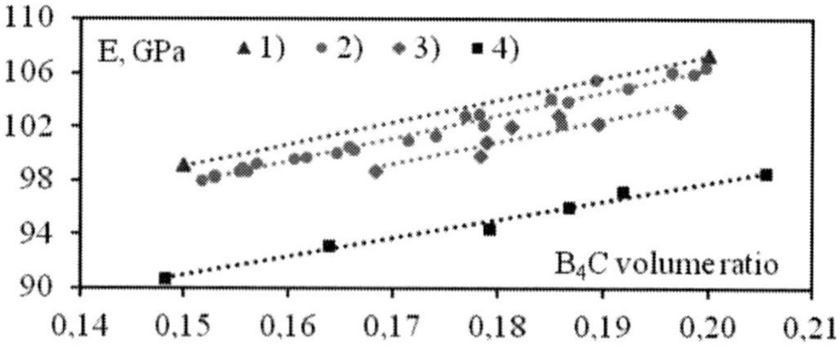

Figure 2. Dependence of Young's modulus on B_4C volume ratio: 1) experiments, 2) 3D model of the real structure, 3) 3D model with ellipsoidal inclusions, 4) 2D model of the real structure

As one can see in figure 2, for the B_4C volume ratio range from 0.14 to 0.21, the dependence of the effective Young's modulus on the boron carbide volume ratio is linearly approximated with a fairly good accuracy for 3D and 2D models. Young's modulus values for a 3D real structure and a 3D model with ellipsoids differ from the experimental values by no more than 1–1.5% and 2–3%, respectively, at the same B_4C volume ratio (figure 2). Thus, 3D models with ellipsoidal inclusions can be used for evaluating the effective elastic moduli, at least for the inclusion volume ratio up to 0.20. The values of Young's modulus for a 2D real structure are 6–8% lower than the experimental values at the same B_4C volume ratio. Hence, it can be concluded that the use of 3D models is clearly preferable to 2D models.

We used 3D models of the real microstructure and 3D models with randomly oriented ellipsoidal inclusions to study the stress-strain state of the B_4C/2024Al composite at the microlevel in the uniaxial tension and the stress concentration near boron carbide inclusions of various shapes [5]. The equivalent Von Mises stress concentrations C_{Mises} are considered as

$$C_{Mises} = \sigma_{MaxMises}/\sigma_{AvMises}, \quad \sigma_{Mises} = \sqrt{[(\sigma_1 - \sigma_2)^2 + (\sigma_2 - \sigma_3)^2 + (\sigma_1 - \sigma_3)^2]/2},$$

Here, $\sigma_1, \sigma_2, \sigma_3$ are the principal stresses, σ_{Mises} is the equivalent Von Mises stress, $\sigma_{MaxMises}$ and $\sigma_{AvMises}$ are its maximum and average values, respectively. Table 1 shows the results of calculating the stress concentration on a mesh of nearly 2 million nodes [5].

Table 1. The equivalent Von Mises stress concentration in computational uniaxial compression experiments

	Model of the real structure		Model with ellipsoids	
Components	B_4C	2024Al	B_4C	2024Al
Number of samples	14	14	4	4
Mean	13.78	4.03	4.78	2.62
Minimum	9.34	2.72	4.10	2.40
Maximum	29.19	6.19	5.65	2.77
Standard deviation	5.3517	0.9956	0.7354	0.1662

As can be seen from the table 1, the maximum concentration is observed in models of the real structure in boron carbide inclusions. In B_4C inclusions, the von Mises stress concentration for the real structure reaches 29.19, and for the model structure with ellipsoids—only 5.65. In a less rigid aluminum matrix, the maximum von Mises stress concentration is significantly lower (6.19 for a real structure and 2.77 for a model with ellipsoids). Thus, studying the concentration of stress intensity, it is important to take into account the real structure of dispersed composites.

2.2 Elastoplastic properties of B_4C/2024Al composite under uniaxial tension

The elastoplastic behavior of the B_4C/2024Al composite was investigated in uniaxial tension computational and real experiments. Calculations were carried out taking into account the progressive failure in the matrix for 3D models of real (figure 3(a)) and model (with ellipsoids) (figure 3(b)) structures. Modeling the progressive damage in the material was performed as follows. In our problem, B_4C inclusions are assumed to be non-destructive. At each load step the whole finite element in the matrix is considered to be damaged if the equivalent strain in it reaches 8%. The equivalent strain in the elements is calculated as the average value over all Gaussian points. In this case, the stiffness of each destroyed finite element is set to be zero. Then the next load step is carried out.

Figure 3(a) shows that for the model of the real structure accounting the progressive damage ensures the coincidence of the calculated and experimental data. As for the model with ellipsoidal inclusions (figure 3(b)), progressive damage does not lead to the coincidence of the calculated and experimental diagrams. On the other hand, the diagrams calculated without progressive failure technique included coincide with each other for both real and model inclusions but not with the experimental diagram. It should be mentioned that the numerical solution procedure with an embedded progressive damage algorithm becomes unstable at some stage of loading (figure 3(b)).

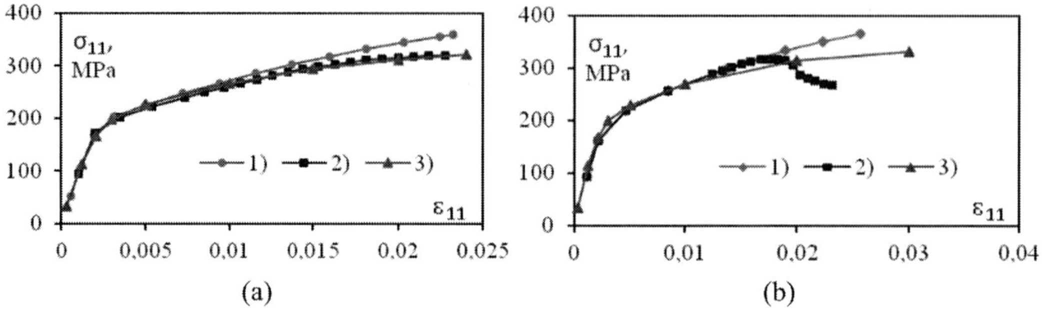

Figure 3. Dependence of stress σ_{11} on strain ε_{11} computed for 3D real structure (a) and 3D model structure with ellipsoidal inclusions (b) in uniaxial tension test: 1) calculations without progressive failure technique, 2) calculations with included progressive damage technique, 3) experiments. B_4C volume ratio equals 0.18 (a) and 0.16 (b)

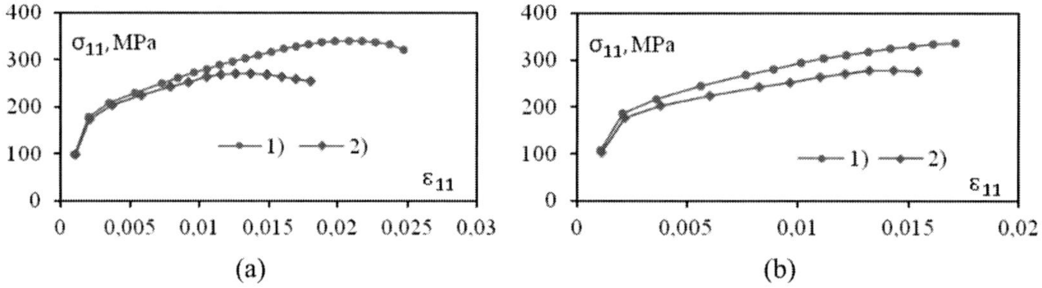

Figure 4. Dependence of stress σ_{11} on strain ε_{11} computed for 1) 3D real structure and 2) 3D model structure with ellipsoidal inclusions in uniaxial tension test with included progressive damage technique. B_4C volume ratio equals 0.15 (a) and 0.20 (b)

Figure 4 shows the deformation diagrams for the real structure model and the model with ellipsoidal inclusions under uniaxial tension, taking into account the progressive damage of the matrix (for B_4C volume ratio of 0.15 (figure 4(a)) and 0.20 (figure 4(b))). It can be noted that the real and ellipsoidal inclusion structures behave completely differently in the plastic region.

3 Determination of effective properties of porous rocks

Asymptotic averaging was also used to determine the effective elastic moduli and Biot's parameter of porous rocks based on 2D and 3D models of the real structure obtained using X-ray tomography. For the calculations, two types of porous rocks were considered: limestone and hyaloclastite. Calculated Young's moduli were compared with experimental data. Calculated Biot's coefficients were verified by the well-known formula, which utilizes bulk moduli of the porous rock and skeleton material. The dependences of Young's modulus and Biot's coefficient on porosity and pore shape were studied.

3.1 Theoretical aspects and solution method

Biot's parameter (pore pressure transfer tensor) is included in the formula for calculating effective stresses [15, 16]:

$$\langle \sigma_{ij}^{eff} \rangle = \langle \sigma_{ij}^{full} \rangle + \alpha_{ij} \langle p \rangle.$$

Here $\langle p \rangle$ is the fluid pressure (positive under compression). Angle brackets indicate averaging over the RVE. For isotropic rocks, Biot's parameter is a scalar coefficient that varies from 0 to 1 and depends on the rock properties (porosity, shape and orientation of pores). In [17, 18], an empirical formula was proposed for calculating the isotropic pore pressure transfer coefficient α:

$$\alpha = 1 - K^{eff}/K_s, \tag{1}$$

where K_s is the bulk modulus of the skeleton material, K^{eff} is the effective bulk modulus. A rigorous mathematical derivation of formula (1) is presented in [19]. The pore pressure transfer tensor (Biot's parameter) is difficult to determine experimentally. Moreover, the experimental determination has not been yet developed for anisotropic rocks. Therefore, the computational method for determining this parameter is relevant and of great practical importance. The asymptotic averaging method used to determine Biot's parameter is similar to the procedure for determining the effective elastic moduli and leads to the solution of local problems in the RVE of the porous rock [6–8].

To determine the effective elastic moduli of a porous medium in the RVE, it is necessary to solve a set of local boundary-value problems with special boundary conditions in the form of linear functions of coordinates on the outer RVE boundary Σ_{RVE} and zero pore pressure p on both pores boundaries Σ_{int} and Σ_{RVE} [6–8]:

$$\left(C_{ijkl}(\xi)\,u^1_{k,l}(\xi)\right)_{,j} = 0, \quad \xi \in \mathrm{V_{RVE}}; \quad u_k = \varepsilon\varepsilon^0_{kq}\xi_q = \varepsilon^0_{kq}x_q, \quad \xi \in \Sigma_{\mathrm{RVE}},$$
$$C_{ijkl}(\xi)\,u^1_{k,l}(\xi)\,n_j = 0, \quad \xi \in \Sigma_{\mathrm{int}}; \quad p = 0, \quad \xi \in \Sigma_{\mathrm{RVE}} \cup \Sigma_{\mathrm{int}}.$$

Finally, elastic moduli are calculated by the formulas:

$$C^{eff}_{ij\alpha\beta} = \frac{\langle\sigma_{ij}\rangle}{\langle\varepsilon_{\alpha\beta}\rangle} = \frac{\langle C_{ijkl}u^1_{k,l}\rangle}{\varepsilon^0_{\alpha\beta}}; \quad \langle\sigma_{ij}\rangle = \frac{1}{\mathrm{V_{RVE}}}\int_{\mathrm{V_{RVE}}} \sigma_{ij}\,dV.$$

If the material is isotropic, effective Young's modulus E^{eff}, Poisson's ratio ν^{eff} and bulk modulus K^{eff} are calculated using the formulas of elasticity theory (the superscript "eff" is omitted for ease of notation):

$$C_{\alpha\alpha\alpha\alpha} = \lambda + 2\mu, \quad C_{\beta\beta\alpha\alpha} = \lambda, \quad \nu = \frac{\lambda}{2(\lambda+\mu)}, \quad E = 2\mu(1+\nu), \quad K = \lambda + \frac{2\mu}{3}.$$

Here λ and μ are the elastic Lame moduli.

For determining Biot's parameter, a boundary value problem is formulated in the RVE as follows. A constant fluid pressure is set on the pore boundary Σ_{int}, while the outer RVE boundary Σ_{RVE} is fixed in the perpendicular direction [6–8]:

$$\left[C_{ijkl}(\xi)\,u^2_{k,l}\right]_{,j} = 0, \quad \xi \in V_{\mathrm{RVE}}; \quad C_{ijkl}(\xi)\,u^2_{k,l}n_j = -p_0 n_i(\xi), \quad \xi \in \Sigma_{\mathrm{int}}; \quad u^2_k = 0, \quad \xi \in \Sigma_{\mathrm{RVE}}.$$

The last boundary condition replaces the periodicity condition and is rigorously correct, if the RVE has a symmetric shape. The components of Biot's tensor α_{ij} are calculated by the formula:

$$\alpha_{ij} = \frac{\langle\sigma_{ij}\rangle}{p_0} = \frac{\langle C_{ijkl}u^2_{k,l}\rangle}{p_0}.$$

Certainly, this method is supposed to be implemented using a finite element code.

3.2 Examples of calculations

The calculation of the effective elastic moduli and components of Biot's tensor was carried out for two types of rocks such as dolomitic limestone and hyaloclastite.

Dolomitic limestone core samples were taken in the south-eastern part of Moscow from wells in the depth interval from 15 to 38 m. Dolomitic limestone consists of calcite (92%), dolomite (6.5%), quartz and clay minerals (< 2%). Limestone has a microcrystalline structure, a homogeneous porous texture (figure 5(a)). The pores are evenly distributed, the pore size is from 0.1 to 1–2 mm. In the calculations, the material of the skeleton was considered homogeneous with elastic properties: $E = 50$ GPa, $\nu = 0.33$, given according to literature data.

To study the stress-strain state of rocks at the microlevel and calculate the effective properties, 3D (figure 5(b)) and 2D (figure 5(c)) models of the real microstructure were used, obtained using a Yamato TDM-1000H-II X-ray computer microtomograph (Lomonosov Moscow State University) and digitization in the VG MAX 3.3 program in OOO Sovtest-Service (Kursk). Two-dimensional calculations were performed for a plane strain state.

Figure 6(a) shows the dependence of Young's modulus on porosity for dolomitic limestones based on the results of calculations and experiments. As one can see, the results obtained using 3D models are close to the experimental data, the difference is less than 5%. Young's modulus values calculated using 2D models differ from the results of 3D numerical calculations and experimental data by 20–30% at porosity from 7 to 28%. At low values of porosity (less than 5%), the values of Young's modulus for 2D and 3D models are close to each other.

(a) (b) (c)

Figure 5. Photo of a thin section (a), 3D model (b) and 2D model (c) of dolomitic limestone

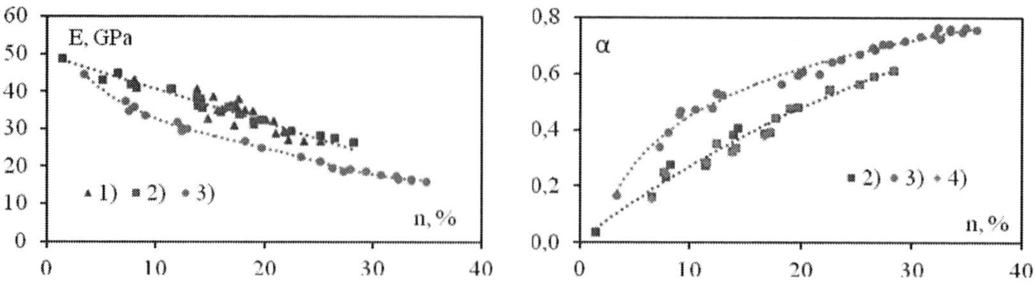

(a) (b)

Figure 6. Dependence of Young's modulus (a) and Biot's coefficient (b) on porosity for the dolomitic limestone: 1) experiments, 2) 3D model, 3) 2D model, 4) calculation results using formula (1)

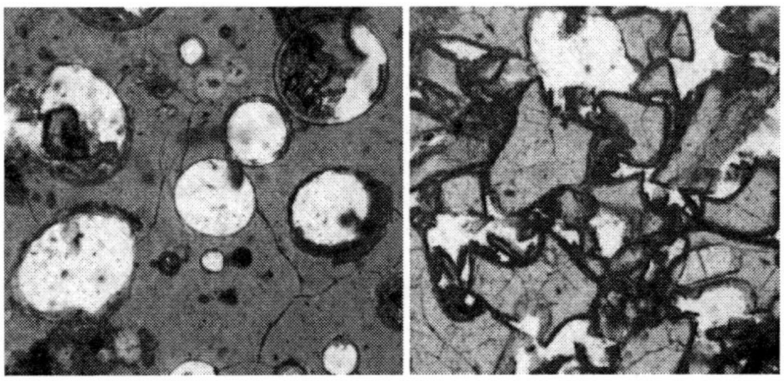

(a) (b)

Figure 7. Photographs of thin sections of hyaloclastite—with round (a) and angular (b) pores

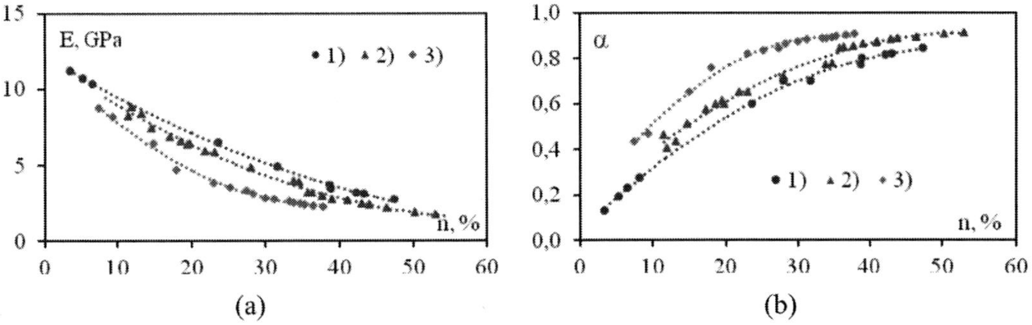

Figure 8. Dependence of Young's modulus (a) and Biot's coefficient (b) on porosity and pore shape for hyaloclastite based on the results of 2D calculations: 1) round pores, 2) mixed pores, 3) angular pores

Figure 6(b) shows the results of calculations of Biot's parameter for limestone. It can be noted (figure 6(b)) that the results of 2D and 3D calculations by the averaging method are close to each other at low porosity (up to 5%). The values of Biot's coefficient for 2D models are 20–35% higher than for 3D models at porosity from 5 to 20%. The discrepancy becomes smaller and amounts to 10–15% with a porosity of more than 20%. As one can see (figure 6(b)), the results obtained by 3D calculations and given by the formula (1) coincide.

Hyaloclastites are volcanogenic-sedimentary rocks obtained from the southern and southwestern regions of Iceland. They were formed in the process of solidification and crystallization of magmatic melt during volcanic eruptions in underwater conditions [20]. Slightly altered hyaloclastites consist of fragments of volcanic basalt glass. Hyaloclastites have got pores of a round (figure 7(a)) or angular (figure 7(b)) shape due to the peculiarities of the origin. Therefore, such geomaterials can be used to study the effect of pore shape on effective properties. The following elastic properties of the skeleton material were set in the calculations: $E = 12.3$ GPa, $\nu = 0.35$.

We investigated the effect of pore shape on Young's modulus (figure 8(a)) and Biot's parameter (figure 8(b)) using 2D models of hyaloclastites. It turned out that for the same porosity in models with round pores, Young's modulus is higher (figure 8(a)), and Biot's coefficient, on the contrary, is lower (figure 8(b)) than in models with elongated angular pores. Probably, this is due to the fact that the contact area of the pores and the solid phase for round pores is smaller than for elongated pores.

4 Conclusion

The so-called local problems related to asymptotic averaging were numerically implemented in application to B_4C/2024Al composite and porous geomaterials. Numerical implementation uses 3D structure images obtained by means of X-ray tomography. Commercial code VG MAX 3.3 was used for FE mesh generation and following porting to FE software. The FE calculations were also done for a model structure in which B_4C inclusions are modeled by ellipsoids. A significant discrepancy in the computed effective stress/strain diagrams in the plastic strain range was revealed between the real structure of B_4C/2024Al and the model structure with ellipsoids, if the progressive destruction of the matrix was taken into account. The calculated stress/strain diagram coincides with the experimental data quite well, if the image-based composite structure is exploited.

The results of calculating the effective elastic properties of the dispersed composite and porous rocks showed that 2D models can be used to determine the Young's modulus of B_4C/2024Al composite, and for porous rocks 2D models are unsuitable and 3D models should be used instead.

The averaging method provides a general way to calculate Biot's tensor of porous rocks. Knowledge of the accurate assessment of the pore pressure transfer tensor is necessary for the correct determination of effective stresses. Adequate determination of Biot's tensor of porous water-saturated rocks by the asymptotic averaging method is possible only with the use of 3D models. A significant influence of the pore shape on the effective Young's modulus and Biot's parameter of porous rocks was revealed.

This research was supported by the Russian Foundation for Basic Research (project 20-01-00431) and the Interdisciplinary Scientific and Educational School of Moscow University "Fundamental and Applied Space Research".
In this work, a Yamato TDM-1000H-II X-ray computer microtomograph (Japan) obtained within the framework of the Development Program of Lomonosov Moscow State University was used.

References

[1] A. Bensoussan, J.-L. Lions, G. Papanicolaou, *Asymptotic analysis for periodic structures* (North-Holland Pub. Co., Amsterdam, New York, 1978)
[2] E. Sanchez-Palencia, *Non-homogenous media and vibration theory (Lecture Notes in Physics)* (Springer-Verlag, New York, 1980)
[3] N.S. Bakhvalov, G.P. Panasenko, *Homogenization: averaging processes in periodic media* (Kluwer Academic Publishers, Dordrecht, Boston, London, 1989)
[4] S.V. Sheshenin, Zhang Qiang, N.B. Artamonova, F.B. Kiselev, M.A. Volkov, AIP Conf. Proc., **2216(1)**, 040017-1 (2020)
[5] S.V. Sheshenin, N.B. Artamonova, P.D. Klement'ev, F.B. Kiselev, R.R. Muradkhanov, E.A. Orlov, Zhang Qiang, Mech. Compos. Mater., **57(1)**, 19 (2021)
[6] S.V. Sheshenin, N.B. Artamonova, Yu.V. Frolova, V.M. Ladygin, Moscow University Geology Bulletin, **70(4)**, 354 (2015)
[7] N.B. Artamonova, A.Zh. Mukatova, S.V. Sheshenin, Mech. Solids, **52(2)**, 212 (2017)
[8] N.B. Artamonova, S.V. Sheshenin, Yu.V. Frolova, O.Yu. Bessonova, P.V. Novikov, Mech. Compos. Mater., **55(6)**, 715 (2020)
[9] R. Hill, J. Mech. Phys. Solids, **11**, 357 (1963)
[10] R. Hill, J. Mech. Phys. Solids, **13**, 213 (1965)
[11] K. Wang, X. Li, Q. Li, G. Shu, G. Tang, Mater. Sci. Eng. A, **696**, 248 (2017)
[12] C. Karthikkumar, R. Baranirajan, I. Premnauth, P. Manimaran, Int. J. Eng. Res. Gen. Sci., **4(2)**, 405 (2016)

[13] P. Zhang, Y. Li, W. Wang, Z. Gao, B. Wang, J. Nucl. Mater., **437**, 350 (2013)
[14] *VG Studio MAX 3.3. Reference Manual. Volume Graphics GmbH* (printed in Germany, Heidelberg, 2019)
[15] M.A. Biot, J. Appl. Phys., **12(2)**, 155 (1941)
[16] Y. Gueguen, M. Bouteca, *Mechanics of fluid-saturated rocks* (Elsevier Acad. Press, 2004)
[17] M.A. Biot, D.G. Willis, J. Appl. Phys., **24**, 594 (1957)
[18] J. Geertsma, Trans. AIME, **210**, 331 (1957)
[19] A. Nur, J.D. Byerlee, J. Geophys. Res., **76(26)**, 6414 (1971)
[20] Yu.V. Frolova, Moscow University Geology Bulletin, **65(2)**, 104 (2010)

MATEC Web of Conferences **362**, 01030 (2022) https://doi.org/10.1051/matecconf/202236201030
CMMASS 2021

To the dynamic reconstruction of sliding controls

Nina Subbotina[1,2,*] and *Evgenii* Krupennikov[1,2]

[1]N.N. Krasovskii Institute of Mathematics and Mechanics of the Ural Branch of the Russian Academy of Sciences (IMM UB RAS), 16, S. Kovalevskaya st., Yekaterinburg, 620108, Russia
[2]Ural Federal University named after the first President of Russia B.N. Yeltsin, 19, Mira st., Ekaterinburg, 620002, Russia

Abstract. The paper is devoted to the problem of dynamic control reconstruction for controlled deterministic affine systems. The reconstruction has to be carried out in real time using known discrete inaccurate measurements of an observed trajectory of the system. This trajectory is generated by an unknown measurable control with values in a given compact set. A correct formulation of the reconstruction problem for the case of non-convex control restriction set is given. An approach to solving this problem is suggested. This approach is based on auxiliary variational problems with non-classical convex-concave Tikhonov-regularized integral cost. A numerical method for solving dynamic control reconstruction problem is suggested. It reduces the reconstruction problem to solving systems of linear ordinary differential equations. Matching conditions for the approximation parameters (accuracy and step of the known measurements and a Tikhonov regularizing parameter) such that the constructed approximations converge to the solution are obtained. Results of numerical simulation are exposed to illustrate the theory.

1 Introduction

1.1 Inverse problems

Inverse problems play an important role in the theory and applications of dynamical systems. While forward optimal control problems for dynamical systems aim at constructing control laws that optimize given quality indexes, inverse problems aim at reconstructing unknown input control by known output—an observed movement of the system. The term dynamic reconstruction means that the reconstruction is carried out in real time when a new piece of information is received from the observations. Such tasks are actual for many modern applied control problems (for example, in mechanics, medicine and economics).

1.2 Modern approaches to solving inverse problems

There exists a variety of modern approaches to solving inverse problems. We offer a brief review of some of them. A classical approach to solving inverse problems, based on the least squares method, is suggested in [1]. In work [2] inverse problems are solved by analyzing the response of the dynamical system to special controls. In [3], a regularized filtering function

*e-mail: subb@uran.ru

is used. Some of the suggested methods are based on algebraic approaches, such as the method described in [4], that is based on the Lie groups theory. Another method, developed in [5], relies on an optimization approach based on gradient-type methods. In this approach inverse problems are formulated as operator equations and then are reduced to minimizing the corresponding residual functionals. Adjoint variables are used to compute the gradients. The authors of [6] consider effective iterative methods, based on generalized Golub–Kahan bidiagonalization, which allows automatic regularization of the solution. Similar problems are considered in [7], where an approach, based on using Fourier series, is proposed.

Let us also note two approaches that rely on the Tikhonov regularization [8] procedures. The first one [9] suggests an optimization method based on discrete linear operator equations that reduces inverse problems to minimization of Tikhonov functionals in which an additional penalty term leads to smoothness of the minimizer.

Another well-known approach to solving the dynamic control reconstruction problems was proposed in [10]. This approach consists of minimizing a Tikhonov-regularized discrepancy between the dynamics and measurements of the observed states of the system. It uses the extreme aiming procedure, which has roots in the works of the school of N.N. Krasovskii on the theory of positional optimal control [11].

More detailed reviews of modern methods for solving inverse problems are presented in [12, 13].

1.3 The new approach

This work develops a new variational approach to solving dynamic control reconstruction problems, which was proposed by N.N. Subbotina, E.A. Krupennikov, T.B. Tokmantsev [14–16]. This approach uses auxiliary variational problems on minimum of a nonclassical integral convex-concave Tikhonov-regularized residual functional. A feature of the proposed approach is that it does not imply direct solving of the auxiliary problem and uses only linearized necessary conditions for an extremum. This method reduces dynamic control reconstruction problem to integration of Hamiltonian systems of linear ordinary differential equations.

In this paper, we consider deterministic dynamical controlled systems affine in controls. It is supposed that discrete inaccurate measurements of the observed trajectory are known. This trajectory is called the basic trajectory. Reconstruction is carried out in real time as new measurements are received. Admissible controls are bounded measurable functions.

Note that other approaches to solving inverse problems, that rely on auxiliary optimization problems, use convex residual functionals, as a rule (see, for example, [9]). The new approach uses convex-concave functionals. This innovation ensures that the obtained approximated movements oscillate around the real movement of the dynamical system. This character of the residual is well-suited for averaging and provides *-weak convergence of the constructed control approximations to the desired solution.

A numerical algorithm based on the new approach is proposed. It reduces the dynamic control reconstruction problem to integrating systems of linear ODEs. The variety of existing numerical methods for solving ODEs ensures the efficiency of the algorithm.

1.4 New Results

Justification of the new approach is based on the results obtained in [14–16]. These results are obtained for the case of convex control restriction set, which describes geometrical constrains on the controls. In this work a generalization of this case is introduced. Namely, controls with values in non-convex sets are considered. This type of restrictions, in particular, includes controls with sliding regimes.

Note that the control reconstruction problem is ill-posed because one basic trajectory can be generated by different admissible controls. To regularize the problem, the definition of normal control is introduced. That is the admissible control that generates the basic trajectory and has the least possible L^2-norm. It has been shown in [16] that under some general assumptions there exists a unique normal control in the case of convex control restriction set.

Now we need to complement the set of admissible controls by generalized controls, which are measurable functions with values in the set of regular probability measures on the given set of geometrical restrictions. It allows us to include sliding controls in this set of admissible controls (for references on generalized controls see [17]). But it implies that we need to redefine the notion of the unique normal control to make the problem well-posed.

Let us remark that for each generalized control there exists a unique averaged control [17]. That is a measurable function with the values in the convex hull of the set of the geometric control restrictions. We will say that it is equivalent to the corresponding generalized control in the sense that they generate the same trajectory of the dynamical system. One averaged control can be equivalent to a set of generalized controls. We will not differ generalized controls that are equivalent to the same averaged control. Hence, we will consider the set of averaged controls as the set of admissible generalized controls.

So, the set of control restrictions is now convex, as in [14–16]. According to [16], there exists a unique generalized normal control. That is the admissible generalized control that generates the basic trajectory and has the least possible L^2-norm. So, we can consider the well-posed problem of control reconstruction as the problem of approximation of the generalized averaged normal control. The generalized averaged normal control will be further called the normal control. The results obtained in [15, 16] can be applied to the introduced regularization of the problem for the non-convex case.

Remark that there may exist many admissible controls that are equivalent to the normal control. There is no means to find which one of them generated the basic trajectory in reality.

2 Problem statement

2.1 Dynamics

Dynamical controlled systems of the following form are considered:

$$\begin{gathered}\dot{x}(t) = G(t, x(t))u(t) + f(t, x(t)), \quad t \in [0, T],\\ x(\cdot)\colon [0,T] \to R^n, \quad u(\cdot)\colon [0,T] \to R^m, \quad m \geq n,\\ G(\cdot)\colon [0,T]\times R^n \to R^{n\times m}, \quad f(\cdot)\colon [0,T]\times R^n \to R^n,\end{gathered} \tag{1}$$

where $u(\cdot)$ is the vector of the controls (the input of the system) and $x(\cdot)$ is the state variables vector (the output of the system). Admissible controls are measurable functions with values in a compact set $U \subset R^m$:

$$u(t) \in U \text{ a.e. on } [0,T]. \tag{2}$$

Note that the set of restrictions U is not necessarily convex. In the case of non-convex restrictions there can appear controls with sliding regimes. We introduce generalized controls to describe the motions of system (1) generated by such controls. Generalized admissible controls are measurable functions with values in

$$u(t) \in \overline{\text{co}}U \text{ a.e. on } [0,T]. \tag{3}$$

2.2 Measurements

It is supposed that we observe a trajectory of system (1) that is generated by an unknown generalized admissible control. This trajectory is called the basic trajectory $x^*(\cdot)\colon [0,T] \to R^n$. Discrete inaccurate measurements $\{y_i^\delta\}$ of the basic trajectory have the error $\delta > 0$ and are received with the time step $h^\delta > 0$:

$$\left\{y_i^\delta,\ i = 0, \ldots, N\right\}, \quad \left\|y_i^\delta - x^*(t_i)\right\| \le \delta, \quad t_i = ih^\delta, \quad t_N = T. \tag{4}$$

We introduce the following assumptions:

Assumption 1. There exist parameters $d_0 > 0$, $\delta_0 > 0$, $h_0 > 0$ and a compact $\Psi \subset R^n$ such that for any $\delta \in (0, \delta_0]$, $h^\delta \in (0, h_0] \bigcup_{i=\overline{0,N}} B_{d_0}[y_i^\delta] \subset \Psi$, where $B_r[x]$ is a ball of the radius r and the center in x.

Assumption 2. In dynamics (1) the elements of the matrix function $G(\cdot)$ and the vector function $f(\cdot)$ are Lipschitz continuous in all variables for $(t, x) \in D_0 = [0, T] \times \Psi$.

Assumption 3. The rang of the matrix $G(t, x)$ equals $\min\{m, n\}$ for $(t, x) \in D_0$.

2.3 Well-posed dynamic control reconstruction problem

Generally speaking, the problem of reconstruction of the unknown control is ill-posed because the same basic trajectory can be generated by a set of different admissible generalized controls. It can be illustrated on the following example:

$$\dot{x}(t) = u_1(t) + u_2(t), \quad x \in R, \quad u \in \{-1, 1\} \times \{-1, 1\}, \quad t \in [0, 1],$$
$$x^*(t) \equiv 0, \quad U^* = \left\{(u_1(\cdot), u_2(\cdot))^{\mathrm{T}} : u_1(t) + u_2(t) = 0\right\},$$

where U^* is the set of admissible generalized controls generating the basic trajectory $x^*(\cdot)$.

To state a well-posed problem the definition of normal control is introduced:

Definition. The normal control $u^*(\cdot)\colon [0, T] \to R^m$ is the measurable control that generates $x^*(\cdot)$ and has the least possible norm in the L^2 space:

$$u^*(\cdot) \in U^* \colon \|u^*(\cdot)\|_{L^2} = \min_{u(\cdot)\in U^*} \|u(\cdot)\|_{L^2}, \tag{5}$$

It was proved in [16] that if assumptions 1–3 hold for $x^*(\cdot)$, there exists a unique normal control $u^*(\cdot)$.

Assumption 4. The normal control satisfy condition (3). Therefore, it is a generalized admissible control.

So, the following *dynamic control reconstruction problem* can be stated: for any measurements set (4) obtained for a $\delta \in (0, \delta_0]$, $h^\delta \in (0, h_0]$ to find a measurable control $u^\delta(\cdot)\colon [0, T] \to R^m$ such that:

1. The controls $u^\delta(\cdot)$ are bounded essentially and uniformly by δ.
2. Each control $u^\delta(\cdot)$ generates a trajectory $x^\delta(\cdot)$ of system (1) such that these trajectories converge uniformly to $x^*(\cdot)$ in $C([0, T]; R^n)$, as $\delta \to 0$.
3. The controls $u^\delta(\cdot)$ converge *-weakly to the normal control $u^*(\cdot)$ in $L^1([0, T]; R^m)$, as $\delta \to 0$.

So, we can approximate in this way a real admissible sliding control that is equivalent to the generalized normal control. Since a sliding regime can't be approximated in the L^2 space, *-weak convergence is used in the problem statement.

3 Algorithm for solving the dynamic control reconstruction problem

A step-by-step algorithm for solving the dynamic control reconstruction problem is suggested. Each i-th step ($i = 1, \ldots, N$) is performed when the measurement point y_i^δ from (4) is received. On each i-th step the solution is constructed on the interval $[t_{i-1}, t_i] = [(i-1)h^\delta, ih^\delta]$, $i = 1, \ldots, N$.

The construction consists of 3 sub-steps:

3.1 Interpolation

The function $y^\delta(\cdot)\colon [0, t_i] \to R^n$ is extended on the interval $[t_{i-1}, t_i]$. It is a third-degree spline of the first-order smoothness which satisfies the conditions

$$y^\delta(t_{i-1}) = y_{i-1}^\delta, \quad y^\delta(t_i) = y_i^\delta, \quad \dot{y}^\delta(t_{i-1}) = f(t_{i-1}, y_{i-1}^\delta), \quad \dot{y}^\delta(t_i) = f(t_i, y_i^\delta),$$

3.2 Solving auxiliary variational problem

The constructed interpolation $y^\delta(\cdot)$ is used in the following auxiliary variational problem on the minimum of the functional

$$I(x_i(\cdot), u_i(\cdot)) = \int_{t_{i-1}}^{t_i} \frac{\|x_i(t) - y^\delta(t)\|^2}{2} + \alpha^2 \frac{\|u_i(\cdot)\|^2}{2}\, dt \to \min, \tag{6}$$

where α is a small Tikhonov-regularizing [11] parameter. The functional is considered on the set of continuously differentiable functions $x_i(\cdot)\colon [t_{i-1}, t_i] \to R^n$, $u_i(\cdot)\colon [t_{i-1}, t_i] \to R^m$ with the boundary conditions

$$i = 1\colon \quad x_1(t_0) = y_0^\delta, \quad \dot{x}_1(t_0) = \dot{y}^\delta(t_0), \quad i = 2, \ldots, N\colon \quad x_i(t_{i-1}) = y_{i-1}^\delta, \quad \dot{x}_i(t_{i-1}) = \dot{x}_{i-1}(t_{i-1}).$$

The necessary optimality conditions can be written in the form of a Hamiltonian system of ODEs [18]. The equations in this system are non-linear. They are linearized by "freezing" the coefficients

$$G(t, x(t)) \xrightarrow{to} G_i = G(t_{i-1}, y_{i-1}^\delta), \quad f(t, x(t)) \xrightarrow{to} f_i = f(t_{i-1}, y_{i-1}^\delta)$$

to obtain a system of linear ODEs:

$$\begin{gathered} \dot{x}_i(t) = -\alpha^{-2} G_i G_i^T s_i(t) + f_i, \quad \dot{s}_i(t) = x_i(t) - y^\delta(t), \\ t \in [t_{i-1}, t_i], \quad i = 1\colon \quad x_1(t_0) = y_0^\delta, \quad \dot{s}_1(t_0) = 0, \\ i = 2, \ldots, N\colon \quad x_i(t_{i-1}) = y_{i-1}^\delta, \quad \dot{s}_i(t_{i-1}) = \dot{s}_{i-1}(t_{i-1}), \end{gathered} \tag{7}$$

where $s_i(\cdot)$ is the adjoint variables vector corresponding to the i-th step.

3.3 Construction of the solution

System (7) is numerically integrated by any method and its solution $s_i(\cdot)\colon [t_{i-1}, t_i] \to R^n$ is used to construct the approximation $u_i^\delta(t)$ of the solution of the dynamic control reconstruction problem on this step:

$$u_i^\delta(t) = -\alpha^{-2} G_i^T s_i(t), \quad t \in [t_{i-1}, t_i].$$

This formula is one of the Euler equations in the auxiliary variational problem (6). The piecewise-continuous function

$$u^\delta(t) = u_i^\delta(t), \quad t \in [t_{i-1}, t_i], \quad i = 1, \ldots, N \tag{8}$$

is suggested as the approximation of the solution of the dynamic control reconstruction problem.

More detailed description and justification of the algorithm is exposed in [16].

The following theorem is proved [15, 16]:

Theorem. If assumptions 1–4 hold for the input data (1), (4), then functions (8) are an approximation of the normal control, which is the solution to the dynamic control reconstruction problem if the following agreement of the parameters holds:

$$h^\delta = h(\delta) \xrightarrow{\delta\to 0} 0, \quad \frac{\delta}{h^\delta} \xrightarrow{\delta\to 0} 0, \quad \alpha = \alpha(\delta) \xrightarrow{\delta\to 0} 0, \quad \frac{\alpha}{(h^\delta)^2} \xrightarrow{\delta\to 0} K_0 > 0. \tag{9}$$

4 Example

As an example, let us consider a model of controlled flight used in the problem of maintaining the flight altitude [19]:

$$\dot{v}(t) = \frac{\beta(t)}{m(t)} - \frac{q(t)v(t)}{m(t)}, \qquad \dot{m}(t) = -\beta(t), \tag{10}$$

$$\beta(t) \in \{0; 1\}, \quad Q(t) \in [0, 1], \quad t \in [0, 1], \quad T = 1, \quad v(0) = -1, \quad m(0) = 1. \tag{11}$$

The state variables are the vertical velocity $v(\cdot)$ and the mass $m(\cdot)$. The unknown control parameters are the fuel consumption $\beta(\cdot)$ and the aerodynamic drag coefficient $q(\cdot)$. Note that the bounding set $\{0; 1\}$ for β (11) is not convex.

To simulate the measurements that are input data in the dynamic control reconstruction problem, the following generalized controls have been considered:

$$\beta^*(t) = \begin{cases} 0.8, & t \in [0, 0.3), \\ 0.5, & t \in [0.3, 0.6), \\ 0, & t \in [0.6, 1], \end{cases} \qquad q^*(t) = 0.1 + 0.2 \sin t.$$

Since the generalized admissible control $\beta^*(\cdot)$ does not satisfy the restriction (11), we introduce an admissible control $\bar{\beta}^*(\cdot)$ that is an approximation of a sliding control that is equivalent to $\beta^*(\cdot)$ in the meaning that it generates the same trajectory as $\beta^*(\cdot)$. This approximation has the form of a function that can switch between the bounds $\{0; 1\}$ with the frequency 10^{-3}. The admissible controls $\bar{\beta}^*(\cdot)$ and $q^*(\cdot)$ were substituted in the dynamics (10) to calculate the basic trajectory. Then, it was perturbed in discrete points in steps $h^\delta = 10^{-3}$ with random error not exceeding $\delta = 10^{-4}$. The obtained set of "measurements" is considered as the input data for the dynamic control reconstruction problem.

The suggested algorithm has been applied to the obtained data set. The results are exposed on figure 1 (the reconstructed trajectories $v^\delta(t), m^\delta(t)$) and figure 2 (the reconstructed controls $\beta^\delta(t), q^\delta(t)$) for the parameters $\delta = 10^{-4}$, $h^\delta = 10^{-3}$, $N = 100$, $\alpha = 10^{-4}$.

As a numerical experiment, the reconstructed controls were averaged step-by-step by the formula

$$\hat{\beta}_j^\delta(t) = \frac{1}{H^\delta} \int_{t_{j-1}}^{t_j} \beta^\delta(t)\, dt, \quad t_j = jH^\delta, \quad j = 0, \ldots, K, \quad t_K = T, \quad H^\delta = 4h^\delta.$$

The results of averaging are exposed on figure 3.

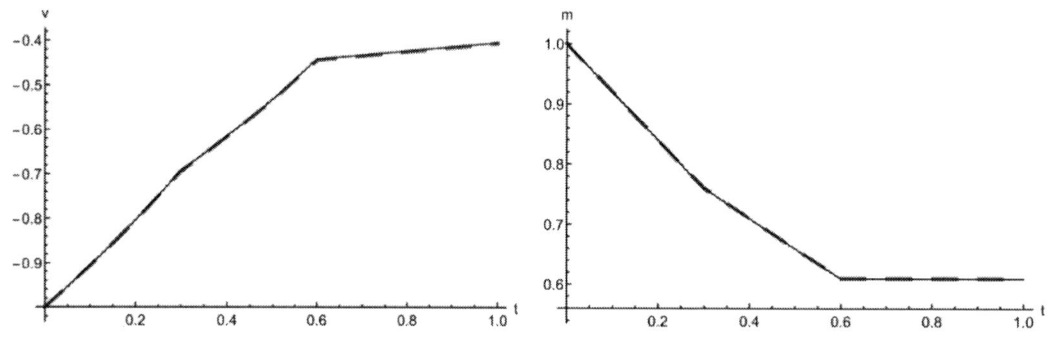

(a) (b)

Figure 1. Reconstructed trajectories: (a) is $v^\delta(t)$, (b) is $m^\delta(t)$. ——reconstruction, – – –—the base trajectory

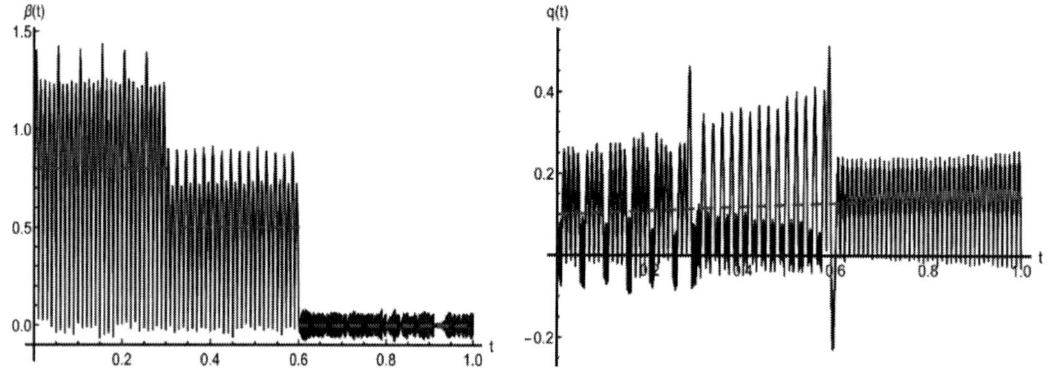

(a) (b)

Figure 2. Reconstructed controls: (a) is $\beta^\delta(t)$, (b) is $q^\delta(t)$. ——reconstruction, – – –—generalized unknown control

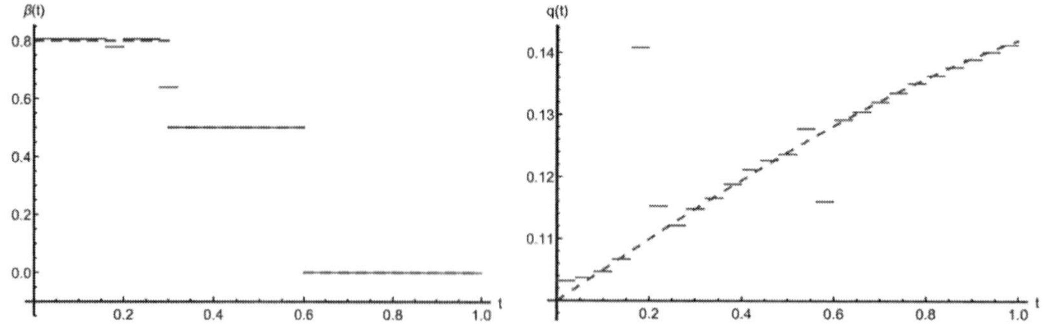

(a) (b)

Figure 3. Averaged reconstructed controls: (a) is $\hat{\beta}^\delta(t)$, (b) is $\hat{q}^\delta(t)$. ——averaged reconstruction, – – –—unknown control

5 Conclusion

A new approach for solving dynamic control reconstruction problems, suggested in [14–16], is developed. A generalization for the case of non-convex control restrictions set has been suggested. A new definition of the generalized normal control is introduced. It allows, in particular, to consider dynamical systems with controls that have sliding regimes.

In the future, it is planned to develop and justify a procedure that allows to construct numerical approximations of admissible sliding controls with the help of measurable controls with values in non-convex sets of geometrical restrictions. The approximations have to converge *-weakly to the generalized normal control.

The work on the first two sections of the paper (problem statement) is supported by the Russian Foundation for Basic Research (project no. 20-01-00362). The work on the last two sections (algorithm development and numerical simulations) is performed as a part of research conducted in the Ural Mathematical Center with the financial support of the Ministry of Science and Higher Education of the Russian Federation (Agreement number 075-02-2022-874).

References

[1] D.V. Ngoc, H. Marcelo, Ang Jr., *Dynamic model identification for industrial robots*, Acta Polytech. Hungarica, **6(5)**, 51 (2009)

[2] Y.R. Sturz, L.M. Affolter, R.S. Smith, *Parameter identification of the KUKA LBR iiwa robot including constraints on physical feasibility*, IFAC PapersOnLine, **50(1)**, 6863 (2017)

[3] C. Ren, N. Wang, Q. Liu, Ch. Liu, *Dynamic force identification problem based on a novel improved Tikhonov regularization*, Math. Probl. Eng., **2019**, 6095184 (2019)

[4] Y.C. Liu, Y.W. Chen, Y.T. Wang, J.R. Chang, *A high-order Lie groups scheme for solving the recovery of external force in nonlinear system*, Inverse Probl. Sci. Eng., **26(12)**, 1749 (2018)

[5] S.I. Kabanikhin, O.I. Krivorotko, *Identification of biological models described by systems of nonlinear differential equations*, J. Inverse III-Posed Prob., **23(5)**, 519 (2015)

[6] J. Chung, A.K. Saibaba, M. Brown, E. Westman, *Efficient generalized Golub–Kahan based methods for dynamic inverse problems*, Inverse Probl., **34**, 024005 (2018)

[7] M.C. D'Autilia, I. Sgura, B. Bozzini, *Parameter identification in ODE models with oscillatory dynamics: a Fourier regularization approach*, Inverse Probl., **33(12)**, 124009 (2017)

[8] A.N. Tikhonov, *Solution of Incorrectly Formulated Problems and the Regularization Method*, Soviet Mathematics, **4(5)**, 1035 (1963)

[9] U. Schmitt, A.K. Louis, C. Wolters, M. Vauhkonen, *Efficient algorithms for the regularization of dynamic inverse problems. Inverse Theory*, Inverse Probl., **18(3)**, 659 (2002)

[10] A.V. Kryazhimskij, Yu.S. Osipov, *Modelling of a control in a dynamic system*, Engineering Cybernetics, **21(2)**, 38 (1983)

[11] N.N. Krasovskii, A.I. Subbotin, *Game-theoretical control problems* (New York, Springer, 1987)

[12] N. Schuster, M. Burger, B. Hahn, *Dynamic inverse problems: modelling-regularization-numerics. Preface*, Inverse Probl., **34**, 040301 (2018)

[13] P.C. Sabatier, *Past and future of inverse problems*, J. Math. Phys., **41(6)**, 4082 (2000)

[14] N.N. Subbotina, E.A. Krupennikov, *Hamiltonian systems for dynamic control reconstruction problems*, Minimax Theory Applications, **5(2)**, 439 (2020)

[15] N.N. Subbotina, E.A. Krupennikov, **-weak approximations of the solution of the dynamic reconstruction problem*, Trudi Instituta Matematiki i Mehanici UrO RAN, **27(2)**, 208 (2021)

[16] N.N. Subbotina, E.A. Krupennikov, **-weak solution of the dynamic reconstruction problem*, Trudi Instituta im. V.A. Steklova, **315** (2021)

[17] R.V. Gamkrelidze, *Principles of optimal control theory* (Springer, New York, 1978)

[18] A.D. Ioffe, V.M. Tihomirov, *Theory of extremal problems* (North-Holland Publishing Company, New York, 1979)

[19] A.M. Letov, *Dynamics of flight and control* (Nauka, Moscow, 1969)

About the development of applied software for mechatronic systems of SLM technology

Olga Tretiyakova[1,2,*] and *Andrey* Molotkov[1,2,**]

[1]Moscow Aviation Institute (National Research University), 4, Volokolamskoe shosse, Moscow, 125993, Russia

[2]LLC "PROMIS LAB" group of companies SPC "Lasers and Apparatus TM", 5 building 1, Georgievsky prospect, Moscow, 124498, Russia

Abstract. The article describes the structure of the created software package for mechatronic systems that implements the technology of selective laser melting (SLM). One of the part of this software package is a software platform for solving machine vision tasks. Examples of several problems solved using the software platform are given. The solution to define the geometry and location of a metal sheet is considered in detail. The developed algorithm is used to solve the task of determining the geometry of the melted layer, the results of which will later be used for layer-by-layer reconstruction of the geometry of the manufactured part. The presented software platform was developed to implement analysis and control functions within the application software for implementing in SLM technology, however, as a result, the created software platform has received a wider range of applications.

1 Introduction

SLM (Selective Laser Melting) is an innovative technology to manufacture of products of complex shapes and structures from metal and ceramic powders using mathematical models of automated design systems. This process consists in sequential layer-by-layer melting of a powder material using powerful laser radiation and is used, for example, in medicine [1], aircraft construction [2], and mechanical engineering.

The mechatronic system for high-precision control of industrial equipment of the ML6 series, created by the group of companies "Lasers and Apparatus TM" for the implementation of selective laser melting process, combines mechanical elements, electrical parts, electronics, microprocessors and software. During development of software for mechatronic systems of SLM technology, the following program blocks have been created to date:

1. Calculation of the temperature distribution during the manufacturing process.
2. Visualization of the selective laser melting process [3].
3. Calculation of the stress-strain state of the part and minimization of the resulting stresses.

*e-mail: tretiyakova_olga@mail.ru
**e-mail: karacerr@gmail.com

4. Automation of the process of selecting the parameters of the technological process.

5. Creation of a control program for an installation with numerical control (CNC—Computer Numeric Control).

6. Analysis of visual data and control of the manufacturing process using machine vision algorithms.

In this work, we will make point in details on the consideration of the software platform that implements machine vision algorithms in the development of SLM technology.

Usage of software for CNC machines shows that it is often required to solve control and analysis problems using machine vision methods. Moreover, these are often problems, the solution of which allows the use of explicit linear algorithms. Examples of such tasks are following: finding a QR code or barcode, determining the position, number and quality of holes after perforation of material, determining the geometry of the working blade, controlling the quality of the seam, determining the geometry of flat products or products made by selective laser fusion SLM.

To reduce labor costs for solving a class of similar problems, we set a goal to create a software platform that would collect previously implemented image filters, and image analysis tools, allowing them to combine into solving specific problems. Such a platform should enable process engineers to solve machine vision problems without requiring programming skills, as well as provide a wide range of tools for software developers.

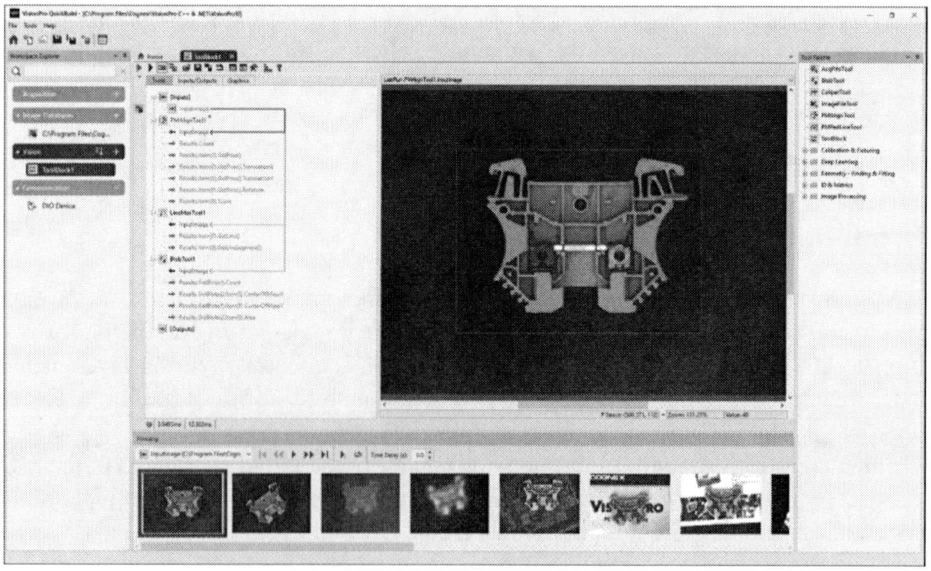

Figure 1. Cognex VisionPro

Such software tools are well in use for today, many foreign companies provide their solutions, such as Microsoft with the Azure Computer Vision product, Google with the Google Cloud Vision API, IBM with Watson Visual Recognition. Most of these solutions provide a programming interface such as the OpenCV libraries [4–6] for use in third-party software solutions. This makes it very difficult to use such tools for specialists without the necessary programming skills. A slightly different approach is taken by Cognex Corporation, a US manufacturer of machine vision systems, software and sensors used in automated manufacturing to inspect and identify parts, detect defects, verify product assembly, and guide assembly robots. This company offers not only separate solutions for machine vision, but also a visual development environment (figure 1) for using these tools in their own solution. We set ourselves the goal of implementing a similar environment with an emphasis on accessibility and ease of use.

2 Software platform and common approach

The software platform we are developing has already been partially implemented (figure 2).

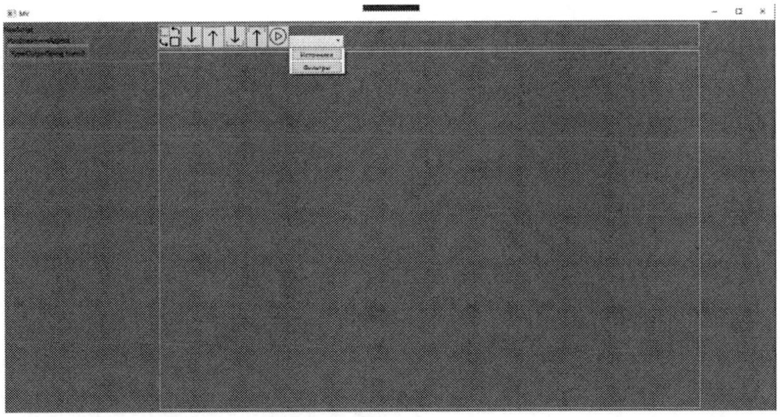

Figure 2. User interface of developed machine vision platform

Using this software platform user can collect data from various sources and perform processing sequence using various image-processing filters (figure 3), conduct analysis using prepared software tools.

Figure 3. Sequence of image processing

The images contain a lot of debugging information, but the general principle of operation is already visible. Using the function of establishing a connection, a connected directed graph is formed, along which the system subsequently transmit data from sources to filters and further along the sequence specified by the user. The absence of cycles in the decision graph is only requirement for this construction to work.

3 Examples of platform operation. Videorecording

One of the capabilities implemented by the platform is interaction with various camera models. In particular, the use of Basler matrix cameras (figure 4) was tested, which are mainly used in industrial automation, traffic monitoring, retail, as well as medicine and biological sciences.

Further, the data from the camera used for subsequent analysis or video recording of the studied processes (figure 5). The use of machine vision cameras allows you to monitor processes in environments inaccessible to the observer.

Figure 4. Matrix camera Basler

Figure 5. Video recording of process

One of the most common task in machine vision is object recognition. One of the common case of such task in industry is finding a sheet in the working field of the machine. Let's consider one of the options for solving a similar problem using the capabilities of the platform under consideration. We use a Basler matrix camera as a data source. Sheets can be located in different places of the working area, have different shapes and consist of different materials, so it is supposed to start by looking for abstract objects in the image. To do this, various image comparison filters are possible to use. The idea behind the approach is simple: get an image of an empty workspace, then get a new image of the same area with the sheet already positioned and identify its location by comparing it with the original image (figure 6).

Distortion is clearly visible in both photographs. Distortion caused by using a wide angle lens. It is often not possible to select lenses and position the aperture diaphragm in the optical scheme. In this case, mathematics comes to our aid and another distortion correction filter was implemented in the platform (figure 7). The filter is based on an expression of the form:

$$\vec{R} = b_0\vec{r} + F_3 r^2\vec{r} + F_5 r^4\vec{r} + F_7 r^6\vec{r}, \tag{1}$$

where $\vec{R}$—coordinate vector after transformation; $\vec{r}$—coordinate vector before transformation; $\vec{b}_0$—linear magnification factor; F_3, F_5, F_7—distortion coefficients.

To maintain image dimensions, the linear magnification factor is calculated automatically, thus it is not necessary to define it manually. The distortion correction is seen on the straightened pallet blades (figure 7).

Next, using one of the filter options that determine the difference between the two images, compare the area with a sheet and the same area without a sheet and find the desired object in the image. For color images, there are several similar approaches, which are based on

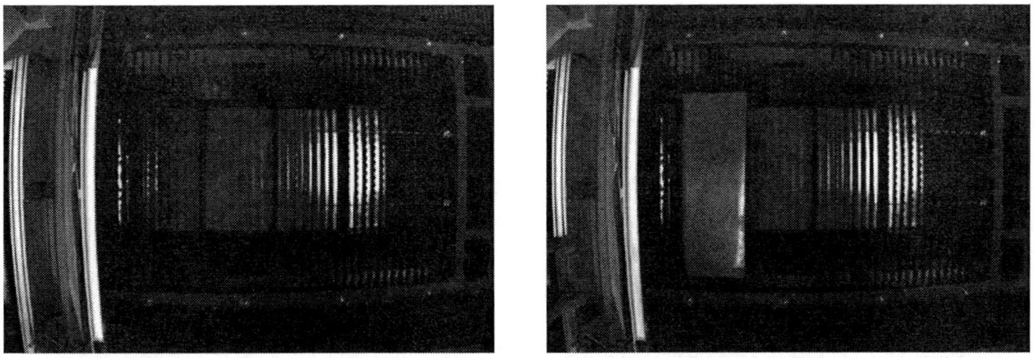

(a)   (b)

Figure 6. Images of working area: (a) without metal sheet; (b) with metal sheet

Figure 7. Working area after distortion correction

different metrics for calculating the color distance between pixels. The most commonly used metrics in this case are the Euclidean metric

$$d(p,q) = \sqrt{(p_1 - q_1)^2 + (p_2 - q_2)^2 + \ldots + (p_n - q_n)^2} = \sqrt{\sum_{k-1}^{n} (p_k - q_k)^2} \qquad (2)$$

and the Minkowski metric

$$d_1(p,q) = \|p - q\|_1 = \sum_{k=1}^{n} |p_k - q_k|. \qquad (3)$$

Obviously, both metrics give the same distance for a grayscale image. The value of threshold for filtering pixels is selected using the materials of objects and lighting.

Further, we find the angles in the image using various implementations of the algorithms for determining the angles [7–10]. The resulting angles will define the geometry of the desired object. The most linear and sensitive angle detection algorithms are recommended to use there.

Now we collect our corners into geometry contours; for this, the platform implements a clockwise and counterclockwise traversal algorithm. The idea of this algorithm is simple: we take the upper left unused corner from the list of defined ones. If the top pixel is dark, then we go counterclockwise, if it is light, then we go clockwise. Thus, we divide the contours into external and internal (figure 8).

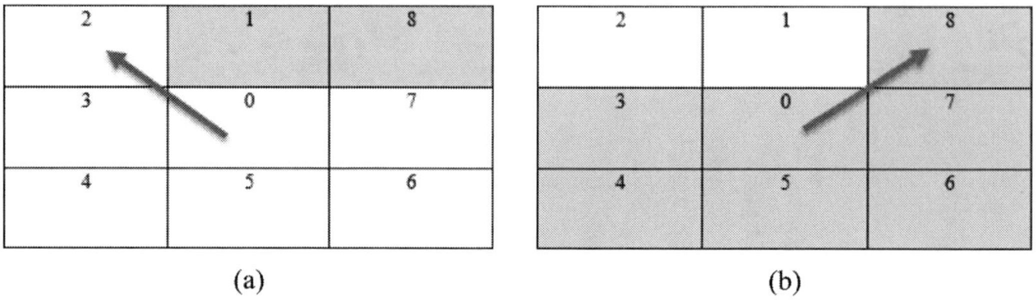

Figure 8. Traversal algorithm for finding edge: (a) inner boundary traversal; (b) outer boundary traversal

The traversal continues until we come back to begin, that means contour is closed. All pixels from the list of corners encountered along the way make up the found contour (figure 9).

The quality of the obtained result depends on the quality of the image, lighting and the accuracy of the selection of processing parameters.

Figure 9. The desired contour highlighted in yellow from the list of all contours

The same approach is in use to find the geometry of the melted layer in the SLM process. The preview image is a photo of the previous layer. The new image is a photo of the next melted layer (figure 10).

The further recognition process is identical to that described earlier. In fact captured images acquired using parsing of video stream. This recording was made using regular camera. Neither special lenses nor matched lighting were used. The camera was not fixed during

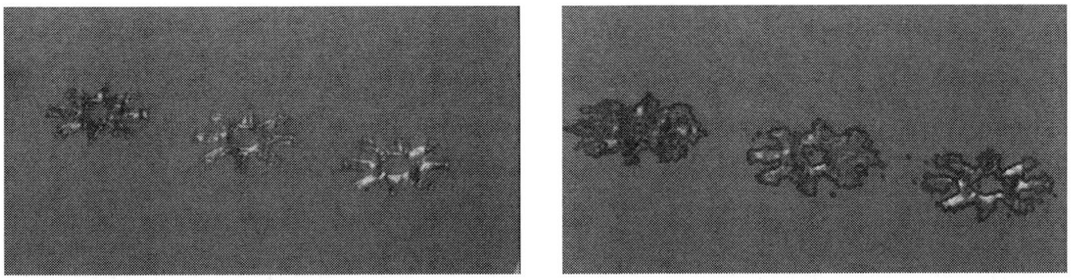

(a) (b)

Figure 10. Recognition of layers by image frame of a video stream: (a) initial image; (b) recognized image

shooting. Even under all the above conditions, it was possible to successfully recognize the contours of the product.

4 Conclusion

The basis of a new software platform to solve machine vision problems has been made. Using this software platform several tasks have been solved, for example, the tasks of recognizing a flat object, providing video recording of the production process, determining the holes on the workpiece have been solved. The task of recognizing the geometry of the product obtained using SLM process has been partially solved. The presented software platform created to implement the analysis and control functions within the application software for mechatronic systems of equipment that implements SLM technology, but as a result, the resulting platform found a wider range of applications. Our work continues on the development of a software platform and a software package to support SLM technology, in the process of solving tasks of tracking the layer and autofocusing of optical systems.

The authors thank management and colleagues from the group of companies "Lasers and Apparatus TM" for their support of this work.

References

[1] K. Munir, A. Biesiekierski, C. Wen, Y. Li, *Selective laser melting in biomedical manufacturing*, in Metallic Biomaterials Processing and Medical Device Manufacturing, 235 (Woodhead Publishing, Duxford, United Kingdom, 2020)

[2] W. Chen, Z. Li, *Additive manufacturing of titanium aluminides*, in Additive Manufacturing for the Aerospace Industry, 235 (Elsevier, 2019)

[3] A.A. Molotkov, O.N. Tretiyakova, *On possible approaches to visualizing the process of selective laser melting*, in Scientific Visualization 11.4 (2019)

[4] Z. Zhu, Y. Cheng, *Application of attitude tracking algorithm for face recognition based on OpenCV in the intelligent door lock*, Comput. Commun., **154**, 390 (2020)

[5] S. Sivkov, L. Novikov, G. Romanova, A. Romanova, D. Vaganov, M. Valitov, S. Vasiliev, *The algorithm development for operation of a computer vision system via the OpenCV library*, Procedia Computer Science, **169**, 662 (2020)

[6] C. Dominguez, J. Heras, V. Pascual, *IJ-OpenCV: Combining ImageJ and Open CV for processing images in biomedicine*, Comput. Biol. Med., **84**, 189 (2017)

[7] A. Semma, Y. Hannad, I. Siddiqi, C. Djeddi, M. El Y. El Kettani, *Writer Identification using Deep Learning with FAST Keypoints and Harris corner detector*, Expert Syst. Appl., **184** (2021)

[8] K. Rangarajan, M. Shah, D. Van Brackle, *Optimal corner detector*, Computer Vision, Graphics, and Image Processing, **48(2)**, 230 (1989)

[9] G. Olague, B. Hernandez, *A new accurate and flexible model based multi-corner detector for measurement and recognition*, Pattern Recognit. Lett., **26(1)**, 27 (2005)

[10] S. Teng, R. Md. N. Sadat, G. Lu, *Effective and efficient contour-based corner detectors*, Pattern Recognit., **48(7)**, 2185 (2015)

An analytical solution of boundary problems of shell deformation mechanics

Yuri Vinogradov[1] and *Bulat* Kashfutdinov[1,*]

[1]Bauman Moscow State Technical University, 5, 2nd Baumanskaya st., 105005, Moscow, Russia

Abstract. This paper proposes an algorithm for the analytical (with a controllable error) solution of boundary-value problems of shell deformation mechanics. Decomposition of partial differential equations is per-formed. The obtained ordinary differential equations in matrix form are solved in the form of converging matrix series in the corresponding coordinates. Atthe point of intersection of the coordinates of ordinary differential equations, the algebraic equations of their connection are satisfied. The error control of the solution of the boundary value problem is carried out by comparing the results when the grid of the de-composition of equations is refined. The study solved the problems of deformation a hinged cylindrical shell and a shallow shell fixed on the edges loaded with external pressure.

1 Introduction

The development of new constructions for transport infrastructure for safe and regular delivery of goods to Mars is one of the tasks of the development of cosmonautics for the period after 2030 [1]. The development of novel promising calculation methods for finding the optimal required characteristics of various structures is an integral part of the development of new technology [2]. As shown in [3], the use of analytical algorithms for solving boundary problems in the mechanics of deformation of thin shells can significantly increase the accuracy of calculations, especially in stress concentrators. The essence of the method is that linear differential equations with partial derivatives written in canonical form, that is, for quantities characterizing the state of the section of the shell, are taken as a mathematical model of the mechanics of shell deformation. Generally, the methods for modeling shell structures in rocket and space, aviation technology are based on FEM [4, 5]. Contrary to them, the proposed method makes it possible to simulate the stress concentration in thin-walled structures with the required accuracy. Partial differential equations [6, 7] are reduced to ordinary differential equations [8, 9], which can be solved analytically [10–12], via decomposition.

2 Materials and Methods

According to the method proposed, the system of p equations of p desired functions $z_1 = z_1(\alpha,\beta), \ldots, z_m = z_m(\alpha,\beta)$:

$$L_i(z_1, \ldots, z_m) = f_i(\alpha,\beta), \quad i = (1 \ldots m). \tag{1}$$

*e-mail: bkashfutdinov@bmstu.ru

With boundary conditions:

$$l_j(z_1, \ldots, z_m) = \varphi_j(\alpha, \beta), \quad j = (1 \ldots r) \tag{2}$$

by representing the operator in the form of h addends $L_i = \sum_{k=1}^{h} L_{ik}$, $i = (1 \ldots p)$ and representing right-hand sides of the equations was a sum of unknown functions $f_i(\alpha, \beta) = \sum_{k=1}^{h} f_i^{(k)}(\alpha, \beta)$ can be reduced to a system of simpler equations and coupling equations:

$$L_{ik}(z_1^{(k)}, \ldots, z_m^{(k)}) = f_i^{(k)}(\alpha, \beta). \tag{3}$$

With boundary conditions:

$$l_j(z_1^{(k)}, \ldots, z_m^{(k)}) = \varphi_j(\alpha, \beta). \tag{4}$$

And coupling equations:

$$z_1^{(1)}(\alpha, \beta) = z_i^{(k)}(\alpha, \beta). \tag{5}$$

Via decomposition, the system of partial differential equations is reduced to system of equations [13] written in the following form:

$$\begin{gathered} F\left(\alpha, \beta, z_i^{(k)}, \frac{\partial z_i^{(k)}}{\partial \alpha}, \frac{\partial^2 z_i^{(k)}}{\partial \alpha^2}, \ldots, \frac{\partial^n z_i^{(k)}}{\partial \alpha^n}, f_i^{(k)}\right) = 0, \\ F\left(\alpha, \beta, z_i^{(k)}, \frac{\partial z_i^{(k)}}{\partial \beta}, \frac{\partial^2 z_i^{(k)}}{\partial \beta^2}, \ldots, \frac{\partial^n z_i^{(k)}}{\partial \beta^m}, f_i^{(k)}\right) = 0. \end{gathered} \tag{6}$$

With partial derivatives with respect to only one of the variables in the following form:

$$F\left(\alpha, \beta, z_i^{(k)}, \frac{\partial z_i^{(k)}}{\partial \alpha}, \frac{\partial^2 z_i^{(k)}}{\partial \alpha^2}, \frac{\partial^2 z_i^{(k)}}{\partial \alpha\, \partial \beta}, \ldots, \frac{\partial^r z_i^{(k)}}{\partial \alpha^r}, \frac{\partial^s z_i^{(k)}}{\partial \alpha^t\, \partial \beta^{s-t}}, f_i^{(k)}\right) = 0 \tag{7}$$

and

$$F\left(\alpha, \beta, z_i^{(k)}, \frac{\partial z_i^{(k)}}{\partial \beta}, \frac{\partial^2 z_i^{(k)}}{\partial \beta^2}, \frac{\partial^2 z_i^{(k)}}{\partial \alpha\, \partial \beta}, \ldots, \frac{\partial^u z_i^{(k)}}{\partial \beta^u}, \frac{\partial^v z_i^{(k)}}{\partial \alpha^w\, \partial \beta^{v-w}}, f_i^{(k)}\right) = 0. \tag{8}$$

In these equations, there coupling equations, partial derivatives with respect to a single variable and mixed derivatives.

Mixed derivatives of equations (8) and (9) are considered as partial derivatives with respect to a single variable:

$$\frac{\partial^s z_i^{(k)}}{\partial \alpha^t\, \partial \beta^{s-t}} = \frac{\partial^t \left(\dfrac{\partial^{s-t} z_i^{(k)}}{\partial \beta^{s-t}}\right)}{\partial \alpha^t} \quad \text{or} \quad \frac{\partial^v z_i^{(k)}}{\partial \alpha^w\, \partial \beta^{v-w}} = \frac{\partial^{v-w} \left(\dfrac{\partial^w z_i^{(k)}}{\partial \alpha^w}\right)}{\partial \beta^{v-w}}. \tag{9}$$

Thus, these equations are reduced to the form in (2), (3). Now, equations (2) and (4) can be written in the matrix form:

$$z'_\alpha(\alpha, \beta) = A_\alpha(\alpha, \beta)\, z_\alpha(\alpha, \beta) + f_\alpha(\alpha, \beta). \tag{10}$$

Similarly, equations (2) and (3) are written:

$$z'_\beta(\alpha, \beta) = A_\beta(\alpha, \beta)\, z_\beta(\alpha, \beta) + f_\beta(\alpha, \beta). \tag{11}$$

Evidently, the solution of equations (4) will be as accurate as $\frac{\partial^{s-t} z_i^{(k)}}{\partial \beta^{s-t}}$, while solution of equations (5) will be as accurate as $\frac{\partial^{w} z_i^{(k)}}{\partial \alpha^{w}}$. If differentiated equations (1) are to be used as the coupling equations, then these solutions can be used to satisfy coupling equations. Lesser-order derivatives of functions and the functions themselves do not have to be determined from (4) and (5). They are calculated from (2) and (3).

Such an approach to mixed derivatives implies the need to have many additional boundary functions for the function $\frac{\partial^{s-t} z_i^{(k)}}{\partial \beta^{s-t}}$, $\frac{\partial^{w} z_i^{(k)}}{\partial \alpha^{w}}$ and their derivatives for equations (4) and (5), accordingly. If boundary conditions cannot be expressed naturally, one should satisfy coupling equation in boundary points or in some additional points.

Analytical solutions formulas [14, 15] for ordinary differential equations can also be used for equations similar to (7), (8). As $z(\alpha,\beta)$ is a function of two variables, columns of parameters on the interval boundaries should be interpreted as columns of arbitrary function in formulas [11]. By fixing the value of one of the variables in (7), (8), that is, assuming that $z(\alpha,\beta_1)$, $z(\alpha,\beta_2)\ldots$ and $z(\alpha_1,\beta)$, $z(\alpha_2,\beta)\ldots$, one can solve the differential equations along the corresponging coordinate line for the function of one variable by using [16]. So, one can construct a mesh of solution lines by intermittently fixing one coordinate, then the other (and considering them as parameters of the corresponding equations).

In the intersection points, coupling equations should be applied. By solving the coupling equations for unknown functions $f(\alpha,\beta)$, we get the expressions for their averaged values in the following form:

$$C_{\alpha_i} z_\alpha(\alpha,\beta_i) + C_{\beta_j} z_\beta(\alpha_j,\beta) + C_{ij} q, \tag{12}$$

where C_{α_i}, C_{β_j}, C_{ij}—coefficient matrices, q—external load column. By plugging them in formulas [16] instead of f_i, we can write a system of algebraic equations. Thus, the boundary problem is finally solved by solving the obtained system of algebraic equations.

3 Results

A series of analysis are carried out to test the algorithm [16]. The circumferential and meridian bending moments are calculated in a cylindrical panel (figure 1) along two lines (red in the figure 1). The panel is simple supported.

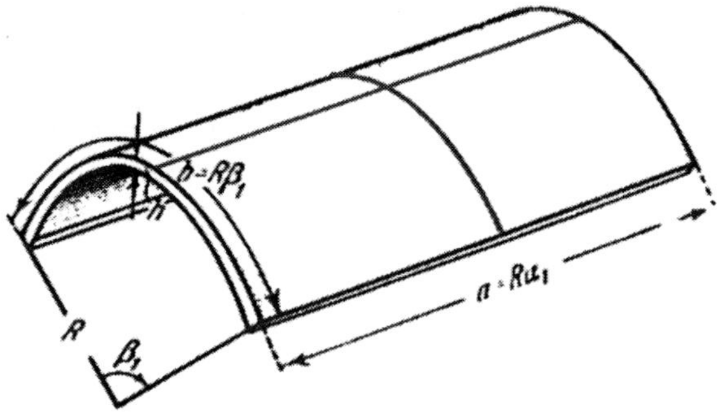

Figure 1. Cylindrical plate

The dimensions of the panel are: in the direction α is 1.5 m, the length of the arc in the direction β is 1.5 m. The radius are varied from 0.48 m (half-cylinder) to 50 m (almost flat plate). The bending moments M_1—meridional and M_2—circumferential are calculated using double rows according to the eq. (13)

$$M_{1k} = \frac{16\mu}{\pi^4} \sum_{m=1}^{\infty} \sum_{n=1}^{\infty} \frac{m^3(m^2 + \nu\lambda^2 n^2) \sin\left(\frac{m\pi\alpha}{\alpha_1}\right) \sin\left(\frac{m\pi\beta}{\beta_1}\right)}{n(m^2 + \lambda^2 n^2)^2[(m^2 + \lambda^2 n^2)^4 + \mu m^4]},$$
$$M_{2k} = \frac{16\mu}{\pi^4} \sum_{m=1}^{\infty} \sum_{n=1}^{\infty} \frac{m^3(\lambda^2 n^2 + \nu m^2) \sin\left(\frac{m\pi\alpha}{\alpha_1}\right) \sin\left(\frac{m\pi\beta}{\beta_1}\right)}{n(m^2 + \lambda^2 n^2)^2[(m^2 + \lambda^2 n^2)^4 + \mu m^4]}. \quad (13)$$

The diagrams below show the moment diagrams: in figures 2(a) and (b) in the range of radius from 0.48 m to 1.20 m, in figures 3(a) and (b) up to 50 m.

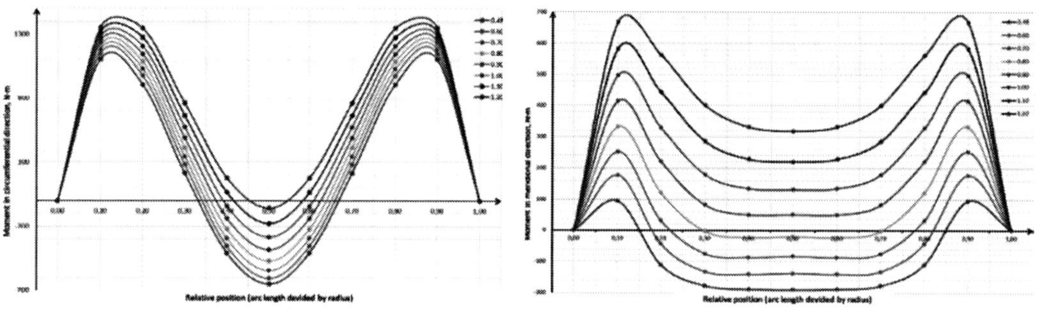

(a) (b)

Figure 2. Result for cylindrical simply supported panel: (a) moment in circumferential direction; (b) moment in meridional direction

As shown for some values of the radius, for example 1.0 m, moments become negative, this means that values gave different sign at center and near boundary condition zone. As the radius increases, the panel straightens into a square plate and the moment diagrams become the same in the center part.

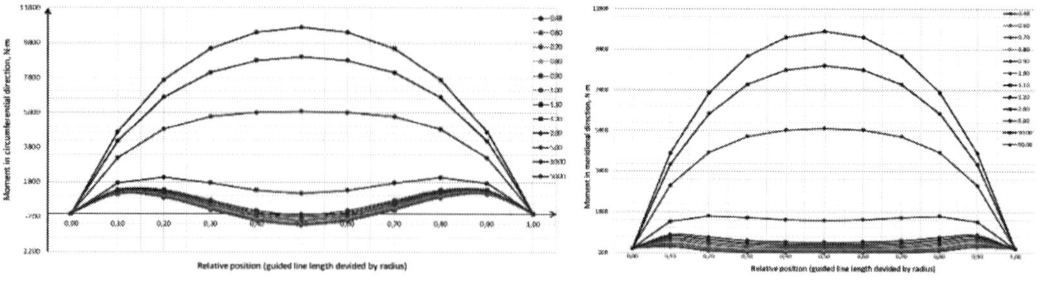

(a) (b)

Figure 3. Result for cylindrical simply supported panel: (a) moment in circumferential direction; (b) moment in meridional direction

For a cylindrical panel, the decomposition solution was compared to the solution in double trigonometric series. Table 1 compares the results of calculating deflection and bending moments.

Table 1. Comparison of deflections and bending moments of a cylindrical shell

Column Header Goes Here	Solution in double trigonometric series	Decomposition solution	Difference, %
Deflection, m	$14.23 \cdot 10^{-5}$	$14.19 \cdot 10^{-5}$	2.3
Meridional bending moment, N · m	131.0	130.5	3.5
Circumferential bending moment, N · m	−285.9	−284.8	3.8

Table 2. Results of mesh density influence evaluation

Iteration number	1	2	3	4	5	6	7	8	9	10
Mesh density, meridians × × parallels	5 × 5	7 × 7	9 × 9	11 × 11	13 × 13	17 × 17	19 × 19	21 × 21	23 × 23	25 × 25
Deflection, mm	3.16409	3.16700	3.17151	3.17349	3.17433	3.1747	3.1749	3.17496	3.1749	3.1750
Relative difference, %	0.0919	0.1424	0.0621	0.0267	0.0119	0.0054	0.0025	0.0010	0.0003	0.0000
Bending moment, N	27.315	27.174	27.205	27.238	27.260	27.274	27.284	27.291	27.296	27.300
Relative difference, %	−0.5187	0.1140	0.1196	0.0808	0.0534	0.0364	0.0257	0.0187	0.0140	0.0108

Also, we solved a boundary problem of deformation of a shallow shell fixed at the edges and loaded with external pressure. Displacements w and bending moments M of a double-curvature shell are calculated in its pole. On the shell, a mesh of meridians and parallels was imposed to separate the sides of the shell into intervals. The calculation results are shown in table 2.

Aside from the absolute values of deflections and moments, to evaluate convergence and the solution errors, we also calculated the difference of the calculated value at two subsequent iterations divided by the value obtained at current iteration. In table 2, this quantity is referred to as the relative difference. The proposed approach to solving the problems of shell mechanics and its program implementation enjoys fast convergence of the solution. The solution error can be reduced by increasing the density of the decomposition mesh.

4 Conclusion

The computational algorithm of the proposed analytical method is extremely simple: an analytical solution of ordinary differential equations is determined by formulas in the form of convergent matrix series [7] on intervals with a controlled error, a system of algebraic equations is formed and solved. The problem of calculation stability is overcome by dividing the main (boundary) interval into sub-intervals. A special feature of this algorithm is that no constant integration of differential equations is required.

The efficiency of parametric investigations with variable calculation is ensured by repeated solution of the system of algebraic equation with another load and/or boundary conditions without having to solve the system of differential equations again.

This work was supported by the RFBR grant No. 22-29-00733.

References

[1] V. Igritsky, V. Mayorova, *Promising transport infrastructure for the development of the planet Mars at the construction stage and the start of operation of a permanent base*, in XLIV ACADEMIC SPACE CONFERENCE: Dedicated to the Memory of Academician S.P. Korolev and Other Outstanding Russian Scientists—Pioneers of Space Exploration (2021)

[2] K.A. Neusypin, M.S. Selezneva, *Aircraft measurement complex design using intelligent technologies*, in XLIII ACADEMIC SPACE CONFERENCE: Dedicated to the Memory of Academician S.P. Korolev and Other Outstanding Russian Scientists—Pioneers of Space Exploration (2019)

[3] Y.I. Vinogradov, *Analysis of Stress Concentration with a Controlled Error*, in Thin-Walled Structures (Transport and Launch Container), Mech. Solids, **56**, 230 (2021)

[4] C. Yanyan, V.A. Nelyub, G.V. Malysheva, *An Investigation of the Kinetics of the Heating Process for Parts Made of Carbon Fiber*, Process of Curing. Polymer Science, Series D, **12(3)**, 296 (2019)

[5] V.A. Nelyub, *The stress-strain state evaluation of elementary carbon fibers with metal coatings*, J. Phys.: Conf. Ser., **1990**, 012071 (2021)

[6] I.F. Obraztsov, *Structural mechanics of aircraft: A textbook for aviation specialties of universities* (Mashinostroenie, Moscow, 1986)

[7] V.Z. Vlasov, *Selected Works*, vol. 1 (Publishing house of the Academy of Sciences of the USSR, Moscow, 1962)

[8] A.N. Krylov, *About the calculation of beams lying on an elastic foundation* (Akad. Nauk SSSR, Leningrad, 1931)

[9] A.N. Krylov, *Lectures on approximate calculations* (State. publishing house of technical and theoretical literature, Moscow, 1954)

[10] G. Peano, *Integration par series des equations differentielles lineaires*, Math. Ann., **32**, 450 (1888)

[11] F.R. Gantmakher, *Matrix theory* (Nauka, Moscow, 1988)

[12] R. Bellman, *Introduction to the theory of matrices* (Nauka, Moscow, 1976)

[13] G.I. Pshenichnov. *Solution of some problems of structural mechanics by the decomposition method*, Construction mechanics and calculation of constructions, **4**, 12 (1986)

[14] Yu.I. Vinogradov, *Methods for solving linear ordinary differential equations*, Dokl. RAS, **409(1)**, 15 2006)

[15] Yu.I. Vinogradov, D.B. Molchanov, *Fundamentals of the modified finite element method for solving boundary value problems of structural mechanics of thin-walled structures*, PMM, **82(4)**, 498 (2018)

[16] Y.I. Vinogradov, D.B. Molchanov, V.N. Bakulin, *Analytical matrix solutions of linear ordinary differential equations with constant coefficients*, J. Phys.: Conf. Ser., **1392**, 012077 (2019)

MATEC Web of Conferences **362**, 01033 (2022) https://doi.org/10.1051/matecconf/202236201033
CMMASS 2021

Application of an actuator made of a material with a shape memory effect for a transformable space structure

Vladimir Zimin[1,*], *Aleksey* Krylov[1], *Georgy* Kuvyrkin[1], and *Artur* Shakhverdov[1]

[1]Bauman Moscow State Technical University, 5, 2nd Baumanskaya st., 105005, Moscow, Russia

Abstract. Increase in the operational functionality of space technology leads to the necessity of creating large-sized transformable systems. The development of adequate models which are used in computational experiments to simulate the opening of transformable space structures is of great importance. Significant progress has been made in the field of creating large-sized transformable space antennas: the ratio of the mass of the reflector to its working area has decreased to 0.5–1.5 kg/m^2. Despite the significant progress achieved in the design of such structures, the task of ensuring smooth and reliable deployment of large-sized transformable space structures, while ensuring their subsequent functioning, remains important. An important area of applied research in the field of shape memory materials is the creation of thermomechanical force actuators that deploy large-sized space structures. Actuators with shape memory effect makes the opening process easy to control, ensuring its shock-free nature. As an active element of the force actuator, it is proposed to use a wire made of titanium nickelide material with a shape memory effect. To develop a functional model of the actuator's active element made of a shape memory material, a series of experimental and theoretical studies was carried out.

1 Introduction

The complication of design schemes and the increase in the dimensions of modern large space structures due to the increase in their operational functionality leads to the necessity of creating systems with different configurations in transport and operating states (figure 1).

When the required parameters of working orbits are reached, the process of deploying or transformation of large space structures is carried out [1]. The procedure of opening such space systems occurs under the influence of force actuators, which, generally, use various kinds of springs and electric motors. To date, in the field of space systems, many variants of opening large structures have been developed [2–5]. Shock-loads occur when the operating state is fixed after the completion of the structural transformation process in orbit [6].

In recent years, materials with shape memory effect (SME) are widely used in space technology [7, 8]. It is possible to create large space structures in orbit that use force actuators for their deployment, the active elements of which are made of materials with SME [9–13]. Active elements that use a change in temperature to control their deformation will allow for controlled slow deployment of large space structures. The defining relations for materials with SME, which represent the stress-strain dependences, temperature, and the fraction of

*e-mail: zimin@bmstu.ru

(a) (b)

Figure 1. Large-sized transformable space antennas: (a) umbrella-type antenna reflector with a diameter of 12 m; (b) petal-type antenna with a diameter of 10 m

phase composition, are essentially nonlinear. This causes great mathematical difficulties in calculating the force characteristics (force-displacement relations) of active elements of actuators made of materials with SME [14]. Also, along with the problem of deformation of active elements, it becomes necessary to solve the problem of thermal conductivity [15, 16]. Therefore, the main role in creating such force actuators is currently played by experimental methods [17].

2 Experimental studies

In a force actuator, it is proposed to use a wire made of titanium nickelide material as an active element, which is heated during operation by passing an electric current through it (figure 2).

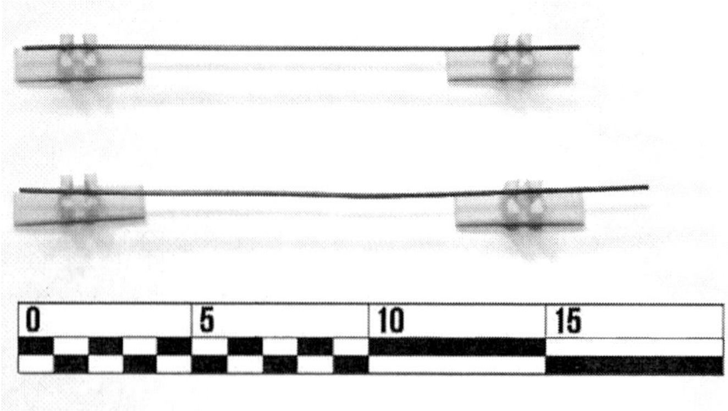

Figure 2. Active elements made of titanium nickelide material

The main characteristics of the active element of a force actuator made of SME material are: developed force, actuation time, maximum displacement or working stroke. To study the main characteristics of active elements, a large series of experimental studies were conducted. In the course of these experimental studies, the following parameters were measured: temperature of active element, electrical resistance, operating stroke, actuation time, and generated force.

The active elements of the force actuator (a wire made of titanium nickelide material with a diameter of 1.5 mm) were previously subjected to various types of heat treatment. Length of the samples was 160 mm. Heat treatment was performed to eliminate possible internal stresses. Then the samples of active elements were preliminarily stretched (deformed) to a relative elongation of about 10% and fixed in this state, the maximum external force was 1.2 kN. After removing the fixation, their length decreased by 1.0–1.2% of the new length. Then the active elements were heated. When the temperature of the beginning of the reverse martensitic transformation was reached, the active elements decreased their length by 4–5%. This decrease in the length of active elements ceased at the temperature of the end of the reverse martensitic transformation. During the repeated test cycle, the active elements were stretched to a relative elongation of about 7% and fixed in this state, while the external force was 1.15 kN. After removing the fixation, their length decreased by 1.2–1.4% of the new length. Further, the active elements were reheated and they reduced their length by 5%.

In the next cycle, the active elements were stretched to a relative elongation of about 7%, with an external force of 1.25 kN. After removing the fixation, their length decreased by 1.4–1.7% of the new length (figure 3).

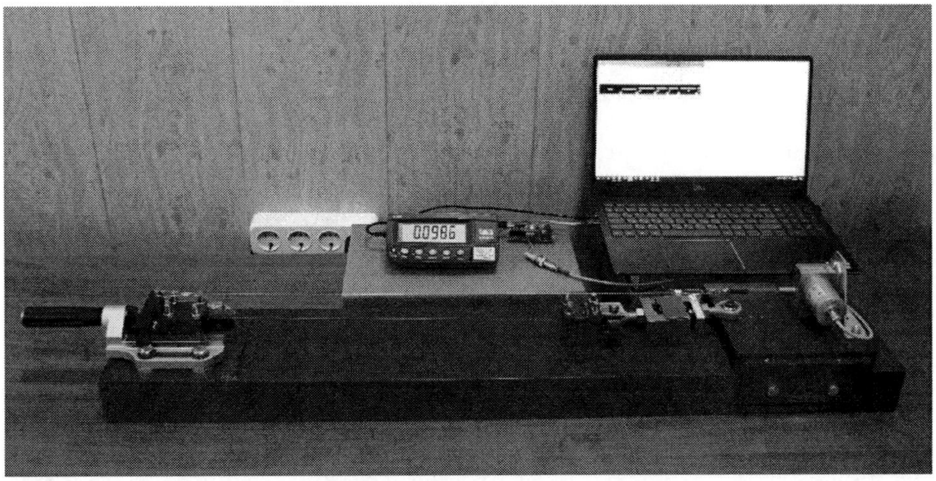

Figure 3. Experimental setup for preliminary deformation of active elements

Then the active elements were heated again and they reduced their length by 5%. The length of the samples after three test cycles was approximately 170 mm. When heated, they developed a longitudinal force of about 0.8–0.9 kN. Tests of active elements of the force actuator made of material with SME were carried out on the experimental setup shown in figure 4.

These tests were carried out to assess the stability of the main characteristics of the active elements of the force actuator, such as the working stroke (reduction of the relative elongation of the active element), the developed force (force generated by the active element) and the actuation time. Figure 5 shows the results of the conducted experimental studies.

Two methods were used to measure the temperature of the active element: contact and non-contact. In the contact method, a set of thermocouple sensors placed along the length of the active element was used. For non-contact measurement, a thermal imager was used that measures the radiation of the active element in the IR range. Using a digital multimeter operating in the ammeter mode, the current was measured. The magnitude of the current changed due to changes in the resistivity of the active element during the test. To measure the force, developed by the active element, a digital dynamometer with an external strain gauge

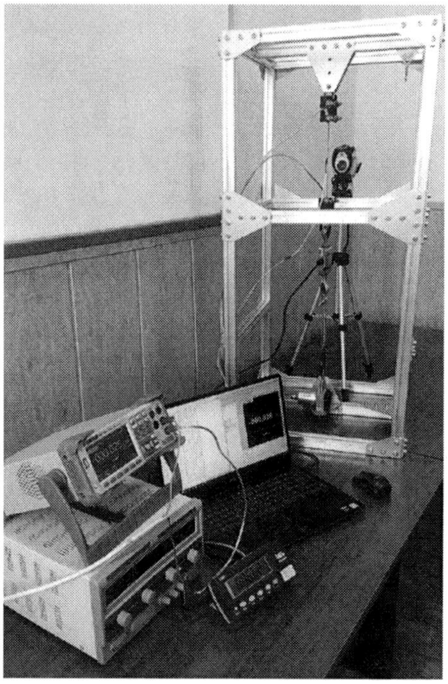

Figure 4. Experimental setup for determining the main characteristics of the active elements of the force actuator

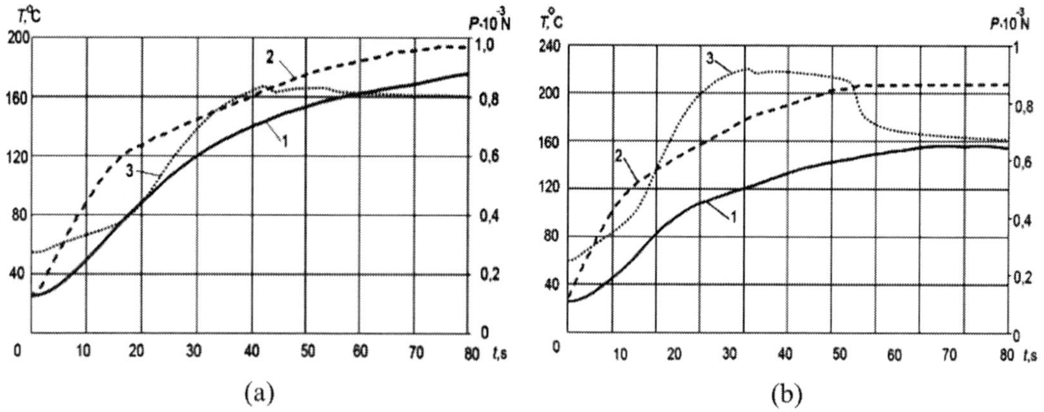

(a) (b)

Figure 5. Experimental results: (a) active element with a length of 174 mm after the first deformation; (b) active element with a length of 180 mm after the second deformation; 1—temperature change when measured by thermocouple; 2—temperature change when measured by a thermal imager; 3—change in the force developed by the active element

was used. During the tests, the displacement of the active element was measured using a special cable sensor, the operation of which is based on the strain effect. This measurement made it possible to determine the change in the relative elongation (working stroke) of the active element with a fairly high accuracy.

When calculating the temperature change of the active elements of the force actuator in ground tests, the heat exchange coefficient was selected. When conducting a numerical experiment, the results obtained correlated well with the data of experimental studies. At the same time, the calculations took into account changes in the resistance and heat capacity of active elements during their heating.

3 Model of functioning of the active element of the force actuator

In the course of ground-based experimental studies, it can be assumed that the change in the thermal energy of the active elements of the force actuator is equal to the amount of heat received due to electrical energy minus heat losses from natural convection

$$cm\frac{dT}{dt} = RI_1^2 - \alpha S(T - T_a), \tag{1}$$

where c—specific heat capacity; m—mass of the active element; T—temperature of the active element; t—time; R—resistance of the active element; I_1—current in the active element during testing; α—heat exchange coefficient; S—heat exchange surface area; T_a—ambient temperature.

When operating in orbit, it can be assumed that the change in the thermal energy of the active element of the force actuator is equal to the amount of heat received due to electrical energy minus thermal losses from radiation in orbit

$$cm\frac{dT}{dt} = RI_2^2 - \varepsilon\sigma_0 T^4 S, \tag{2}$$

where I_2—current in the active element in orbit; ε—emissivity; σ_0—Stefan–Boltzmann constant. Find the equilibrium temperatures from (1) and (2) under the condition $dT/dt = 0$

$$\bar{T}_1 = \frac{RI_1^2}{\alpha S} + T_a, \quad \bar{T}_2 = \sqrt[4]{\frac{RI_2^2}{\varepsilon\sigma_0 S}}. \tag{3}$$

The power supply voltage is determined by the formula

$$U = RI.$$

Then the relation (3) can be represented as

$$\bar{T}_1 = \frac{U_1^2}{\alpha S R} + T_a, \quad \bar{T}_2 = \sqrt[4]{\frac{U_2^2}{\varepsilon\sigma_0 S R}}, \tag{4}$$

where U_1, U_2 is the voltage of power sources on earth and in space accordingly.

To ensure an approximate similarity of processes during ground tests and in orbit, it is necessary that

$$\bar{T}_1 = \bar{T}_2. \tag{5}$$

Then from (5) we obtain

$$\frac{U_1^2}{\alpha S R} + T_a = \sqrt[4]{\frac{U_2^2}{\varepsilon\sigma_0 S R}}. \tag{6}$$

Convert (6) to the form

$$U_2 = \sqrt{\varepsilon\sigma_0 S R}\left(\frac{U_1^2}{\alpha S R} + T_a\right)^2. \tag{7}$$

In the course of conducting experimental studies in ground conditions, the initial length of the active element of the force actuator was equal to 174 mm, the power supply voltage was 2.2 V. The resistance of the active element of the force actuator, obtained during the tests, was 0.09 Ohms. Substituting these values in formula (7), we obtain an approximate value of the power supply voltage U_2 in orbit, which is necessary to fulfill condition (5),

$$U_2 \approx 1.77 \text{ V}.$$

The following values included in relation (7) were used in the calculation:

$$\alpha = 50 \,\frac{\text{W}}{\text{m}^2\,\text{K}}, \quad S = 8.19 \cdot 10^{-4}\ \text{m}^2, \quad T_a = 298\ \text{K},$$
$$\varepsilon = 0.66, \quad \sigma_0 = 5.67 \cdot 10^{-8}\,\frac{\text{W}}{\text{m}^2\,\text{K}^4}.$$

4 Conclusion

The proposed actuator uses an active element, made of a wire of titanium nickelide material with SME, which is heated during operation by passing an electric current through it. As a result of experimental studies, the strain-force characteristics of active elements operating under uniaxial compression were studied. It has been determined that force actuators with active elements made of titanium nickelide material with SME have the advantages of generating significant forces, low weight and low power consumption.

The change in the relative elongation and the force generated by the active element of the force actuator occurred smoothly during experimental studies. When varying the amount of voltage supplied to the active element of the force actuator, the time of its operation varied greatly. However, it is worth noting that despite the significant difference in the actuation speed of the force actuator, the working stroke of the active element remained constant.

In the course of quantitative and qualitative studies of the active elements of the force actuator, it was established that it ensures very slow movements, which are necessary for a smooth shock-free transformation of large space structures in orbit. The main characteristics of these force actuators will ensure reliable opening of promising transformable space structures.

The development of a model of the functioning of the active element of the force actuator will allow to obtain its main characteristics (developed force, actuation time, working stroke), as well as the power and voltage of the power source. If necessary, using the model, it will be possible to adjust the design parameters of the active element (length, shape and dimensions of its cross-section) and the conditions of its heating.

References

[1] N.V. Banichuk, I.I. Karpov, D.M. Klimov et al., *Mechanics of large space structures* (Factorial, Moscow, 1997)

[2] A.V. Lopatin, M.A. Rutkovskaya, *Review of the designs of modern transformed space antennas (Part 1)*, Bulletin of the Siberian State Aerospace University named after academician M.F. Reshetnev, **2**, 51 (2007)

[3] V. Hasanzade, S.H. Sedighy, M. Shahravi, *Compact deployable umbrella antenna design with optimum communication properties*, J. Spacecr. Rockets, **54(3)**, 782 (2017)

[4] F. Zheng, *Affordable system conceptual structure design of new deployable spaceborne antenna*, in 33rd AIAA International Communications Satellite Systems Conference and Exhibition, 7–10 September 2015, Queensland, Australia (2015)

[5] S.V. Ponomarev, *Transformable reflectors of spacecraft antennas*, Bulletin of the Tomsk State University. Mathematics and Mechanics, **4(12)**, 110 (2011)

[6] V.N. Zimin., Z. Zikun, A.V. Krylov, S.A. Churilin, *Mathematical modelling of the deployment of a large transformable space structure*, AIP Conf. Proc., **2171**, 030002 (2019)

[7] L.W.Y. Feng, P. Martinez, *MEDUSA—Mechanism for Entrapment of Debris Using Shape Memory Alloy*, in 67th International Astronautical Congress, 26–30 September 2016, Guadalajara, Mexico (2016)

[8] E. Im, M. Thomson, H. Fang, J. Pearson, J. Moore, J. Lin, *Prospects of large deployable reflector antennas for a new generation of geostationary Doppler weather radar satellites*, in AIAA SPACE Conference & Exposition, 18–20 September 2007, Long Beach, California, USA (2007)

[9] K.J. De Laurentis, A. Fisch, J. Nikitczuk, C. Mavroidis, *Optimal design of shape memory alloy wire bundle actuators*, in Proceedings 2002 IEEE International Conference on Robotics and Automation, **3**, 2363 (2002)

[10] F. Schiedeck, T. Hemsel, J. Wallaschek, *The use of shape memory alloy wires in actuators*, Solid state Phenomena, **113**, 195 (2006)

[11] C. Liang, C.A. Rogers, *Design of shape memory alloy Actuators*, J. Intell. Mater. Syst. Struct., **8(4)**, 303 (1997)

[12] V.A. Likhachev, A.I. Razov, A.G. Cherniavsky, Yu.D. Kravchenko, S.N. Trusov, *Truss mounting in space by shape memory alloys*, in The First International Conference on Shape Memory and Superelastic Technologies, 7–10 March 1994, California, USA (1994)

[13] A. Riad, A. Ainamany, M. Benzohra, *The shape memory alloy actuator controlled by the Sun's radiation*, Materials Research Express, **4(7)**, 075701 (2017)

[14] I.V. Mishustin, *Three-level model of a shape memory alloy and its application to the description of the effect of reversible shape memory*, Mech. Compos. Mater. Struct., **18(4)**, 463 (2012)

[15] V.A. Barvinok, V.I. Bogdanovich, O.V. Lomovsky et al., *Development of reversible power drives from materials with the effect of memory of form for devices, applied in the knots of check off of space vehicles*, Izvestia of Samara Scientific Center of the Russian Academy of Sciences, **13(4-2)**, 301 (2011)

[16] V.A. Barvinok, V.I. Bogdanovich, A.A. Groshev et al., *Design method of power drives from material shape memory effect for rocket and space technology*, Izvestia of Samara Scientific Center of the Russian Academy of Sciences, **15(6)**, 272 (2013)

[17] V.N. Zimin, A.V. Krylov, A.O. Shakhverdov, *Development of the mathematical model of the force actuator for deployment of large-sized space structures*, J. Phys.: Conf. Ser., **1902(1)**, 012115 (2021)

AUTHOR INDEX